Graphs Everyone Should Know and How to Create Them in Stata

Graphs Everyone Should Know and How to Create Them in Stata

Franz Buscha
University of Westminster

A Stata Press Publication
StataCorp LLC
College Station, Texas

Published by Stata Press, 4905 Lakeway Drive, College Station, Texas 77845
Typeset in LaTeX 2ε
Printed in the United States of America
10 9 8 7 6 5 4 3 2 1

Print ISBN-10: 1-59718-413-6
Print ISBN-13: 978-1-59718-413-7
ePub ISBN-10: 1-59718-414-4
ePub ISBN-13: 978-1-59718-414-4

Library of Congress Control Number: 2024945728

Dedication

To Lena, Felix, Lito, Anne, and Andreas

Contents

Figures

Tables

Preface

The art of data visualization has advanced significantly since the turn of the millennium. In particular, accessibility and ease of use have taken center stage. Beautiful, complex, and sophisticated data visualizations are no longer limited to those with highly technical and specialized coding skills. Many software platforms now make it easy for anyone to start creating graphs and visualizing data. Now a much broader user base can find creative ways to reveal patterns and connections in data and, in turn, weave them into their stories and narratives.

Stata exemplifies both ease of use and powerful capabilities in data visualization. Continuous updates to Stata's graphics features, combined with a large and innovative user base, have transformed it from statistical software with basic graphic functions into a true graphics powerhouse. This evolution allows users to create an extensive array of sophisticated and visually compelling graphs with relative ease, leveraging both built-in functionalities and community-contributed commands.

This book aims to unleash that powerhouse and show users the wide range of statistical graphs that can be created with Stata. It is written in an informal style and contains no equations. The aim of the book is not to explain how each graph and statistical command works "under the bonnet", and there is an assumption that users have a basic understanding of statistics. This book is not intended to be linear and read from beginning to end. It is designed to be flipped through casually, during which readers can identify a particular type of graph or a series of graphs they are interested in. They then focus on that content, either reading the text around the graph before applying the relevant code and idea to their own data or reading the entire chapter for a more comprehensive introduction and discussion of that type of graph.

Each graph is directly preceded by the command that generates it. This will help users quickly adapt the code to their own data and create similar-looking graphs. As much as possible, only Stata datasets are used. These come preinstalled with Stata and will already reside on local hard drives. In other instances, examples use online datasets from Stata's official website that are fast to download. Custom or bespoke data sources are avoided because they can be difficult to access or work with. Users will, therefore, notice repetition of certain datasets such as `auto.dta`, `citytemp.dta`, `nlsw88.dta`, or `uslifeexp.dta`. These datasets often have particular variables or data setups that make them useful for teaching purposes, and their repetition throughout the book will help users better understand how different graph types represent the same data.

The book is split into four major parts, of which three are related to statistical concepts around data visualization. The first part looks at graphs that focus on univariate (one variable) data. Here the emphasis is often on presenting distributions of single variables. The second part examines graphs for bivariate (two variables) data. These graphs often examine how two variables are related and how such data are jointly distributed. The third part presents graphs for multivariate (three or more variables) data, which aim to examine how many variables are jointly related. These are often highly specialized graphs that use some trick to insert additional data dimensions. The final part looks at some useful graphic capabilities that users might encounter in data analysis, including how regression results can be easily visualized or how maps can be created.

Each chapter first demonstrates a basic graph, after which a series of important options is highlighted. This book does not aim to present and explain the many options available for each graph command. However, some options matter more than others, and this book tries to highlight how selected options can significantly change a graph. Also note that many graphs have access to generic graph options that change marker colors, line shapes, aspect ratios, axis titles, etc. These options are not explained in detail, because it is assumed that users have some basic knowledge of the graphic capabilities in Stata. However, these options are highlighted and explained when used.

A vital feature of this book is that it goes beyond the standard, official Stata installation. This book aims to represent the wide range of graph commands written by other Stata users and available at the Statistical Software Components Archive. Each chapter will clarify what additional commands or packages need to be installed to be able to follow along with that chapter. All chapters are standalone, and the code will work from top to bottom. The chapters are designed so that readers can execute all the relevant code without worrying about previous or later chapters.

I hope you enjoy this book and keep it close when working with data. Stata learning and teaching have come a long way since I started learning many years ago, and I am glad to provide you with a friendly new resource to augment Stata's extensive documentation. Of course, you are always encouraged to read the relevant help file and examine Stata's manuals in depth for all the nitty-gritty details. However, to rapidly accumulate high-level knowledge of many different processes and abilities, you cannot beat the plethora of Stata Press books now available for users.

Best wishes,
Franz Buscha

Acknowledgments

This journey into the world of Stata graphs would not have been complete without the invaluable insights and guidance from a few exceptional individuals. I extend my heartfelt gratitude to Chuck Huber and Lisa Gilmore, whose constructive feedback, keen eye for details, and knowledge of Stata were vital in making this book a reality. To both, thank you for your support and help. I also thank Kristin MacDonald, Adam Crawley, and Deirdre Skaggs, who were instrumental in creating, editing, and shaping this book. Finally, a special thanks to Stephanie White, who translated all my thoughts into reality.

How to use this book

This book is aimed at users who are somewhat familiar with Stata: it assumes the reader knows how to input data, use previously created datasets, create new variables, run regressions, and perform other basic Stata functions.

This book is highly modular. Users can immediately narrow down a particular graph they want using the index and then read the relevant section. Or they can follow a "textbook" approach, reading the book from front to back, following each step and each piece of code. Many users will likely follow an intermediate path.

Each chapter is set up similarly:

1. The chapter title is the type of graph on which the chapter will focus.
2. Following the title is a list of advantages and disadvantages, providing a rapid overview of each graph.
3. Next, there is a list of community-contributed commands that must be installed to enable users to follow along with the entire chapter.
4. Then, there is an introduction to the graph and basic examples.
5. Finally, advanced examples and permutations are presented.

Each chapter will consist of two different types of code:

- Code related to data preparation, manipulation, recoding, or statistical estimation is presented on a white background, just like a log file. An example is shown below.

```
. * Load data
. sysuse citytemp
(City temperature data)

. summarize tempjan

    Variable |        Obs        Mean    Std. dev.       Min        Max
-------------+---------------------------------------------------------
     tempjan |        954    35.74895    14.18813        2.2       72.6
```

- Code that produces graphs is presented in a box. An example is shown below.

Creating figure X

```
. * A basic histogram
. histogram tempjan
```

Boxed code will be of primary interest to users. It lets users identify which piece of code leads to a graph. However, note that users must first load (and possibly manipulate) the relevant data in all cases. This means that code on white backgrounds must be executed before any code in boxes can be executed.

Users should also understand how code is often presented in Stata books such as this one. I use the `typewriter` font to refer to commands, syntax, and variables. Readers of this book should understand the following pieces of syntax:

- Each new line of code starts with a dot prompt (`.`), indicating that you can type verbatim what is displayed after the dot to replicate results.
- Long code is wrapped at the end of a line with a line-join indicator (`///`), which indicates that the code continues on the following line.
- Each line of wrapped code begins with a `>` and not a dot prompt.
- An asterisk (`*`) at the beginning of a line denotes a comment. It indicates to Stata that the line should be ignored.

When working through examples, one should use the associated do-files that come with this book. However, where users are manually copying code or even copying and pasting from electronic versions, they must remove any starting `.` or `>` of each line for the relevant code to work. Copying commented code denoted by an `*` is optional.

Many chapters require users to install community-contributed packages from the Internet. Users will need to contact local IT administrators if they are behind a firewall or do not have read and write permissions. Those working on secure computers or with sensitive data should contact appropriate data security administrators.

Part I

Graphs for univariate data

Univariate data analysis describes the analysis of one characteristic or attribute in a dataset. In practice, this means that univariate analysis tends to focus on each variable separately by examining the value range and central tendency of the value. Numerical analysis of such variables often results in descriptive statistics such as means, medians, standard deviations, minimums and maximums, skewness, and kurtosis. Graphically, such analysis will often focus on visualizing complete distributions and providing users with a sense of where data are located, what shape they take globally and locally, and whether they conform to any known prior distributions.

Univariate analysis is not isolated only to one variable. Multiple variables can be analyzed together. However, a key condition is that such variables are analyzed independently; no joint distributional properties are examined.

Univariate analysis is common when data are first explored. Modern datasets are complex and often contain hundreds of different variables. These variables are often analyzed numerically using commands such as `describe` and `summarize`. Numerical analysis is often preferred when users must scan hundreds of new variables. However, once the focus starts to sharpen on a select few variables, graphical analysis becomes highly beneficial. Key data variables benefit from closer inspection to identify distributional properties such as modalities, outliers, and value ranges. Often, graphical analysis is a speedy way to inform further data manipulation, such as taking the log of long-tailed variables or recoding top-to-bottom coding value ranges.

In general, I would caution against relying only on numerical descriptive statistics when working with data; graphing key variables early will often lead to fewer coding errors, better data context, and improved multivariate modeling.

1 Histograms

Advantages

1. Are easy to understand and widely used
2. Are versatile—bin-width choices can generate different visualizations
3. Have fewer parameters to choose from than kernel density plots
4. Make it easy to identify the most frequent bin (tallest)

Disadvantages

1. Use density for default display, which is less intuitive for general users
2. Require careful bin-width selection for proper data analysis
3. Do not provide granularity in small local regions of data
4. Provide minimal statistical summary information such as median, mean, or quartiles

Required community-contributed packages

To install the community-contributed commands that are required to follow along with this chapter completely, type

```
. ssc install eqprhistogram
. ssc install hist3
. ssc install bihist
. ssc install stripplot
```

1.1 Introduction

Histogram plots are the basic "go-to graphs" when plotting univariate distributions for single continuous variables. They are well known, used widely across the corporate and scientific worlds, and taught early at school. Their primary advantage is their simplicity in conveying how one variable is distributed across a range of values. Their use of bars to display the number of observations in a given data range is an effective method to visually present statistical information. In histograms, both the width and the length of the bar carry numerical information, while in traditional bar charts (which are discussed in chapter 9), the width is irrelevant. This two-dimensional property of histograms (compared with the single dimension of bar charts) is what makes them powerful while also retaining accessibility; they carry more information but not so much to overwhelm general users.

The key to using histograms is bin-width management. Larger bin widths result in bars encapsulating a larger data range, leading to less detail in the graph. Very large bins often result in blocky graphs with little variation and detail. Smaller bins lead to more detail. However, too much detail can result in spiky graphs with large variances and many void areas. One must understand what granularity of data analysis is required and adjust the bin width accordingly.

Another disadvantage of histograms is that they provide minimal statistical summary information. For example, users cannot identify where mean or median values lie on the distribution. In addition, the concept of density may confuse some readers because it refers to relative frequency instead of actual frequency. In basic use cases, one should change the y-axis scale to frequency because it will be more intuitive to nontechnical users.

1.2 A basic histogram

Basic histograms can be created via the `histogram` command in Stata. Only one variable can be specified in this command, so the traditional command for a default histogram is written in the form of `histogram` *varname*. An example is shown in figure 1.1. It uses the city temperature dataset (`citytemp.dta`), which contains temperature information on 956 US cities. The variable `tempjan` is a continuously measured variable of interest.

```
. * Load data
. sysuse citytemp
(City temperature data)

. summarize tempjan

    Variable |        Obs        Mean    Std. dev.       Min        Max
-------------+---------------------------------------------------------
     tempjan |        954    35.74895    14.18813        2.2       72.6
```

Creating figure 1.1

```
. * A basic histogram
. histogram tempjan
```

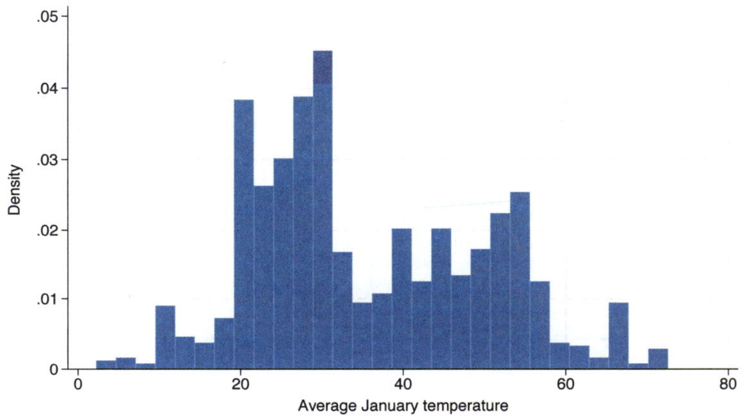

Figure 1.1. A basic histogram

Figure 1.1 presents a basic histogram of average January temperatures of various US cities. The graph suggests that the average temperature in January is possibly bimodal. There are peaks at around 30 degrees and 55 degrees. Few cities experience very cold or very hot days. The y axis is presented as a density by default. Densities are scaled so that the areas of all bins sum to 1. Overall, a user gets an excellent impression of how the numeric temperature values are distributed.

1.2.1 Custom bin widths

The most important options in the **histogram** command relate to adjusting the bin width. This can be done with the **bins()** option, where a numeric value indicates the number of bins on the graph. The **width()** option sets the width of each bin to a numeric value. The **discrete** option is available if data are discrete; this option creates a bin for each unique value in the variable. Examples of each of these options are shown below in figure 1.2. The option **name()** is used to store graphs, and the command **graph combine** is used to merge stored graphs into one graph.

Creating figure 1.2

```
. * Options that change bin sizes
. histogram tempjan, bins(10) name(g1, replace)
. histogram tempjan, bins(50) name(g2, replace)
. histogram tempjan, discrete name(g3, replace)
. histogram tempjan, width(10) name(g4, replace)
. graph combine g1 g3 g2 g4
```

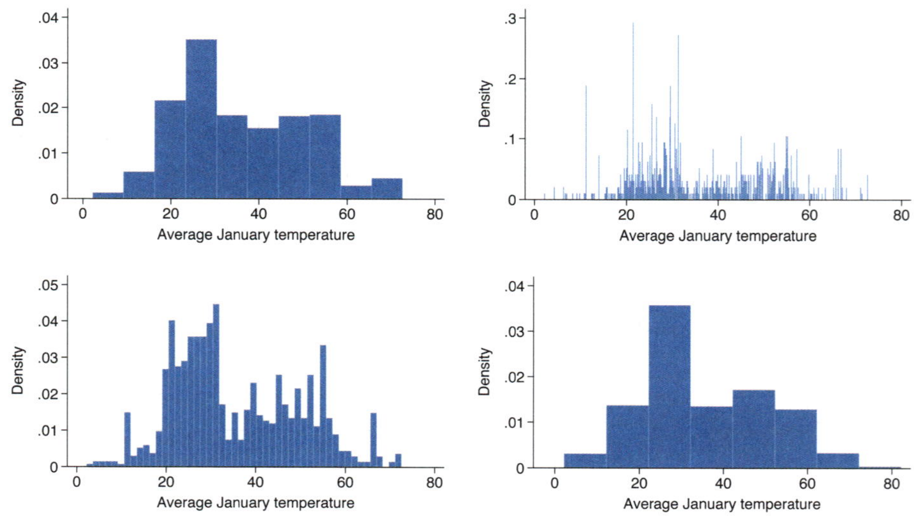

Figure 1.2. Histogram with different bin widths

1.2.2 Changing the y axis

The default y axis of `histogram` is density. Density is a measure of relative frequency, which is frequency divided by the sum of frequencies and then divided by the bin width. This can make interpretation of the y axis difficult for laypeople and those not versed in statistical language. It is possible to transform this axis; for example, multiplying the bin width by the bin height results in proportions, but doing this manually is cumbersome. Luckily, `histogram` has various options that change the default y axis to a different format. The options `density` (default), `fraction`, `frequency`, and `percent` allow users to present density, fractions, frequencies, or percentages on the y axis. Figure 1.3 exemplifies all four options.

Creating figure 1.3

```
. * Options that change the y axis
. histogram tempjan, density name(g1, replace)
. histogram tempjan, fraction name(g2, replace)
. histogram tempjan, frequency name(g3, replace)
. histogram tempjan, percent name(g4, replace)
. graph combine g1 g3 g2 g4
```

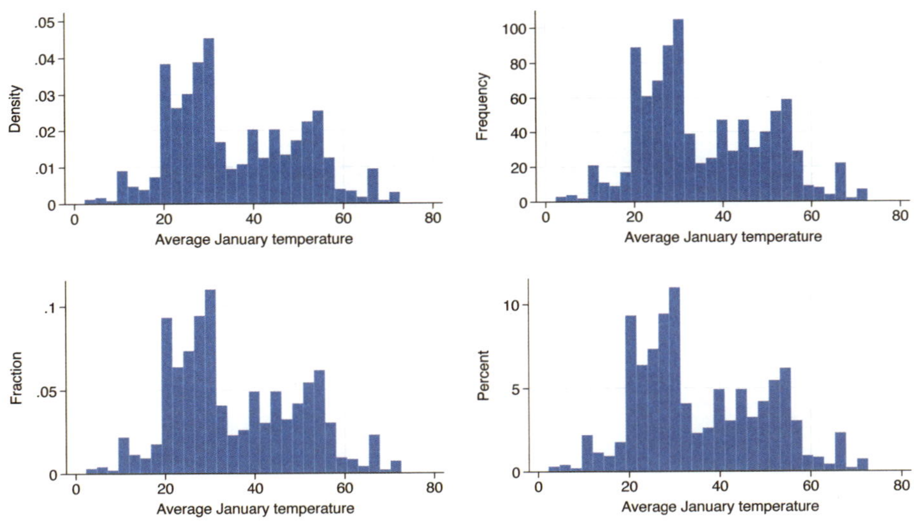

Figure 1.3. Histogram with different y axes

Figure 1.3 highlights the four different y-axis options available to `histogram` users. The graphs look identical, but the y axes contain different values that may be helpful to end users.

1.2.3 Bar labels

Those looking to add more detail to their histograms may want to use the option **addlabels**, which adds height labels to bars and makes individual bar-to-bar comparison substantially easier, especially when bars are of similar sizes. To affect the rendition of the bar labels, the option **addlabopts()** can be used, which takes any options from the *marker_label_options* syntax (see `help marker_label_options`). Figure 1.4 below presents an example of a histogram with marker labels that are angled at 45 degrees, formatted to three decimal places, and very small.

Creating figure 1.4

```
. * Adding bar labels
. histogram tempjan, addlabels       ///
> addlabopts(mlabangle(45) mlabformat(%9.3f) mlabsize(vsmall))
```

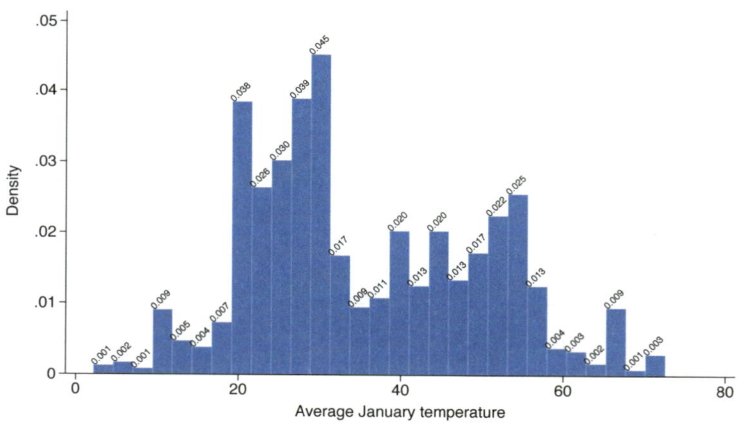

Figure 1.4. Histogram with bar labels

The usefulness of this option will depend on the level of detail required by users. In most cases, histograms are designed to provide general distributional overviews. Individual bar-to-bar comparisons that require bar labels are relatively rare.

1.2.4 Added density plots

Kernel density plots can be added to basic histograms generated with the **histogram** command. The option **normal** adds a normal density to the histogram, while the option **kdensity** adds a kernel density plot to the histogram. The **normal** option is useful for quickly checking whether the histogram mimics a normally distributed variable. The **kdensity** option is useful because kernel density plots provide a smoother visualization of the distribution. Those wishing to have the simplicity of a histogram and the smoothness of a kernel density plot can have both by using this option.

When adding a kernel density plot, the option **kdenopts()** allows for kernel density options to be imposed. Without going into detail, there are many options, of which the most important is **bwidth()**. This option defines the bandwidth of the kernel smoother. One should use bandwidths that are similar to the bin width in the histogram to give some comparability of both plottypes. Figure 1.5 shows an example of this.

Creating figure 1.5

```
. * Adding kernel density plot
. histogram tempjan, width(5) kdensity kdenopts(bwidth(5))
```

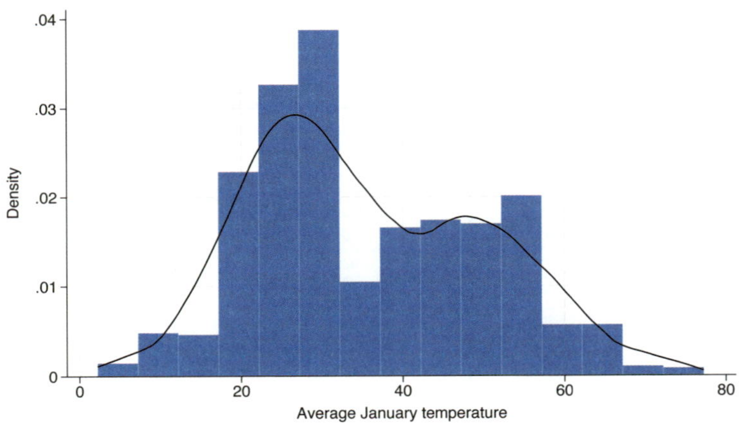

Figure 1.5. Histogram with density plots

1.3 Spike plots

For those wishing to reveal the very fine structure of a data distribution, `spikeplot` offers a convenient alternative to specifying a histogram with a very high bin count (or tiny bin width), for example, `histogram` *varname*`, bins(200)`. A spike plot is essentially a histogram that uses a very high bin count by default, resulting in a highly detailed histogram that reveals the fine structure of an underlying data distribution. An example is shown in figure 1.6.

Creating figure 1.6

```
. * A basic spike plot
. spikeplot tempjan
```

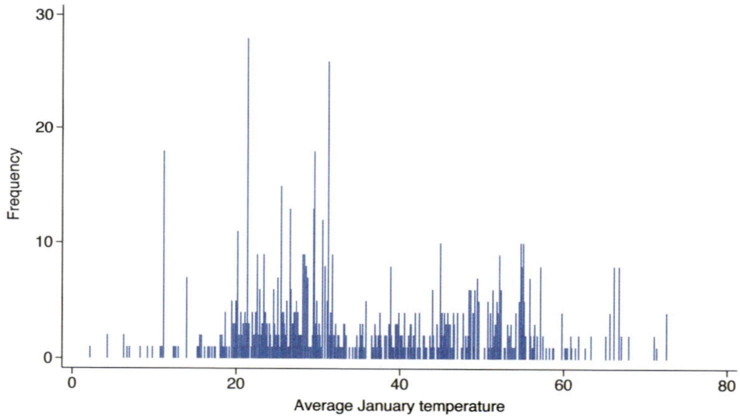

Figure 1.6. A basic spike plot

Figure 1.6 is a default `spikeplot`. It excels at revealing minute data concentrations. However, there are now noticeable gaps in the distribution where no observations occur. Like `histogram`, the `spikeplot` command has a bin-width option that allows users to modify the widths of the spike bars. This can be done via the `round(`#`)` option, which rounds the variable to the nearest multiple of #. However, unlike `histogram`, the bar widths are unchanged. In other words, the thin nature of the bars in `spikeplot` remains under all circumstances, which can result in spike plots that show gaps between bars, while data in each gap are actually represented in each spike. Figure 1.7 shows an example of this.

Creating figure 1.7

```
. * Spike plot with custom bin width
. spikeplot tempjan, round(1)
```

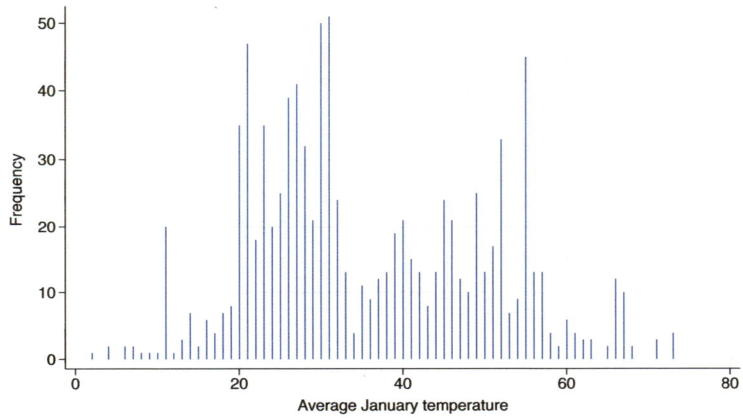

Figure 1.7. Spike plot with custom bin width

Figure 1.7 presents a spike plot that uses integer bins. The gapped nature of this histogram can be useful when seeking to identify which integer values have no temperature observations. However, take care that small gaps are not interpreted as data voids.

It is possible to increase the drawn bar width of spikes in `spikeplot` via the `lwidth()` option. Below is an example that uses a large number in the `round()` option and uses the `fraction` option to change the y axis to fractions. However, one generally should not use `spikeplot` in such a fashion; a normal histogram drawn using the `histogram` command will perform much better unless there is a specific need for blank gaps in a histogram.

Creating figure 1.8

```
. * Spike plot with custom bin width and custom line width
. spikeplot tempjan, round(5) fraction lwidth(*10)
```

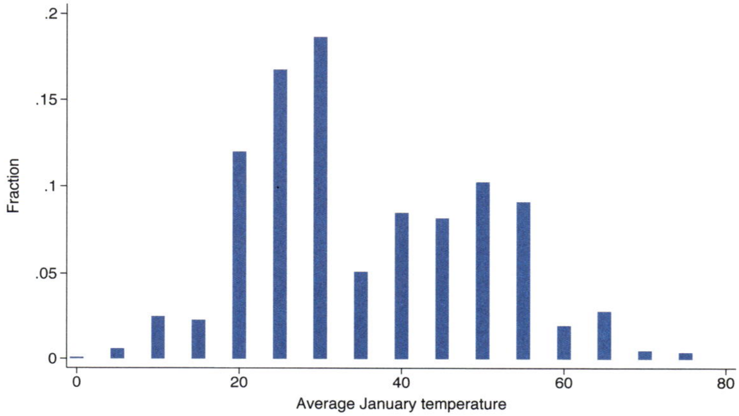

Figure 1.8. Spike plot with custom bin width and custom line width

1.4 Histograms with varying bin widths

Standard histograms have constant bin widths. However, this condition can be relaxed via two community-contributed commands that allow users to modify individual bin widths. The community-contributed command `eqprhistogram` (Cox 2003a) creates equal probability histograms where the widths of the bars vary so that all bars contain equal proportions of the data. This allows users to visualize the percentile distribution in conjunction with the underlying global data distribution. The `bin()` option indicates the number of bins. The default is `bin(8)`, indicating 12.5% of data in each bin. An example with five bins, where each histogram bin captures 20% of the total observations, is shown below in figure 1.9.

Creating figure 1.9

```
. * Equal probability histogram with 5 bins
. eqprhistogram tempjan, bin(5)
```

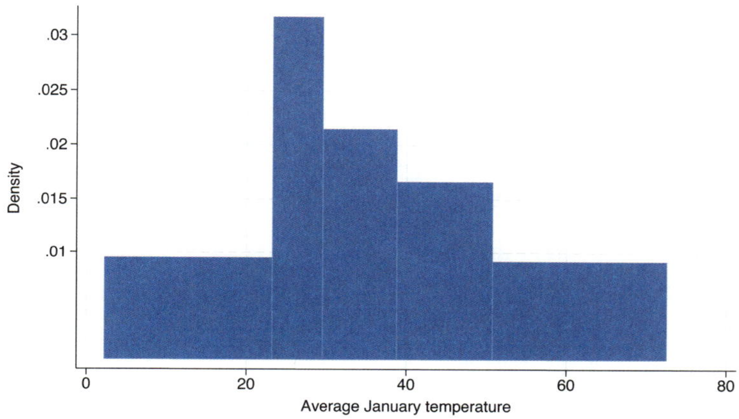

Figure 1.9. Equal probability histogram

The advantage of an equal probability histogram is that it marries distributional analysis with percentile analysis, giving users a better overview of how a distribution is weighted over its observation range. A disadvantage is that areas with little data have large bin widths, denying users details in that particular region. As with all histograms, users should vary the bin size when creating such graphs.

Alternatively, rather than specifying equal probability bins, users can specify custom bins that highlight specific ranges of data. For example, such custom bins might have particular meaning because they belong to an important categorical range group in the underlying continuous data. Such plots can be created with the community-contributed command `hist3` (Kohler and Kuehnel 2000). Unfortunately, this is a relatively old community-contributed command and has not been updated for some time, causing its graphics within Stata to default to a black background with yellow line colors and blue text. However, when these graphs are exported, they default to a black-and-white color scheme. The option `values()` sets the bin cutpoints; note that the start and end cutpoints must be specified. Figure 1.10 presents an example with five custom bins.

Creating figure 1.10

```
. * Histogram with custom bin sizes
. hist3 tempjan, values(2.2, 5, 20, 25, 50, 73)
```

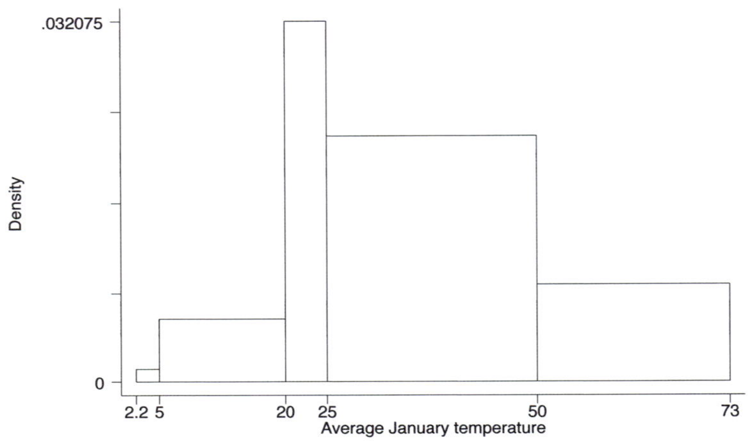

Figure 1.10. Custom bin histogram

Custom bin-width histograms are helpful if users are interested in particular value ranges of the distribution. In this example, a user may be interested in temperatures between 20–25 and 25–50 degrees specifically. However, custom bin-width histograms are generally not often used and require very specific research questions.

1.5 Multiple histograms

Often, users wish to compare multiple distributions. This comparison can be across different continuous variables or subgroups of the same continuous variable. However, histograms are not naturally designed to offer easy comparisons on a single graph, because histograms use significant area shading that causes problems when overlaying multiple histograms. Even when area shading is suppressed, bar lines often continue to overlap, making it hard to identify which histogram belongs to which variable or group. A good rule of thumb is that a kernel density plot should be the first choice when multiple distributions are to be graphed.

Nevertheless, various ways to create multiple histograms exist. The first is via the `by()` option in the official command `histogram`. Users must specify a categorical variable, and Stata will then draw repeated histograms for each subgroup and automatically combine the subgraphs into one graph. An example is shown below in figure 1.11.

Creating figure 1.11

```
. * Multiple histograms with by()
. histogram tempjan, by(region)
```

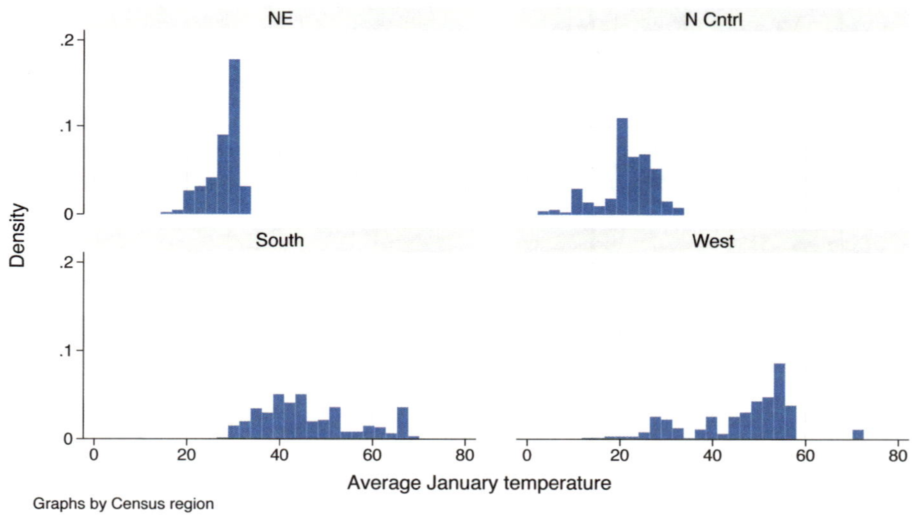

Figure 1.11. Multiple histograms side by side

Figure 1.11 exemplifies use of the `by()` option, which is convenient for a quick overview of how data are distributed across different categorical subgroups. However, side-by-side comparisons are generally useful only for approximate comparison or where there are significant global differences in the distributions. Detailed comparison of local differences is difficult because the graphs are not overlaid.

If detailed local comparisons are required, then the **twoway histogram** command can be used. This command is similar to **histogram** but can be used within the **twoway** wrapper command, which allows multiple Stata graphs to be overlaid on top of each other. The basic command takes the form **twoway (histogram** *var1*) **(histogram** *var2*).

Unfortunately, overlaying multiple **twoway histogram** plots does not automatically change histogram colors. This means that multiple two-way histograms will often visually "block" each other, resulting in a near-useless graph. The trick is to use the **color()** option in each two-way graph and 1) customize histogram colors individually (opposing colors are recommended) and 2) apply opacity via the `%` modifier. A value of `100%` fully hides the background, and a value of `0%` means that the color is fully transparent. An example is shown below in figure 1.12 that uses custom colors from the *colorstyle* list (**help colorstyle**) and 30% opacity. This code uses the **if** qualifier to examine subgroups of the same variable, although different variables could also be used. Note that the legend is not formatted in this graph.

Creating figure 1.12

```
. * Multiple two-way histograms
. twoway (histogram tempjan if region == 1, color(edkblue%30))  ///
> (histogram tempjan if region == 2, color(erose%30))           ///
> (histogram tempjan if region == 3, color(stone%30))           ///
> (histogram tempjan if region == 4, color(emerald%30))
```

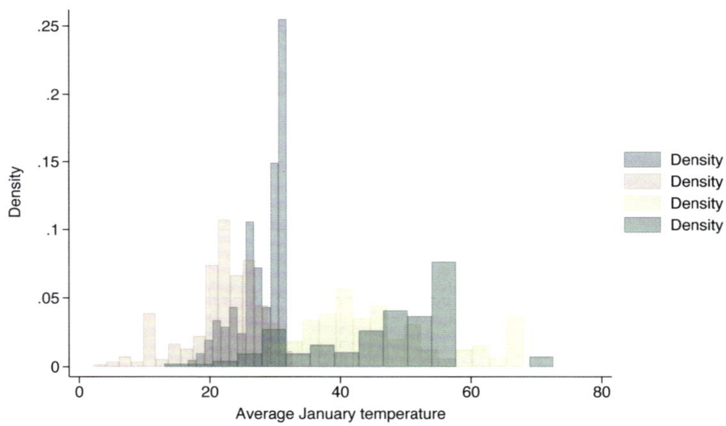

Figure 1.12. Multiple overlaid histograms

1.6 Mirrored histograms

Another approach is to use mirrored histograms. This works only when two histograms must be compared and involves placing one histogram upside down directly below the other histogram. One can manually achieve this by creating two individual histograms and using the `yscale(reverse)` option for one of them. The graphs can then be merged with the **graph combine** command using the `cols(1)` and `xcommon` options, which will result in a mirrorlike merge.

The resulting visual is not perfect, requiring further options to remove margins and align axes. The community-contributed command **bihist** (Nichols 2008) is a more user-friendly method that produces a mirror histogram. The option `by()` must be specified and requires a dichotomous group variable. The **frequency**, **fraction**, and **density** options allow users to choose between different y-axis scales. Figure 1.13 below demonstrates an example with a dichotomous variable and density on the y axis.

```
. * Two histograms with bihist
. generate south = (region==3)

. tabulate south

      south |      Freq.     Percent        Cum.
------------+-----------------------------------
          0 |        706       73.85       73.85
          1 |        250       26.15      100.00
------------+-----------------------------------
      Total |        956      100.00
```

Creating figure 1.13

```
. * Mirrored histogram
. bihist tempjan, by(south) density
```

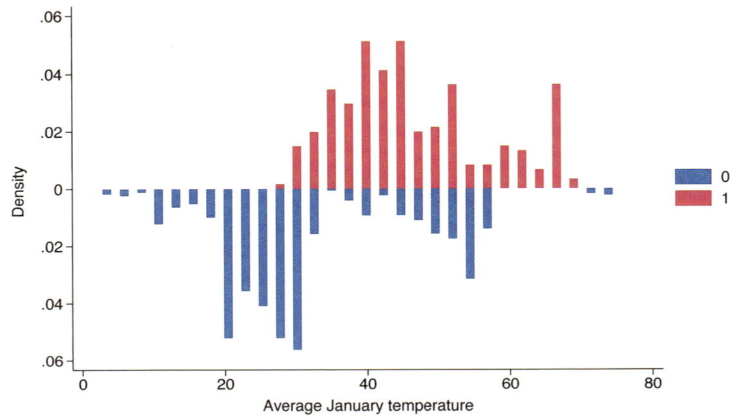

Figure 1.13. Mirrored histogram

Figure 1.13 demonstrates a mirrored histogram. These are particularly useful for comparing two distributions because the distributions do not overlap but still occupy the same graph space and share one x axis, making local comparison fast and easy. A disadvantage of the `bihist` command is that there is no easy way to adjust bin widths. If adjustments are needed, users may need to resort to creating a manual mirrored histogram.

1.7 Ridgeline histograms

If more than two histograms are required, then one can use a ridgeline histogram plot. Rather than merging two histograms onto a common y axis and mirroring one histogram, ridgeline plots graph multiple histograms on separate rows across the y axis. Essentially, the y axis serves the dual purpose of identifying both numeric scale (density, frequency, etc.) and categorical identification. While there is a specific community-contributed command for kernel density ridgeline plots, no such command exists for histograms. However, with some option adjustments, the community-contributed command `stripplot` (Cox 2003d) can be used to create horizontal ridgeline histograms. Alternatively, the official command `dotplot` can be used to create ridgeline plots but only in a vertical orientation. Therefore, `stripplot` is the preferred command to create ridgeline plots.

The `stripplot` command graphs data by placing several marks along a single magnitude axis and can be considered an advanced version of the official command `dotplot`. It can take either multiple variables (for example, `stripplot` *var1 var2*) or one variable when used with the `over()` option (for example, `stripplot` *var1*, `over(`*var2*`)`; this graphs *var1* over *var2* subgroups). Note that the `over()` variable should contain categorical values. The key to using `stripplot` for multiple histograms is to use the `stack`, `width()`, and `msymbol()` options. The `stack` option specifies that data points with identical values are to be stacked, while the `width()` option specifies that values are to be rounded in classes of specified width. The `msymbol(S)` option is used to display square markers. Combining it with the `stack` option results in many stacked squares that are ultimately depicted as vertical bars that identify frequency or density. This is demonstrated in figure 1.14 below. The `msize()` option is used to adjust bar sizes in the graph, although this is only a visual effect.

Creating figure 1.14

```
. * Ridgeline histogram using stripplot command
. stripplot tempjan, over(division) stack msymbol(S) msize(small) width(1)
```

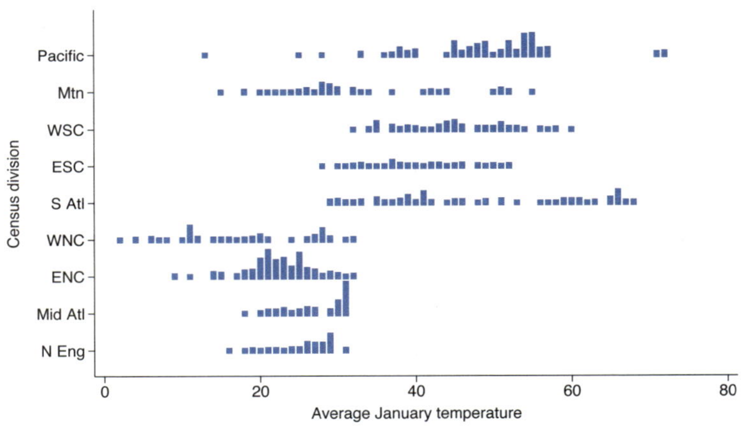

Figure 1.14. Ridgeline histogram using `stripplot`

Figure 1.14 demonstrates how several histograms can be plotted on a single graph. While individual detail in the y axis is lost, global and local distributional comparisons across the different categories are easy and fast. For example, it is easy to identify that the last four regions in the above graph have much colder January temperatures than the first five.

For those unable to download and install community-contributed commands, the official command `dotplot` is an alternative method to produce ridgeline histograms. `dotplot` is based on the idea of showing a point symbol for each data value. The operation of `dotplot` is similar to that of `stripplot`; users specify a continuous variable (*var1*) and use the `over()` option to display a column dot plot for each value of the group variable (*var2*). The `msymbol()` option is used to convert standard circle markers to square markers, and the `msize()` option is used to adjust marker sizes for the graph. The `ny()` option is analogous to the `bins()` option in `histogram`. Figure 1.15 highlights an example with 50 bins.

Creating figure 1.15

```
. * Ridgeline histogram using dotplot command
. dotplot tempjan, over(division) msymbol(S) msize(small) ny(50)
```

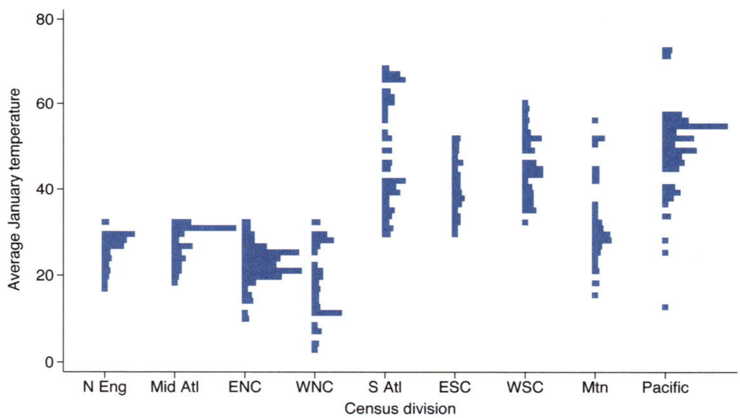

Figure 1.15. Ridgeline histogram using `dotplot`

Figure 1.15 presents an alternative method to create ridgeline histograms. It is similar in style and code to figure 1.14 but can be created using the official command `dotplot`. As mentioned earlier, `dotplot` only creates a vertical graph; it is not possible to rotate the graph and make it horizontal. When using `stripplot`, however, one can simply add the `vertical` option to rotate the graph.

Remember that figures 1.14 and 1.15 are created by careful manipulation of the marker size and bin-width options. Both options must be adjusted synchronously to create an image that has no overlapping bars or wide empty spaces between bars. Some trial and error is required to find the "right" values for both options.

2 Kernel density plots

Advantages

1. Have smoother curves that can depict the general distribution more effectively than histograms
2. Are visually appealing, especially with area shading
3. Highlight peaks and multimodality well
4. Plot multiple distributions better than histograms

Disadvantages

1. Are more complex to understand than histograms
2. Tend to show data where none exist (bias in boundaries is common)
3. Have shapes that can significantly change depending on bandwidth and kernel selections
4. Make it difficult to examine very local areas of distribution
5. Contain no information on summary statistics

Required community-contributed packages

To install the community-contributed commands that are required to follow along with this chapter completely, type

```
. ssc install distplot
. ssc install joyplot
```

2.1 Introduction

Like histograms, kernel density plots (also known as density plots or probability density plots) depict the distribution of data over a continuous interval. Such plots are similar in character to histograms but use kernel smoothing to plot values in a given bandwidth (the concept of bandwidth is analogous to a bin width). This empirical process smooths out the staircase-like representation of bars found in a histogram. It presents users with a smoother, curved graphical illustration of the data distribution, where mountains and valleys represent areas of high and low data concentration.

Like a histogram, a kernel density plot is a nonparametric way to illustrate the probability density function of a continuous variable. Furthermore, kernel density plots require user choices that can greatly affect the distribution display. Unlike histograms, kernel density plots often have at least two parameters that can be adjusted: 1) the bandwidth, which selects the width of the data to be smoothed, and 2) the kernel, which selects the method of data smoothing in any given bandwidth.

A vital advantage of a kernel density plot compared with a histogram is the default smoothness of the plot. Kernel density plots tend to be less disruptive to the eye, allowing casual observers to quickly pick out global features in the data. A fashionable trend is to shade the area underneath the density plot. This gives more contrast to the visualization and provides a sharp focus on peaks and valleys in the distribution.

Kernel density plots are also more helpful when graphing multiple distributions on one plot. Overlaid histograms tend to have many problems with color and line overlap, while overlaid density plots have fewer such problems, especially when area shading is forgone.

Kernel density plots are empirically more complex to understand than histograms, and multiple parameter choices can make choosing the right plot more difficult. Moreover, the smoothness of kernel density plots means that they will often not represent actual data values in a small local range. Boundary effects are common, where nonexistent data are shown in extreme parts of the distribution (for example, negative values for positive distributions). These characteristics can make kernel density plots poor choices when minute details of the distribution require examination; in this case, one should use histograms with many bins or spike plots.

As a rule of thumb, in any situation where a histogram is required, that histogram can often be replaced entirely by a density plot without significant loss of information or statistical precision.

2.2 A basic kernel density plot

Kernel density plots show how observations of one variable are distributed across the data range. In this chapter, we will use `citytemp.dta`, which contains temperature information on 956 US cities.

```
. * Load data
. sysuse citytemp
(City temperature data)
. summarize
```

Variable	Obs	Mean	Std. dev.	Min	Max
division	956	5.135983	2.694249	1	9
region	956	2.623431	1.057724	1	4
heatdd	953	4425.533	2199.605	0	10816
cooldd	953	1240.413	937.6679	0	4389
tempjan	954	35.74895	14.18813	2.2	72.6
tempjuly	954	75.05377	5.495036	58.1	93.6

The above is a summary of the dataset. The variables `tempjan` and `tempjuly` are both continuously measured and contain 954 observations on average city temperature in various regions of the United States. Listing or tabulating these observations individually is not a realistic way to understand how these data are distributed. A kernel density plot can help.

Basic kernel density plots can be accessed via the **kdensity** command. Only one variable can be specified, so the command typically is written in the form of **kdensity** *varname*. An example is shown in figure 2.1.

Creating figure 2.1

```
. * A basic kernel density plot
. kdensity tempjan
```

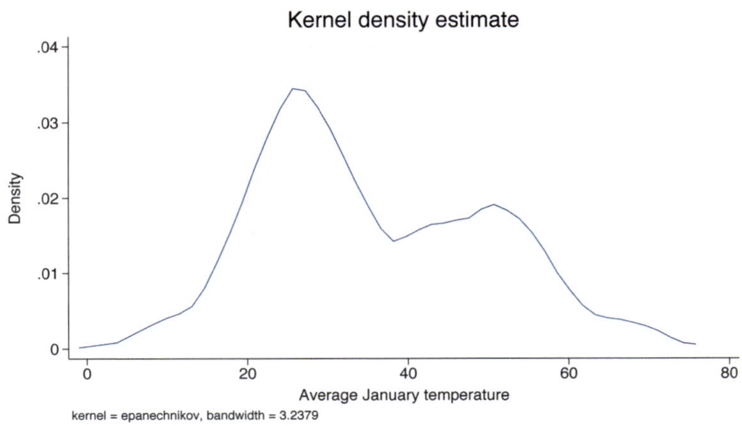

Figure 2.1. A basic kernel density plot

Figure 2.1 presents a basic kernel density plot of average city temperatures in January for various cities in the United States. Instead of the different height bars that a histogram gives, the plot now consists of a continuous, uninterrupted line that goes up and down as it moves along the x axis. High-density values represent higher data concentrations in that particular region of the x axis. As a reminder, density is a concept of relative frequency, and the total area under the graph is 1.

2.2.1 A shaded kernel density plot

Single kernel density plots can look somewhat empty. Most of the graph area will have the default background color, which is often white. It may be useful to shade the area underneath the density estimate. This gives better contrast to the distribution and allows general users to better identify areas of high or low data concentration across the data distribution. Shading can be achieved with the `recast()` option, which is an advanced option (`help advanced_options`) of the two-way plot family, of which `kdensity` is a part. This option determines a new plottype to transform the original graph into. Many plottypes are available (see `help graph_twoway`), and in this case, the `area` plottype will be chosen to shade the area underneath the density plot. This new plottype can use options from the `twoway area` command, and in figure 2.2 below, the `fcolor()` and `lcolor()` options are shown. `fcolor()` determines the fill color of the area shading, and `lcolor()` determines the color of the original density line. Below, a black line color and a light-blue[1] fill color are specified to offer a distinctive contrast.

Creating figure 2.2

```
. * Basic visualization options
. kdensity tempjan, recast(area) fcolor(eltblue) lcolor(black)
```

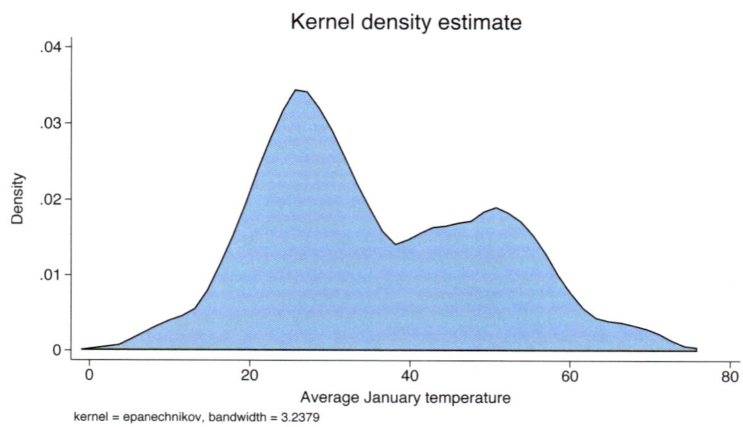

Figure 2.2. A shaded kernel density plot

1. The color scheme here is similar to *The Economist* magazine's.

Figure 2.2 presents a visually appealing alternative to figure 2.1. The shaded area underneath the kernel density line does a great job of showing the distribution of data. Two peaks reveal data concentrations near 30 and 50 degrees Fahrenheit.

2.2.2 Bandwidth choices

Similarly to histograms, the visualization of kernel density plots is primarily a function of the selected bandwidth. Higher bandwidths lead to oversmoothing, while low bandwidths lead to undersmoothing. Both can be problematic. Oversmoothing hides fine detail in local spaces of the distribution and can remove multimodalities, leading users to misidentify the number of peaks in the data. Undersmoothing leads to spiky plots that experience extreme ups and downs. Unlike histograms, which can reveal fine data structure when bin widths are very small, kernel density plots are based on smoothed functions even at small bandwidths. This makes them poor choices for examining small local areas of the distribution because some uncertainty remains about how the data are aggregated even within a small locale.

Bandwidths can be adjusted with the `bwidth()` option, which determines the half-width of the kernel. Figure 2.3 shows four examples of different bandwidth choices. Each individual graph is stored via the `name()` option and then merged together with the `graph combine` command.

Creating figure 2.3

```
. * Density plots with different bandwidths
. kdensity tempjan, bwidth(10) name(g1, replace)
. kdensity tempjan, bwidth(5) name(g2, replace)
. kdensity tempjan, bwidth(1) name(g3, replace)
. kdensity tempjan, bwidth(0.2) name(g4, replace)
. graph combine g1 g2 g3 g4
```

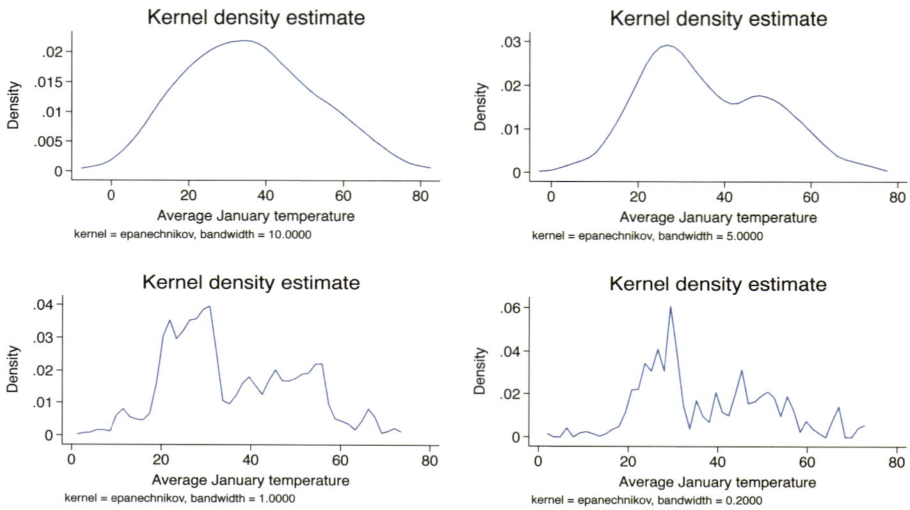

Figure 2.3. The effect of bandwidth choices on kernel density plots

Figure 2.3 clearly demonstrates that the choice of bandwidth matters. Many methods to numerically select bandwidths exist, but ultimately, there is no bandwidth selector that universally fits all situations. Users are strongly advised to generate multiple kernel density plots that vary the bandwidth up and down from its default and then examine the graphs carefully to determine which appears most appropriate for their analytical purposes. In some circumstances, the choice of bandwidth will be obvious. For example, if the running x variable is measured in days, then a good choice for bandwidth would be 7, 30, or 365, representing week, month, or year smoothing, respectively.

2.2.3 Kernel choices

Kernel choice is the second parameter that users can adjust in kernel density plots. The kernel determines how the data are weighted within a given bandwidth. Changes are made via the option `kernel()`, and users can select from the following kernels:

Kernel	Description
`epanechnikov`	Epanechnikov kernel function; the default
`epan2`	Alternative Epanechnikov kernel function
`biweight`	Biweight kernel function
`cosine`	Cosine trace kernel function
`gaussian`	Gaussian kernel function
`parzen`	Parzen kernel function
`rectangle`	Rectangle kernel function
`triangle`	Triangle kernel function

Figure 2.4 below demonstrates all eight kernel choices. Individual kernel plots are combined into one graph by using the `graph combine` command. To reduce clutter, the title is suppressed in the individual graphs via the option `title("")`.

Creating figure 2.4

```
. * Density plots with different kernels
. kdensity tempjan, title("") kernel(epanechnikov) name(g1, replace)
. kdensity tempjan, title("") kernel(epan2) name(g2, replace)
. kdensity tempjan, title("") kernel(biweight) name(g3, replace)
. kdensity tempjan, title("") kernel(cosine) name(g4, replace)
. kdensity tempjan, title("") kernel(gaussian) name(g5, replace)
. kdensity tempjan, title("") kernel(parzen) name(g6, replace)
. kdensity tempjan, title("") kernel(rectangle) name(g7, replace)
. kdensity tempjan, title("") kernel(triangle) name(g8, replace)
. graph combine g1 g2 g3 g4 g5 g6 g7 g8
```

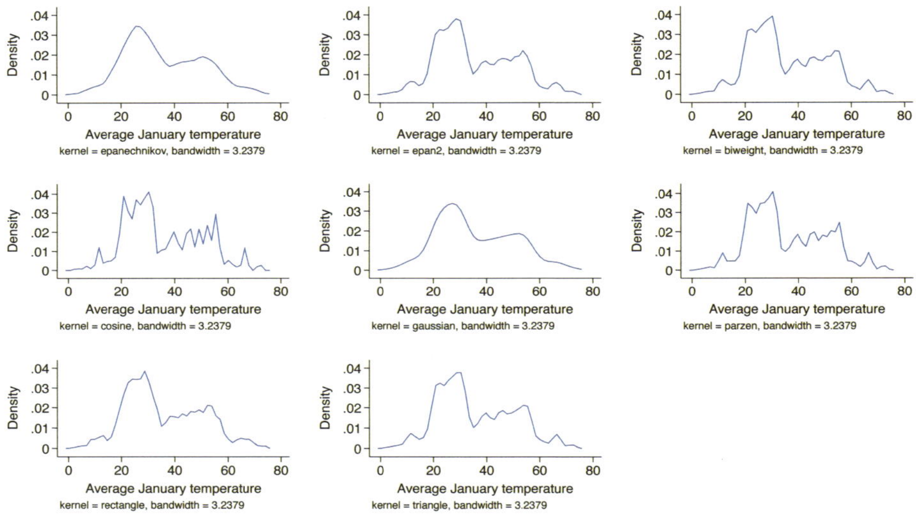

Figure 2.4. The effect of kernel choices on kernel density plots

Figure 2.4 shows how kernel choices affect the rendition of the kernel density plots. Generally, the Epanechnikov kernel (top-left graph) and Gaussian kernel (center graph) result in smoother visualizations. In most cases, the choice of bandwidth is more important than the choice of kernel. Users should keep the default Epanechnikov kernel in most applications, though there can be reasons to use specific kernels. For example, the rectangular kernel gives equal weight to all observations in its entire bandwidth, in contrast to the other kernels, which tend to give less weight to observations further away from the center of the bandwidth. This can be useful if one is examining discrete-time data, for example, the running x-variable increments in seconds or minutes, and one wishes to weigh all values within such discrete time spans equally.

2.3 Cumulative distribution plots

Another way to present distributions is via cumulative distribution plots, which are sometimes known as cumulative distribution functions or empirical cumulative distribution functions. Cumulative distribution plots are the percentage of the sorted numbers plotted over the numbers themselves. They are integrals of standard density plots; rather than showing the fraction of data points at a given value of x, cumulative distribution plots show the sum of data points up to that point of x.

An advantage of cumulative distribution plots is that they offer a direct reading of key quantitative summary information such as quartiles, medians, minimums, and maximums. It is also easy to detect outliers because they will show up as long horizontal bars in the plot. In addition, because these plots emphasize the distribution's global features, they are less sensitive to bandwidth and kernel choices compared with standard kernel density plots. However, they are more complex to interpret because levels and gradients must often be examined jointly, which can be problematic for laypeople or those not well versed in statistics.

It is not possible to create easy cumulative distribution plots by using only one command from official Stata. While cumulative distribution plots can be created via complex usage of the `cumul`, `stack`, and `line` commands, a more convenient way is offered via the community-contributed command `distplot` (Cox 1998a). (A similar community-contributed command, `cdfplot` [Mander 2005a], is also available.)

`distplot` plots the cumulative distribution function or functions for the variables in the variable list. It can take multiple variables at once and contains many useful options. A basic example is shown below in figure 2.5.

Creating figure 2.5

```
. * Cumulative distribution plot
. distplot tempjan
```

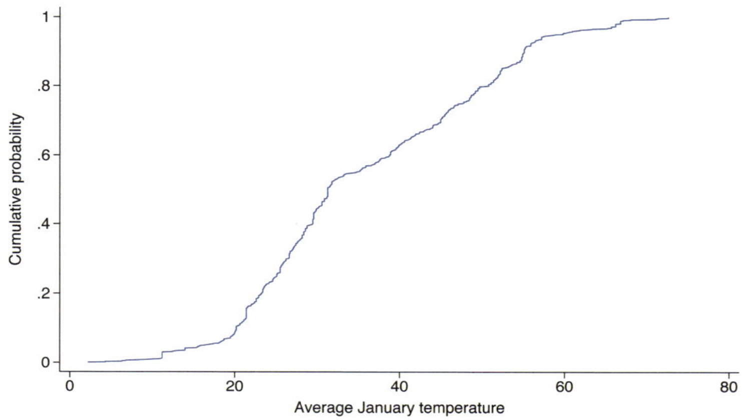

Figure 2.5. Cumulative distribution plot

Figure 2.5 presents a basic cumulative distribution plot. The y axis now measures cumulative distribution, which will always be maximized at 1. Because the cumulative distribution plot is the integral of a probability density plot, its gradients represent the values of a standard density plot. In other words, by examining the gradients along the x axis, users can infer a kernel density plot.

One of the main advantages of the cumulative distribution plot is that information like minimum, maximum, median, and percentiles can be read directly from the graph. For example, the 20th percentile of the average January temperature distribution is about 23 degrees Fahrenheit.

`distplot` has various useful options that can modify the representation of the standard cumulative distribution plot. Figure 2.6 below highlights three useful options in four graphs. The `frequency` option specifies the calculation of cumulative frequency instead of the default cumulative probability, the `reverse` option specifies the plotting of the reverse distribution function, and the `recast()` option offers different visualization styles. The spike and area styles are exemplified below.

Creating figure 2.6

```
. * Cumulative distribution plots with options
. distplot tempjan, frequency name(g1, replace)
. distplot tempjan, reverse name(g2, replace)
. distplot tempjan, recast(spike) name(g3, replace)
. distplot tempjan, recast(area) fcolor(eltblue) lcolor(black) ///
> name(g4, replace)
. graph combine g1 g2 g3 g4
```

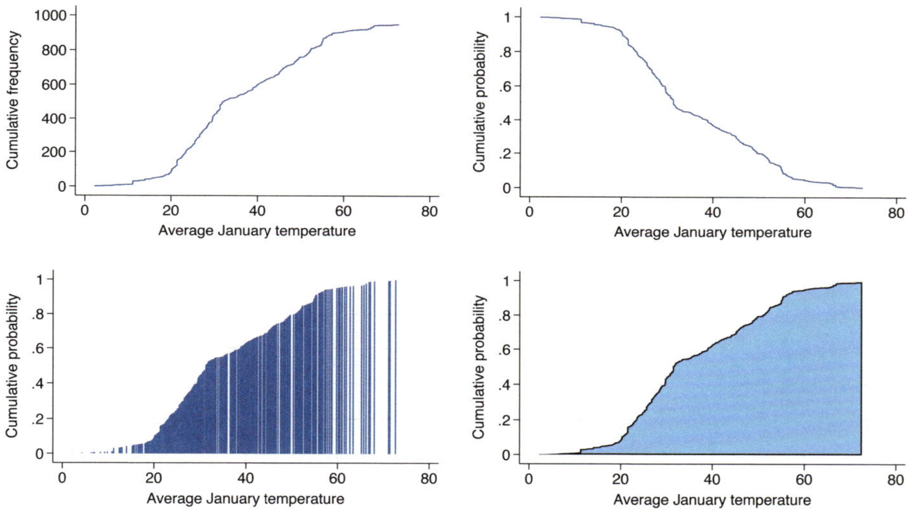

Figure 2.6. Useful `distplot` options

2.4 Multiple kernel density plots

Users may wish to compare multiple kernel density distributions. Thankfully, this is easier to accomplish with kernel density plots than with histograms. Basic kernel density plots tend to have less graphical overlap of density lines than do histogram bin lines, making them more suitable for plotting multiple distributions on a single graph.

There are several ways to create multiple kernel density plots. In Stata, this can be accomplished with the `twoway` version of the `kdensity` command, `twoway kdensity`. The standalone `kdensity` command is not able to draw multiple kernel density plots; the `twoway` command must be used to overlay multiple kernel density plots. Figure 2.7 below demonstrates a basic-use case using the `by()` option to graph multiple kernel density plots over the variable `region` in a side-by-side comparison. Note that variables entered into the `by()` option should be categorical.

Creating figure 2.7

```
. * Multiple kernel density plots
. twoway (kdensity tempjan, by(region))
```

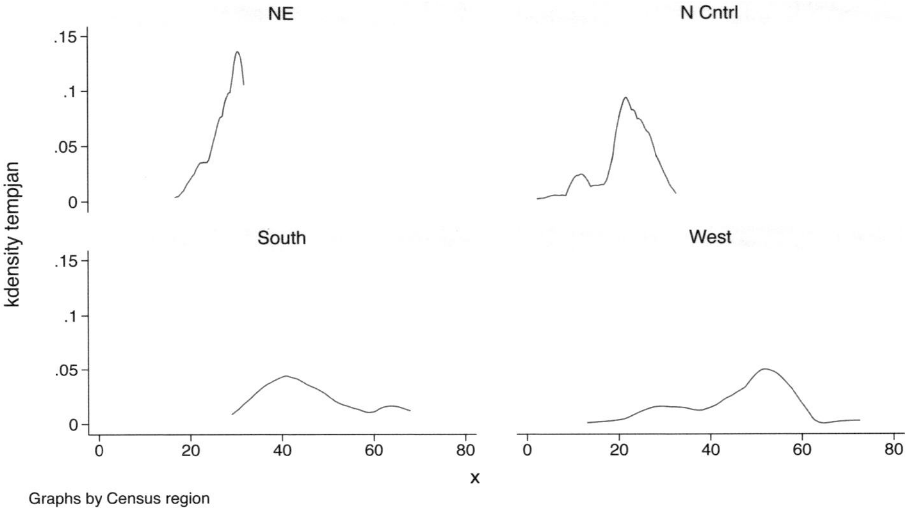

Figure 2.7. Multiple kernel density plots with `by()`

Figure 2.7 presents a basic multiple kernel density plot using the `by()` option. However, detailed local comparison between distributions is made difficult by the side-by-side comparison of `by()`. An alternative visualization method of figure 2.7 is to use `if` qualifiers and overlay multiple `kdensity` plots on each other via the `twoway` wrapper. This will result in code set up in the form `twoway (kdensity` *var1* `if` *groupvar* `== 1)` `(kdensity` *var1* `if` *groupvar* `== 2)`. This is exemplified below in figure 2.8, which uses the four categories in the variable `region` as `if` qualifiers.

Creating figure 2.8

```
. * Multiple kdensity plots with if qualifiers
. twoway (kdensity tempjan if region == 1)  ///
> (kdensity tempjan if region == 2)         ///
> (kdensity tempjan if region == 3)         ///
> (kdensity tempjan if region == 4)
```

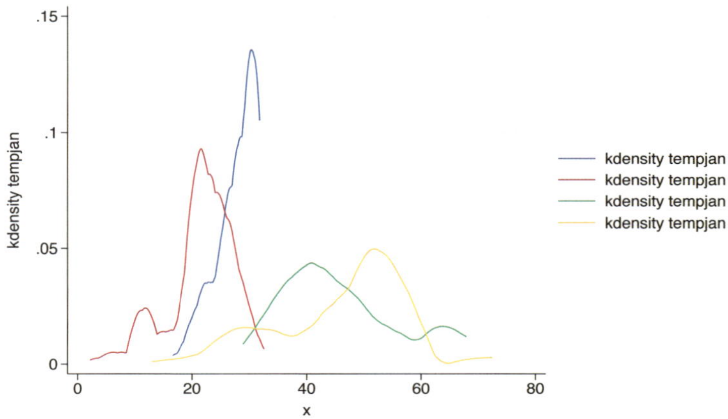

Figure 2.8. Multiple kernel density plots with `if`

Figure 2.8 is an overlaid version of figure 2.7. This graph is more useful in direct comparison of kernel density distributions. Note that legend labeling is not done by default and must be coded or manually edited with the Graph Editor.

The abrupt end of some kernel density functions can look odd. If not too many kernel density graphs are overlaid, coloring can be an interesting option to better emphasize different density plots. Figure 2.9 below demonstrates how the `lcolor()`, `fcolor()`, and `recast()` options can be used to create visually appealing density plots. Colors are made transparent via the `%` modifier.

Creating figure 2.9

```
. * Multiple kdensity plots with if qualifiers and color
. twoway (kdensity tempjan if region == 1, lcolor(black) ///
>     fcolor(ltblue%50) recast(area))                    ///
> (kdensity tempjan if region == 2, lcolor(black)        ///
>     fcolor(erose%50) recast(area))                     ///
> (kdensity tempjan if region == 3, lcolor(black)        ///
>     fcolor(emerald%50) recast(area))                   ///
> (kdensity tempjan if region == 4, lcolor(black)        ///
>     fcolor(sand%50) recast(area))
```

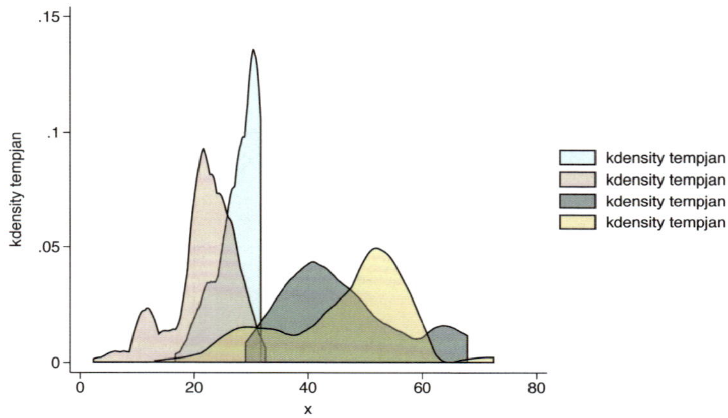

Figure 2.9. Multiple shaded kernel density plots

Figure 2.9 demonstrates how to create overlapping kernel density plots with area shading. Using the % modifier to create transparent colors makes it easier to identify individual plots; however, overlapping too many transparent colors creates new subcolors that can cause clutter and confusion. Moreover, higher levels of transparency reduce the richness of colors, making them (and the relevant distributions) less distinct. Users should experiment with color and opacity choices to find visualizations that suit them.

2.5 Mirrored kernel density plots

When only two kernel density plots need to be compared, a mirror plot is a robust alternative to overlaying plottypes. Mirror plots retain one x axis but split the y axis into top and bottom sections to avoid overplotting. Unfortunately, there is no community-contributed command that creates mirrored kernel density plots. One must thus use various options to customize their graphs and then merge two graphs to create a mirror plot.

In the case of kernel density plots, the code below demonstrates one possible approach. In figure 2.10, two `kdensity` plots are created. Area and color options are used to create an interesting visual style. Moreover, the option `ylabel()` is used to enforce two common y-axis scales. In the first density plot, the option `xscale(alt)` is used to place the x axis at the top. Titles and notes are suppressed. In the second density plot, the option `yscale(reverse)` inverts the plot. Both graphs are stored by using the `name()` option, and these named graphs are then merged with the `graph combine` command. The option `imargin()` sets graph margins to their minimums (left, right, bottom, top), and the option `cols(1)` ensures the second graph is appended vertically. The result is a basic mirrored kernel density plot.

Creating figure 2.10

```
. * Mirrored density plot
. twoway (kdensity tempjan, recast(area) fcolor(ltblue)                ///
> ylabel(0(0.02)0.1) name(g1, replace) xscale(alt) xtitle("")          ///
> title("") note(""))
. twoway (kdensity tempjuly, yscale(reverse) fcolor(erose)             ///
> ylabel(0(0.02)0.1) recast(area) name(g2, replace) xscale(off)        ///
> xtitle() title("") note(""))
. graph combine g1 g2, imargin(0 0 0 0) cols(1)
```

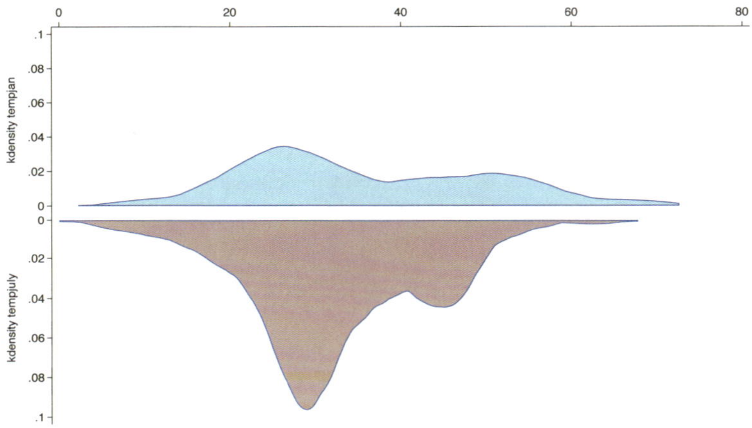

Figure 2.10. Mirrored kernel density plot

Figure 2.10 presents a simple mirrored kernel density plot. Such graphs are useful for comparing two distributions because they do not overlap but occupy the same graph space and share one x axis. This makes local comparison fast and easy. However, they are more complex to create and require much code customization to be produced.

2.6 Ridgeline kernel density plots

If many kernel density plots are required, then a ridgeline kernel density plot (also known as a joy plot) is a possible solution. Rather than merging two density plots onto a common y axis and mirroring one plot, ridgeline plots split each density plot onto a separate y axis. To an extent, ridgeline plots combine elements of both mirrored density plots and overlaid density plots. The y axis now serves the dual purpose of identifying both numeric scale (density, frequency, etc.) and categorical identification.

The primary advantage of a ridgeline plot is its ability to show many different kernel density distributions on a single plot. This, in turn, makes it easy to compare many distributions. A disadvantage is that the y axis now has a dual-purpose role (magnitude of observations and group identifier), making extracting detailed density statistics difficult. Essentially, numerical precision is sacrificed for multiple plots. Ridgeline plots also have a natural group limit. They are useful for plotting about 5 to 20 kernel density plots; beyond that, users will likely experience visual problems because detail is lost or individual plots overlap too much.

The community-contributed command `joyplot` (Naqvi 2023) creates ridgeline density plots. This command has many options, but default usage is relatively easy. Simply specify the `by()` option with a categorical variable that contains the relevant groups. Figure 2.11 demonstrates an example with the variable **`division`**, which contains nine groups.

Creating figure 2.11

```
. * Ridgeline plot
. joyplot tempjan, by(division)
```

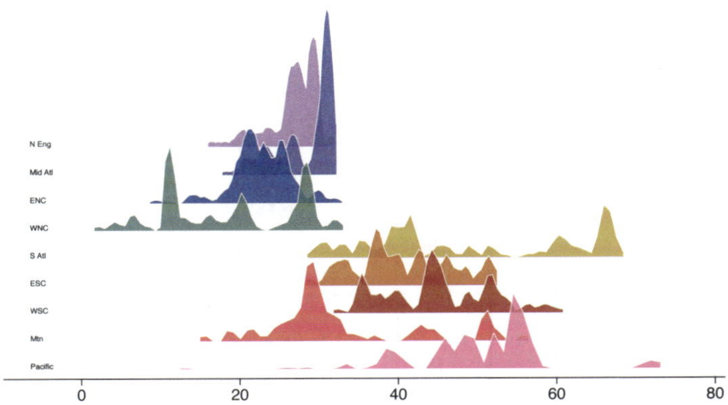

Figure 2.11. Ridgeline kernel density plot

Figure 2.11 presents a ridgeline kernel density plot of average city temperature across nine different geographies in the United States. The default color scheme is `Viridis`, but this can be easily changed (although users will need to install `schemepack` [Naqvi 2021]). Overall, the graph gives a good overview of how temperature is distributed in each geography.

One problem with the default setup of `joyplot` is that there remains a large amount of overlap between the densities. This can be adjusted with the `overlap()` option. The default is `overlap(6)`, and a smaller number will create less overlap. The `bwidth()` option allows users to define custom bandwidths; the default is `bwidth(0.1)`. Also important is the `normalize(global)` option, which normalizes the density heights by the global value and ensures that the heights of each plot are referenced to the overall maximum of the y variable. This makes groups comparable. Using all three additional options results in figure 2.12.

Creating figure 2.12

```
. * Ridgeline plot options
. joyplot tempjan, by(division) bwidth(1) overlap(4) normalize(global)
```

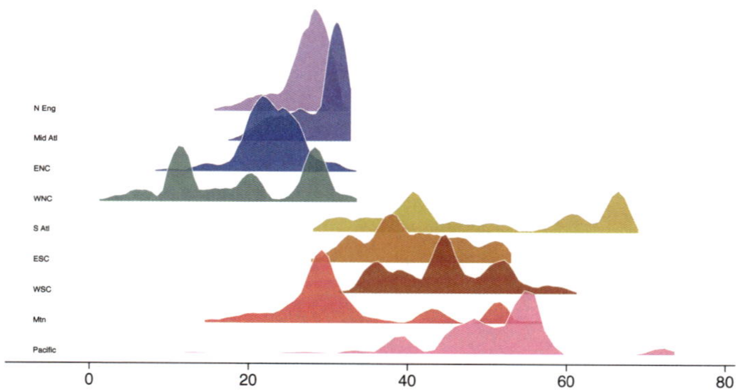

Figure 2.12. Ridgeline kernel density plot with less overlap

Figure 2.12 fixes the visual problems of figure 2.11. The various density plots are now normalized, allowing for easier comparison across each density. In addition, there is less overlap. Ridgeline plots are useful visualizations when dealing with many kernel density plots.

3 Box plots

Advantages

1. Provide a clear and simple overview of the distribution
2. Are great for comparing many distributions
3. Provide summary information on key statistics
4. Highlight outliers
5. Complement histograms and density plots well
6. Show the exact range of data

Disadvantages

1. Lack detail on the shape of the distribution
2. Cannot identify multimodality
3. Do not have standardized outlier calculation
4. Are less useful for single distributions

Required community-contributed package

To install the community-contributed command that is required to follow along with this chapter completely, type

```
. ssc install histbox
```

3.1 Introduction

Box plots, also called box-and-whisker plots, are common visualization techniques that are primarily used to compare multiple univariate distributions. Box plots are distributional plots by design. However, unlike histograms and density plots, they massively reduce the available information and concentrate only on a few select characteristics of the data distribution by reducing the visualization of the distribution to a simple box with emerging whiskers that extend from the box. This visualization provides a straightforward depiction of any given distribution, the primary advantage being simplicity. Many boxes can be stacked next to each other to compare multiple distributions. This stacking advantage means that box plots can often provide visualizations of many more distributions than histograms or kernel density plots on a single graph.

Essentially, box plots are all about quartiles within that box. They graphically display the five-number summary of the distribution. These numbers are the minimum, first quartile (or 25th percentile), median (or 50th percentile), third quartile (or 75th percentile), and maximum. Whiskers often, but not always, represent the minimum and maximum values of the distribution. In some box plots, there are also data markers beyond the whiskers. These are outliers, and most box plots determine that outliers are data points that lie 1.5 times the interquartile range away from the upper or lower quartile. However, there is no consensus around this metric, and other box plots may represent outliers as the last 2% or even not represent outliers. Users should be aware of this and be cautious when using different commands or software programs to create box plots.

A more serious disadvantage of box plots is their inability to display local details of the data distribution. The boxlike nature of visualization does not contain peaks or valleys, making it impossible to identify multimodality in a data distribution. In addition, a minor disadvantage is that, by default, box plots show quartile values, including the median. However, laypeople and nonstatistical users often confuse the median with the mean, especially when it is not noted on the graph.

Box plots are generally about simplicity and being easy to understand. They are excellent choices for comparing many distributions but carry little advantage over histograms or kernel density plots when plotting single data distributions.

3.2 A basic box plot

Box plots illustrate how one variable's observations are distributed across the data range. `citytemp.dta` is used in this chapter. It contains temperature information on 956 US cities. A summary of the dataset is shown below.

```
. * Load data
. sysuse citytemp
(City temperature data)

. summarize

    Variable |        Obs        Mean    Std. dev.       Min        Max
-------------+---------------------------------------------------------
    division |        956    5.135983    2.694249          1          9
      region |        956    2.623431    1.057724          1          4
      heatdd |        953    4425.533    2199.605          0      10816
      cooldd |        953    1240.413    937.6679          0       4389
     tempjan |        954    35.74895    14.18813        2.2       72.6
-------------+---------------------------------------------------------
    tempjuly |        954    75.05377    5.495036       58.1       93.6
```

The variables **tempjan** and **tempjuly** are continuously measured temperature data ranging between 2.2 and 93.6 degrees Fahrenheit. The data are measured across 956 US cities that are spread across 4 regions and 9 divisions. Listing or tabulating these observations individually is not a realistic way to understand how these data are distributed. A box plot can help, especially when comparing these data over geographic areas.

Basic box plots are created with the `graph box` command. Multiple variables can be specified, but to illustrate one variable distribution, the command should be written in the form of `graph box` *varname*. An example is shown in figure 3.1.

Creating figure 3.1

```
. * Basic box plot
. graph box tempjan
```

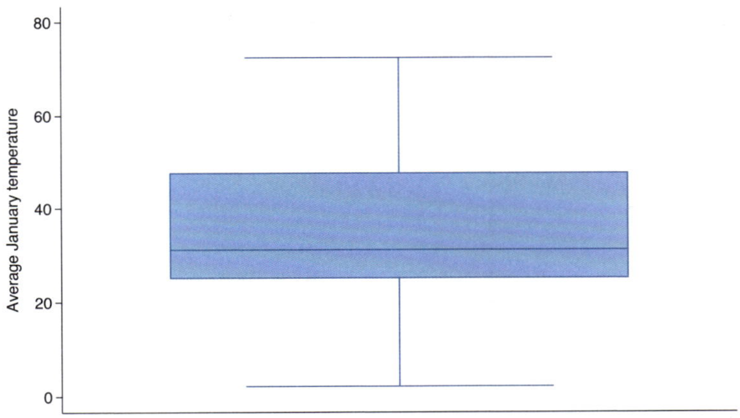

Figure 3.1. A basic box plot

Figure 3.1 presents a basic box plot of average January temperatures. The smallest data value can be found at the bottom whisker. Upward, the actual box begins with the 25th percentile (or first quartile) measurement. Next is the median, followed by the 75th percentile (or third quartile), which ends the box. Finally, the plot finishes with the top whisker, which marks the largest data value. In this case, users can quickly identify that the "middle half" (also called the interquartile range) of the data (the 25th to 75th percentile) is distributed between temperatures of about 25 to 45 degrees Fahrenheit. The median is around 30 degrees.

Not all box plots finish with whiskers; in some box plots, there are data beyond the whiskers. These will be represented as points. Most box plots determine that outliers are data values that lie 1.5 times the interquartile range away from the upper or lower quartile. Figure 3.2 below presents an example with data outliers.

Creating figure 3.2

```
. * Basic box plot with outliers
. graph box tempjan if region == 1
```

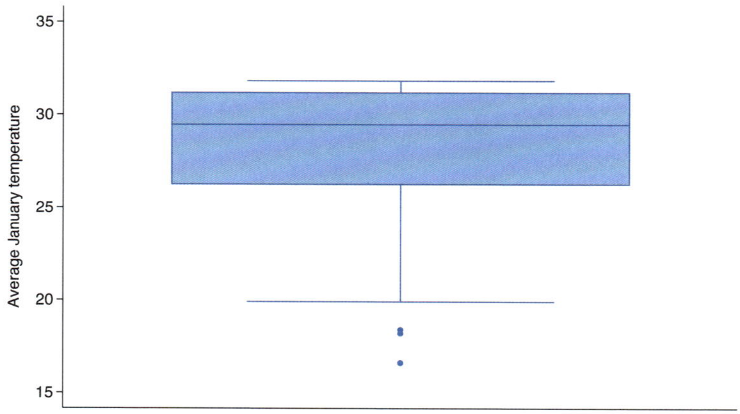

Figure 3.2. Box plot with outliers

Figure 3.2 shows outliers presented as points. Data that are 1.5 times the height of the box away from the box are determined to be outlying data points. In this case, three data values are marked as outliers.

3.2.1 Box plot options

`graph box` has many *boxlook_options* (see `help graph box##boxlook_options`) that allow users to customize their plot. Four common options are presented below. The `medtype()` option defines the method for indicating the median within the box. By default, the median is a line, but it can also be represented as a marker (`marker`) or a custom line (`cline`). The `box(#)` option allows users to specify how the #th box looks using *barlook_options* (see `help barlook_options`). An example with the fill color rose[1] (`fcolor(erose)`) is shown in figure 3.3. The `cwhiskers` option allows users to add whiskers and then customize their visualization by adding the `lines()` option. Below, an example of dashed whiskers is presented.

Creating figure 3.3

```
. * boxlook_options
. graph box tempjan, medtype(marker) box(1, fcolor(erose))  ///
> cwhiskers lines(lpattern(dash))
```

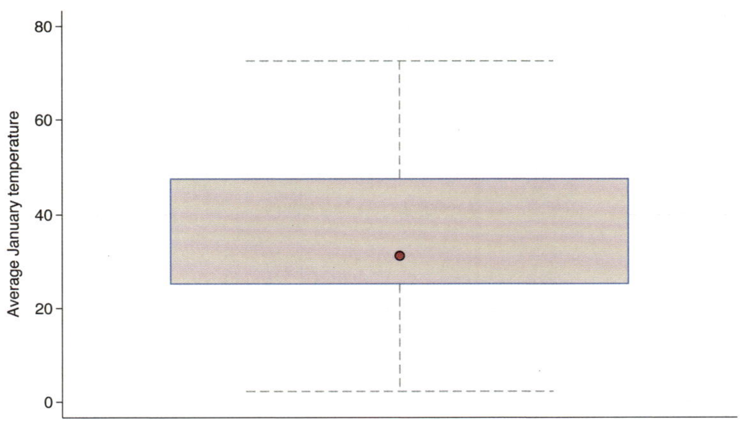

Figure 3.3. Box plot look options

Figure 3.3 is an example of a custom box plot using the *boxlook_options*. Many more customization methods are available.

1. The color scheme here is similar to *The Economist* magazine's.

3.3 Horizontal box plots

The "twin" command `graph hbox` creates horizontal box plots. It works exactly the same as `graph box` and is underpinned by the same options. An example is shown below in figure 3.4.

Creating figure 3.4

```
. * Horizontal box plot
. graph hbox tempjan
```

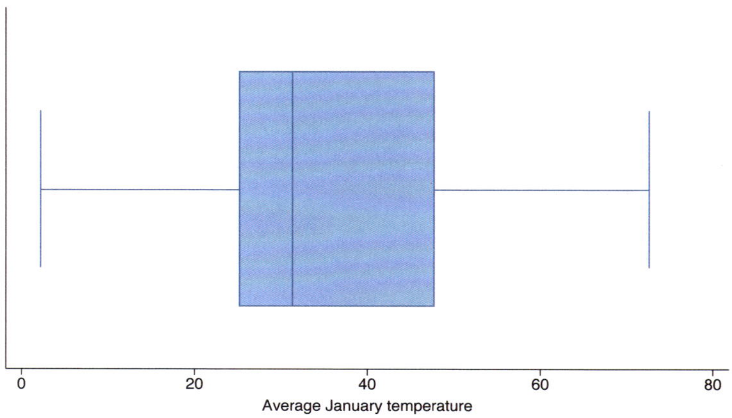

Figure 3.4. Horizontal box plot

When examining single or even multiple distributions, the choice between vertical and horizontal box plots often matters little. However, horizontal box plots gain an advantage when long label names are used. In that case, a horizontal box plot with label names on the horizontal axis allows easier reading (left to right). This avoids users having to stagger or angle their labels on the vertical axis.

3.4 Box plots with histograms

A serious disadvantage of box plots is their inability to signal information about local parts of the distribution, including whether data in the distribution are multimodal. This can be overcome by combining a histogram (or kernel density plot) with a box plot. This combines the advantages of both graphs; histograms are terrible at showing summary statistics, while box plots are poor at showing distributional detail. A simple way of combining them is to generate two graphs and append them vertically by using the `graph combine` command. A basic example is shown below in figure 3.5, which generates a histogram and a horizontal box plot. Both graphs are stored with the `name()` option. In this histogram, y-axis titles are suppressed via `ytitle("")`. The command `graph combine` merges both graphs. The option `cols(1)` ensures that the graphs are appended vertically.

Creating figure 3.5

```
. * Histogram with box plot
. graph hbox tempjan, name(g1, replace)
. histogram tempjan, name(g2, replace) ytitle("")
. graph combine g1 g2, cols(1) xcommon
```

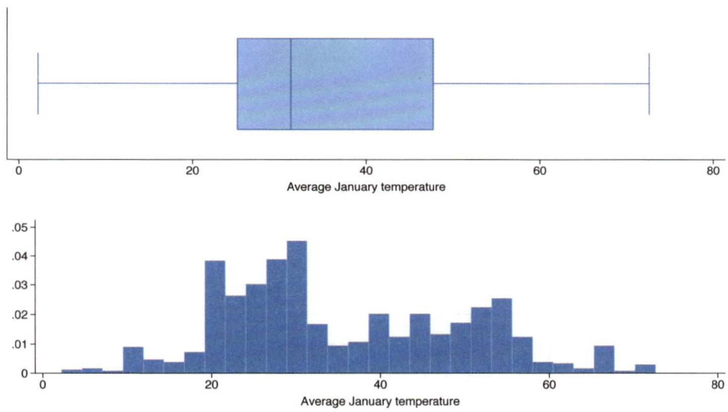

Figure 3.5. Box plot with histogram

Figure 3.5 demonstrates a simple combined plot that allows users to visually identify basic summary statistics and characteristics of the distribution via the box plot and examine local parts of the distribution via the histogram. However, there are two disadvantages to this custom approach. The first is that it requires building and merging multiple graphs, and the second is that the x-axis formatting is not quite identical by default. If users are not specifically interested in the density values, then they can specify the option `ylabel(none)`. This will fix the x-axis formatting in the combined graph at the expense of the loss of some information. Relatively lengthy and complex code is required to fix the x-axis formatting while retaining the density values.

An easier option for combining the graphs is to use the community-contributed command `histbox` (Ender 2002), which creates a basic histogram with a horizontal box plot underneath it. The appeal of `histbox` is its simplicity, having relatively few options. The main option that users will want to specify is `bin()`, which sets the number of bins in the histogram. An example with 30 bins is demonstrated below in figure 3.6.

Creating figure 3.6

```
. * Community-contributed histogram with box plot
. histbox tempjan, bin(30)
```

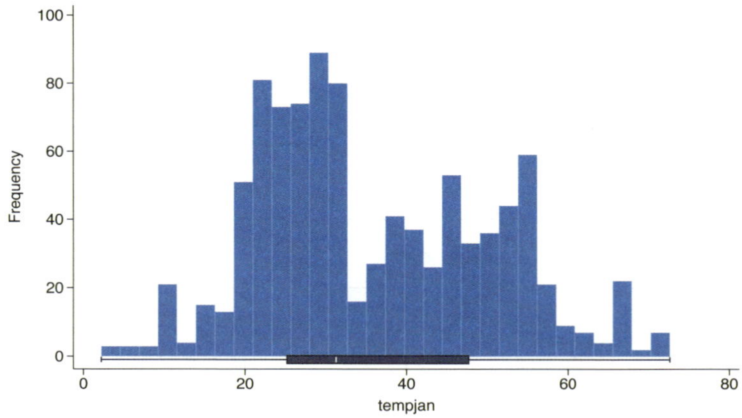

Figure 3.6. Box plot with histogram via `histbox`

Figure 3.6 shows how a histogram can be usefully merged with a box plot to take advantage of both visual techniques. The box plot is deemphasized compared with the histogram but still usefully identifies the interquartile range and percentiles of the distribution. A disadvantage of this technique is that it may be too complex for general users, and some consideration of the technical skills of any audience is required.

3.5 Multiple box plots

The primary purpose of box plots is to plot multiple distributions on a single graph. This is a key advantage of box plots, and any use of box plots should automatically gravitate toward plotting many distributions. There are two methods to create multiple box plots using the `graph box` (or `graph hbox`) command. The first is to specify two or more y variables in the variable list. This will generally result in a command taking the form `graph box` *yvar1 yvar2*. A basic example with the variables `tempjan` and `tempjuly` is shown in figure 3.7.

Creating figure 3.7

```
. * Multiple box plots
. graph box tempjan tempjuly
```

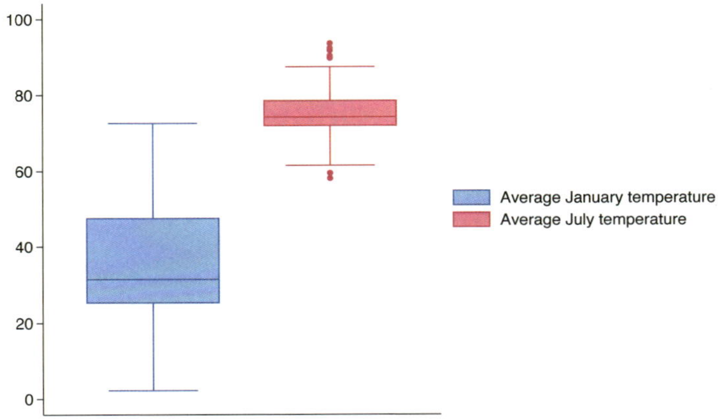

Figure 3.7. Box plot with two distributions

Figure 3.7 demonstrates the power of box plots. It is easy to visually identify that the median average July temperature is higher than the median average January temperature and that the variance in temperature is greater in January than in July. There is some overlap between the two distributions but not much. Most of the July temperatures are located near the maximum of the January temperatures.

Though informative, this graph still does not exemplify the full potential of box plots, which stand on their own when they are used to compare many different groups. The second method to display multiple box plots with the `graph box` command does harness this potential: the `over()` option, which takes a categorical variable. The `over()` option displays a box plot for each value of the categorical variable along with each y variable that was specified. An example with the nine-category variable `division` is shown in figure 3.8.

Creating figure 3.8

```
. * Multiple box plots over a categorical variable
. graph box tempjan tempjuly, over(division)
```

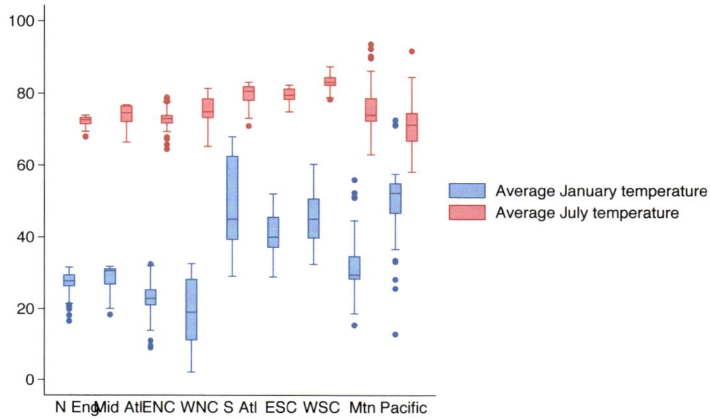

Figure 3.8. Box plot with 18 distributions

Figure 3.8 exemplifies the power of a box plot. It plots 18 different temperature distributions on a single graph and makes comparing them fast and easy. It is relatively straightforward to determine where higher or lower values are located in each distribution and how the spread of values differs by various regional divisions. Discerning this information would not be possible with histograms or kernel density plots and would challenge even ridgeline plots.

When using the `graph box` (or `graph hbox`) command with the `over()` option, an important additional visual option is `ascategory`, which changes the visual style of the plot by treating the first *yvar* as the first `over()` group. This option removes the legend and places the y variables as categories on the x or y axis. This results in a graph that contains a large amount of text because labels are repeated, but it can make group identification faster. This option is particularly useful if the data have hierarchical elements. An example is shown below in figure 3.9. Because of the significant amount of labeling involved, the command `graph hbox` and the option `scale(0.8)` are used to reduce text and marker sizes. Note that this is a general `graph` option and is not specific to the `graph hbox` command.

Creating figure 3.9

```
. * Multiple box plots with ascategory option
. graph hbox tempjan tempjuly, over(division) scale(0.8) ascategory
```

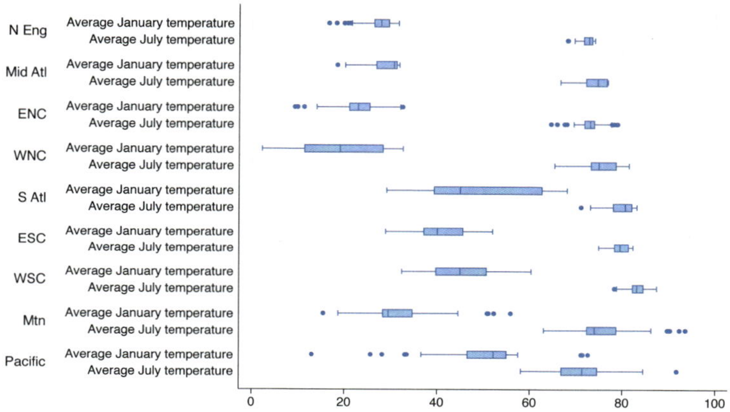

Figure 3.9. Box plot with no color legend

Figure 3.9 shows how the `ascategory` option can suppress color legends of the January and July temperature variables and instead place the relevant information directly on the y axis. This can be a useful way to visualize multiple box plots if monochrome color schemes are required or when users wish to repeat labeling information multiple times.

The `over()` option has access to various suboptions (*over_subopts*) that can be invoked to customize individual (or all) box plots. Users should examine these in detail in the help file (**help graph box##over_subopts**) because they can greatly alter the rendition of the plot. An example with the `sort(#)` suboption is shown in figure 3.10 below. This suboption sorts the various box plots in order of median values and allows users to rank their distributions.

Creating figure 3.10

```
. * Ordering boxes
. graph hbox tempjan, over(division, sort(1))
```

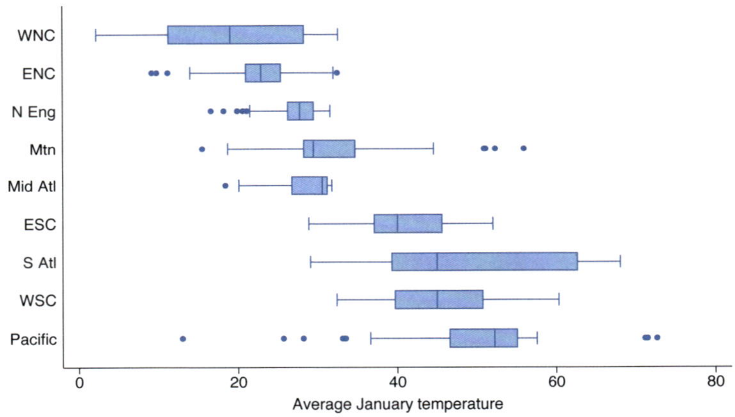

Figure 3.10. Box plot with sorting

4 Violin plots

Advantages

1. Combine density plots with box plots
2. Excel at comparing many distributions
3. Provide summary information on key statistics
4. Provide detail of distributional properties
5. Can identify multimodality

Disadvantages

1. Are complex to understand for general users
2. Require bandwidth and kernel choices
3. Are less useful for a single distribution

Required community-contributed package

To install the community-contributed command that is required to follow along with this chapter completely, type

```
. ssc install vioplot
```

4.1 Introduction

A violin plot is a combination plottype that merges the concept of box plots with a mirrored kernel density plot. A violin plot contains a box plot at its center that is surrounded by two kernel density plots, one on each side. The combination of these graphical elements often creates graphs that look like violins, hence the name violin plot.

Violin plots offer more detailed information compared with standard box plots. They display the full data distribution along with summary statistics, whereas box plots display only summary statistics. Moreover, using a mirrored density plot in the graph emphasizes peaks and valleys in the data distribution, allowing users to quickly identify how the data are distributed.

Violin plots are complex, so they are not frequently used in standard practice and can be more challenging for laypeople to understand. In addition, kernel density plots necessitate understanding bandwidth and kernel tradeoffs because different choices can result in significantly different graphs.

4.2 A basic violin plot

Violin plots provide visualizations of how observations of one continuous variable are distributed across the data range. In this chapter, we will use `citytemp.dta`, which contains temperature information on 956 US cities. A summary of this dataset is shown below.

```
. * Load data
. sysuse citytemp
(City temperature data)
. summarize
    Variable |        Obs        Mean    Std. dev.       Min        Max
-------------+---------------------------------------------------------
    division |        956    5.135983    2.694249          1          9
      region |        956    2.623431    1.057724          1          4
      heatdd |        953    4425.533    2199.605          0      10816
      cooldd |        953    1240.413    937.6679          0       4389
     tempjan |        954    35.74895    14.18813        2.2       72.6
-------------+---------------------------------------------------------
    tempjuly |        954    75.05377    5.495036       58.1       93.6
```

The variables `tempjan` and `tempjuly` measure temperature data in degrees Fahrenheit across 956 US cities. These data are collected over 4 regions and 9 divisions. Listing or tabulating these observations individually is not a feasible approach to understanding how these variables are distributed. A violin plot can help.

Violin plots cannot be created with a dedicated command in official Stata. The community-contributed command `vioplot` (Winter and Nichols 2008) must be installed. The `vioplot` command has syntax similar to `graph box`. One or multiple y variables can be specified, resulting in a basic command syntax that takes the form `vioplot` *varlist*. An example is shown below in figure 4.1.

Creating figure 4.1

```
. * A violin plot
. vioplot tempjan
```

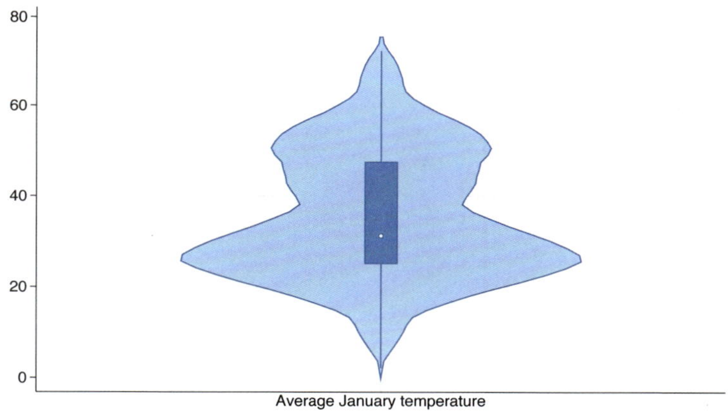

Figure 4.1. A basic violin plot

4.2.1 Custom bandwidths and kernels

Violin plots are composed of kernel density plots and box plots. To modify the density estimation, users can change the default kernel (`epanechnikov`) via the `kernel()` option. The option `bwidth()` allows users to adjust the half-width of the kernel. An example of a Gaussian kernel with a bandwidth of 10 is demonstrated below in figure 4.2.

Creating figure 4.2

```
. * Violin plot with kernel() option
. vioplot tempjan, kernel(gaussian) bwidth(10)
```

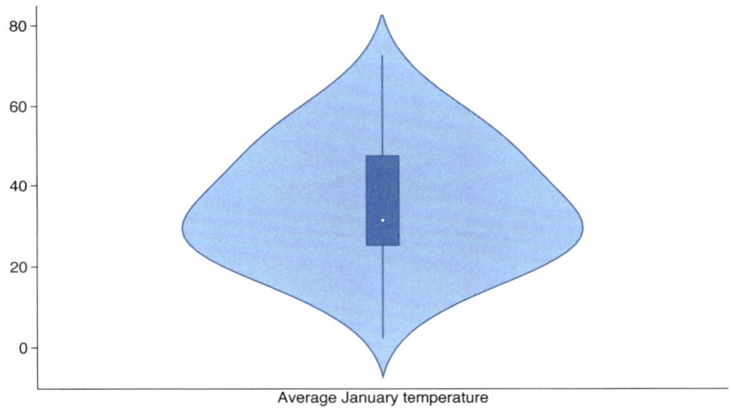

Figure 4.2. Violin plot with custom kernel and bandwidth

Figure 4.2 highlights how kernel and bandwidth choices can influence the visualization of the violin plot and the associated data distribution. Users should have some understanding of how kernel density plots operate and what their limitations are when creating violin plots. The violin plot in figure 4.2 suggests a unimodal data distribution, while figure 4.1 suggested a bimodal data distribution. Care should be taken when using kernel density options, and users should always adjust bandwidth options to explore how different bandwidth sizes affect visual presentation.

4.3 Horizontal violin plot

Like box plots, violin plots can be rotated to create a horizontal plot. This can be useful if long labels are associated with the data. The option `horizontal` creates a horizontal violin plot, as shown in figure 4.3.

Creating figure 4.3

```
. * A horizontal violin plot
. vioplot tempjan, horizontal
```

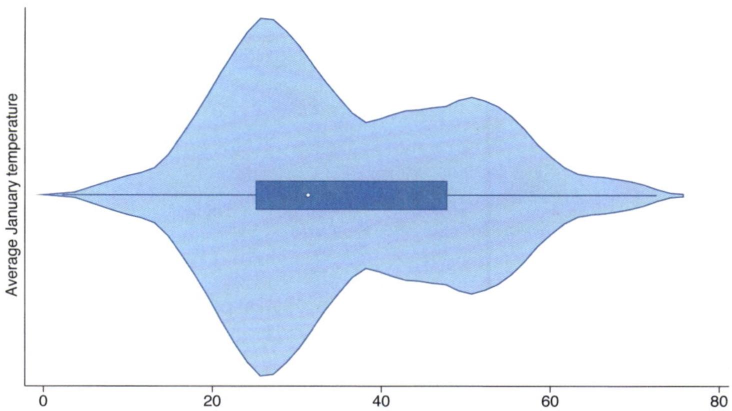

Figure 4.3. A horizontal violin plot

4.4 Multiple violin plots

Similarly to box plots, violin plots shine when they are used to compare many data distributions. Visually, violin plots merge the concept of mirrored ridgeline plots with box plots, giving users a powerful way to present multiple complex distributions. Similarly to the `graph box` command, two methods exist to create multiple box plots using the `vioplot` command. The first method is to add two or more *y* variables to the variable list. This will generally result in a command that takes the form `vioplot` *yvar1 yvar2*. The second method is to specify the `over()` option, in which a categorical variable is specified and violin plots are then repeated for each `over()` group. The two methods can be combined, allowing users to visualize multiple *y* variables over many groups. This is demonstrated below in figure 4.4.

Creating figure 4.4

```
. * Multiple violin plots
. vioplot tempjan tempjuly, over(division)
```

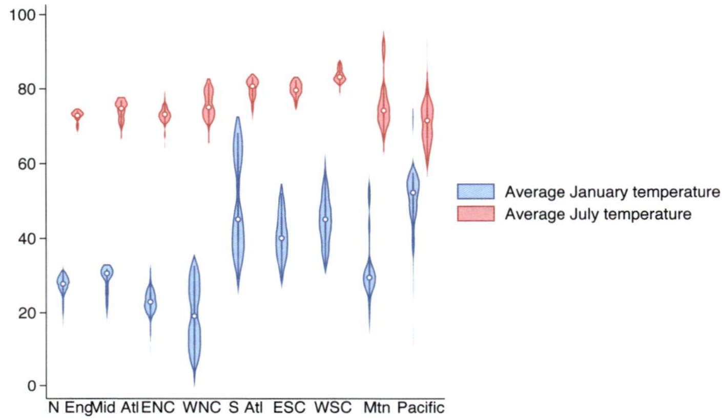

Figure 4.4. Multiple violin plots

Figure 4.4 presents two different temperature distributions for nine different divisions in the United States. While the shapes may look funny at first glance, the multiple violin plots are a compelling way to present complex distributional information. For example, the January temperature distribution in the East South Central (ESC) division is quite different from that in the South Atlantic (S Atl). The first is highly unimodal, and the latter is multimodal. Differences in variance can also be easily observed, with Pacific July temperatures being much more widely distributed than New England (N Eng) temperatures.

Users should remember that violin plots are not as capable of presenting many distributions as box plots. If 50 distributions need to be plotted and compared on a single graph, then a basic box plot is likely the better choice.

4.5 Custom widths

The `vioplot` command help file (`help vioplot`) details many options allowing users to modify the color and other aspects of the box and kernel density plots within violin plots. A useful pair of options is `barwidth()` and `dscale()`, which modify the width of the box plots and the density plots, respectively, as a percentage of the original. For example, a value of 500 in each will lead to a box or density depiction that is 5 times wider than the default. These options can be used to emphasize aspects of the violin plot. An example is shown in figure 4.5 below, where both bars and kernel densities are widened to create a "stockier" appearance.

Creating figure 4.5

```
. * Multiple violin plots with custom widths
. vioplot tempjan tempjuly, over(division) barwidth(500) dscale(200)
```

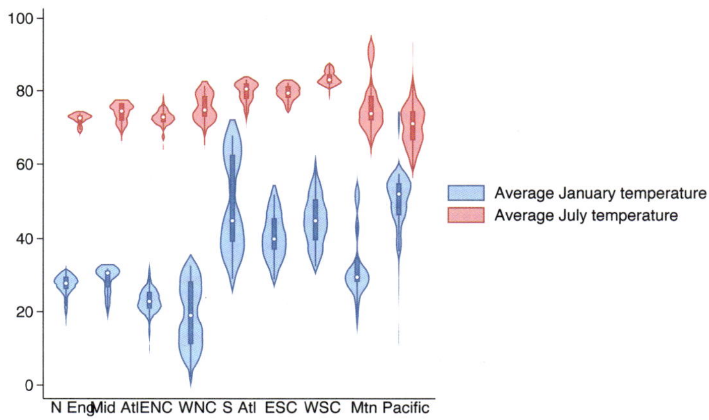

Figure 4.5. Violin plot with custom bar and density width

5 Dot plots

Advantages

1. Are easy to understand
2. Depict each data point
3. Provide a good distributional overview
4. Can identify outliers and clusters of data
5. Conserve numerical information

Disadvantages

1. Are only useful for small-to-medium samples
2. Can change appearance or display "fake" data markers because of binning choices

Required community-contributed packages

To install the community-contributed commands that are required to follow along with this chapter completely, type

```
. ssc install stripplot
. ssc install beamplot
```

5.1 Introduction

Dot plots provide visualizations of the distribution of a continuous variable by plotting data points on a simple scale. They produce a visualization with elements of a box plot, a histogram, and a kernel density plot. They are most closely related to stem-and-leaf plots.

A key feature of dot plots is that they represent every data value in the distribution via a marker. This is both an advantage and a disadvantage. The advantage is that the visualization of every data point provides an easy and intuitive way to understand the plot visualization because the raw data are directly translated into a visual image. In addition, dot plots retain significant numerical information about the underlying dataset because all data values are individually represented. This is not the case with histograms, box plots, or kernel density plots, where data values are hidden behind aggregate visualizations. Simply counting the number of markers in a dot plot reveals the size of the sample. In addition, outliers, gaps, and clusters of data in the distribution are easily identified because they will be represented by a higher or lower density of markers, which can be easily counted.

The disadvantage is that such plots do not work well with large datasets (say, 1,000 or more observations), because too many data points start to clutter the graph. Users should also remember that dot plots maintain the binning aspects of histograms and kernel plots. A bin size, sometimes known as a dot size, regulates how markers are binned and stacked. This implies that markers do not always accurately depict the underlying data point, which may confuse some readers of the graph. For example, integer binning (stacking) will visually suggest that all data values are integers, while most may contain decimal places. This is especially true when small bin sizes are used, resulting in a graph that creates the impression that every data marker has its own unique value, while in reality the visualization is simply tightly binned. Ultimately, the choice of bin width can greatly affect the final visualization, and users should always experiment with different bin-width sizes.

5.2 A basic dot plot

Dot plots visualize how observations of one continuous variable are distributed across the data range. `citytemp.dta` is used in this chapter. It contains the variables `tempjan` and `tempjuly`, which measure temperature data in degrees Fahrenheit across various US cities. The dataset contains 956 observations and also includes multiple region variables that identify 4 unique regions and 9 different divisions. The dataset is summarized below.

```
. * Load data
. sysuse citytemp
(City temperature data)
. summarize
    Variable |        Obs        Mean    Std. dev.       Min        Max
-------------+---------------------------------------------------------
    division |        956    5.135983    2.694249          1          9
      region |        956    2.623431    1.057724          1          4
      heatdd |        953    4425.533    2199.605          0      10816
      cooldd |        953    1240.413    937.6679          0       4389
     tempjan |        954    35.74895    14.18813        2.2       72.6
-------------+---------------------------------------------------------
    tempjuly |        954    75.05377    5.495036       58.1       93.6
```

One can use the official command `dotplot` to create a dot plot. `dotplot` can take a single variable name or multiple variable names, in which case dot plots are plotted over multiple columns in the same graph. Figure 5.1 below demonstrates a basic example with the variable `tempjan` and also includes the option `mean` to plot a horizontal line at the mean value of the distribution. This can be changed to `median` to plot the median value.

Creating figure 5.1

```
. * Basic dot plot
. dotplot tempjan, mean
```

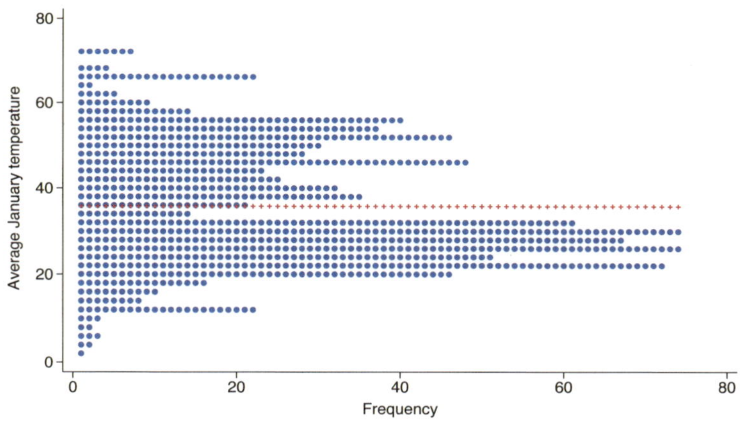

Figure 5.1. A basic dot plot

Figure 5.1 presents a basic dot plot. By default, Stata's dot plot aligns dots horizontally, resulting in the x and y axes flipped compared with traditional density or histogram plots. Frequency (not density!) is plotted on the x axis, while average January temperatures are plotted on the y axis. Each marker represents a unique data value. Markers are binned so that there are 35 individual dot bins by default, implying that each stack represents 1/35 of the total data range. Overall, the visualization in figure 5.1 does a good job of depicting the underlying data distribution. It is easy to identify that temperatures in January are centered around values of 30 degrees Fahrenheit and that there is a second peak near 50 degrees. Moreover, users immediately understand how many data points are in each temperature range because they can count the number of markers in each stack.

5.2.1 Adjusting stack width

Like histograms and kernel density plots, dot plots are binned distributional graphs. Markers are banded into data range groups and stacked on each other. This implies that marker values do not represent the underlying "true" data values. Stack (or bin) sizes can be adjusted via the `ny()` option, which regulates how many overall stacks (bins) there are in the graph. The default is `ny(35)`; numbers larger than this will lead to more detailed graphs, while smaller numbers will lead to less detailed graphs. Figure 5.2 below demonstrates an example with 15 stacks.

Creating figure 5.2

```
. * Dot plot with 15 bins
. dotplot tempjan, ny(15)
```

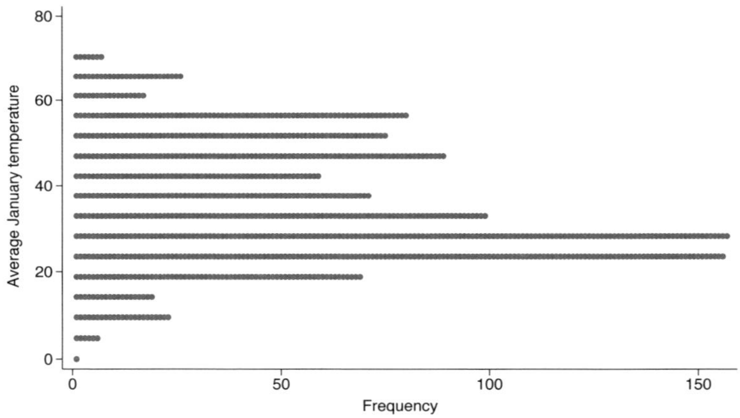

Figure 5.2. Dot plot with 15 stacks

Figure 5.2 demonstrates that stack widths matter. Compared with a histogram, a dot plot with smaller stack or bin widths has two distinct disadvantages. First, the number of individual markers in each stack can quickly become too big, resulting in overlap. This makes it hard to identify individual markers, defeating the purpose of the dot plot. Second, because stacks are based on marker sizes, which do not vary, white empty gaps start to appear between stacks. This is because markers do not adjust to fill the intervening void (unlike histogram bars) automatically. This, in turn, creates the impression of empty data space and uncertainty about which temperature data range each stack represents.

5.2.2 Adjusting marker size

The first problem can be partially fixed if the data size is not too large. In this example, the underlying data contain 954 observations, and by using marker options (see `help marker_options`), users can adjust the size of the individual dots. Figure 5.3 demonstrates an example where the options `msize()`, `msymbol()`, `mlwidth()`, and `mlcolor()` are used to adjust marker size, create hollow dots, reduce the marker line width, and color marker lines black.

Creating figure 5.3

```
. * Dot plot with marker options
. dotplot tempjan, ny(15) msize(*0.4) msymbol(Oh) mlwidth(*0.5)  ///
> mlcolor(black)
```

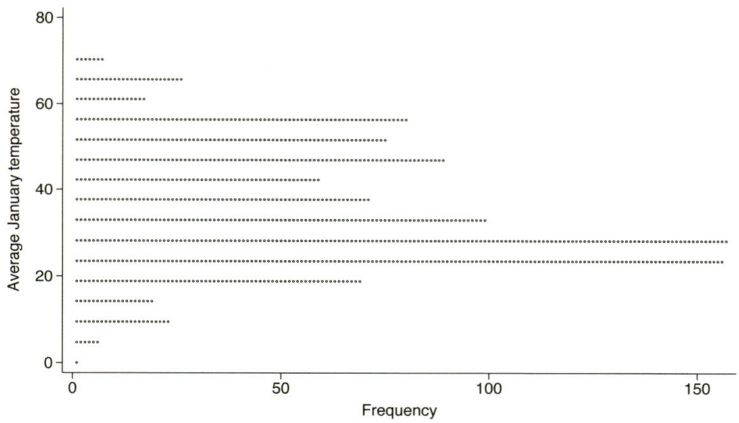

Figure 5.3. Dot plot with 15 stacks and custom markers

The result is a dot plot that can display every data point without overlap. However, a natural limit is quickly reached when observations go into the thousands. Moreover, the smaller marker sizes have exacerbated the white gaps between the dot stacks. The final graph continues to be somewhat unsatisfactory; users should avoid dot plots with few bins.

5.2.3 Dot plots without binning

Users may wonder how dot plots look without bins or, in other words, where each data marker represents its true underlying value. This can be achieved with the `nogroup` option, which provides visualizations of the actual values of the y variable. Stacking still occurs but only for perfectly identical values. The resulting graph is shown in figure 5.4.

Creating figure 5.4

```
. * Dot plot without grouping
. dotplot tempjan, nogroup
```

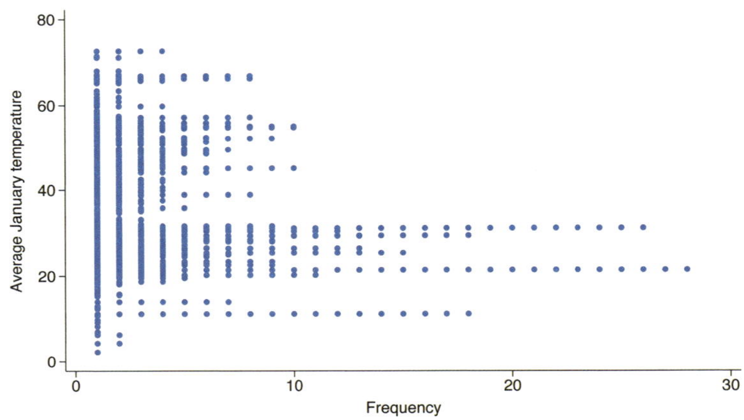

Figure 5.4. Dot plot without binning

Figure 5.4 presents a dot plot with no grouping or binning, meaning that most observations are placed at the 1 value on the frequency x axis. This results in significant overlap, making it hard to discern underlying distributional properties. However, there is some use for this kind of visualization. Any stacks that emerge from the data mass are duplicate observations that contain identical temperature values. While the `nogroup` option should not be used to examine general distributional properties for continuous variables, it can be a helpful tool when identifying duplicate observations is required. In this case, there appear to be certain temperature values that are dominant, such as 21.4 or 31.3. Users could modify this graph with the `mlabel()` option to identify which observations are duplicated.

5.3 Multiple dot plots

`dotplot` can plot multiple dot plots on a single graph, creating quasiridgeline plot structures. It can do this by specifying multiple *y* variables after the main command or via the `over()` option, which creates one dot plot for each value of a categorical variable specified in `over()`. However, the methods are mutually exclusive; it is not possible to combine multiple *y* variables with the `over()` option. A basic example with two *y* variables is shown below in figure 5.5.

Creating figure 5.5

```
. * Dot plots of two variables
. dotplot tempjan tempjuly
```

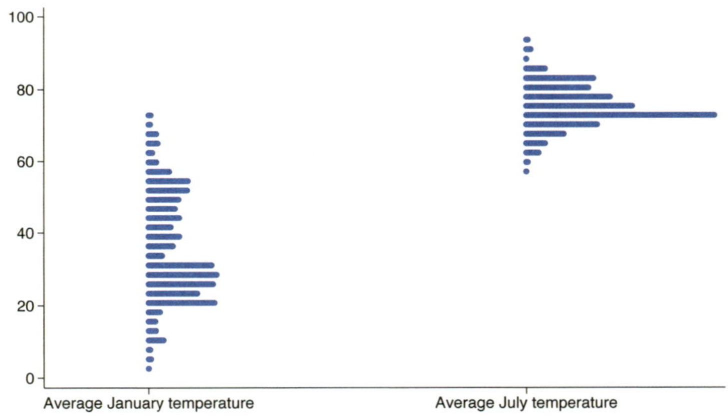

Figure 5.5. Multiple dot plots of two *yvars*

Figure 5.5 exemplifies multiple dot plots. Similarly to ridgeline plots, one axis represents both numerical scale and group identification. These plots are useful for comparing multiple distributions because they provide a simple but effective overview of the distributions. In this example, it is clear that July temperatures are higher than January temperatures and that the spread of January temperatures is greater than the spread of July temperatures.

A disadvantage of using dot plots in such a way is that individual markers quickly become overplotted, so it is not possible to identify individual observations. These plots now largely function as histograms instead of dot plots; if the `msymbol(S)` option were used (which modifies the marker shape into a square), then such plots would indeed look very similar to histograms.

Figure 5.6 demonstrates the `over()` option with the categorical variable `division`. This variable contains nine different geographies, and one dot plot will be created for each group.

Creating figure 5.6

```
. * Dot plots over groups
. dotplot tempjan, over(division)
```

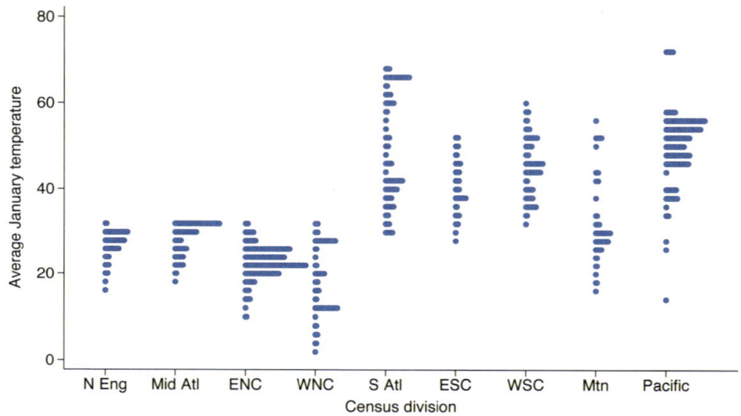

Figure 5.6. Multiple dot plots over nine groups

5.4 Horizontal dot plots

Users who wish to create vertically stacked dot plots, perhaps to mimic histograms better, cannot do so with the `dotplot` command. However, the community-contributed command `stripplot` (Cox 2003d) stacks dots vertically by default. The `stripplot` command displays data as several marks aligned with a single magnitude axis, similarly to `dotplot`. However, `stripplot` has various advanced options that make it a more powerful command than `dotplot`.

The `stripplot` command has syntax similar to `dotplot`. One or more y variables can be specified to create single or multiple dot plots, or the `over()` option can be used to create multiple plots over a categorical variable. (The mutual exclusivity between multiple y variables and `over()` remains.)

An example of multiple dot plots using the `over()` option is shown below in figure 5.7. The key to using `stripplot` as an advanced cousin of `dotplot` is the `stack` option, which specifies that data points with identical values are to be stacked. By default, there is no binning in `stripplot`, leading to dot plots without binning. Thus, users should add a custom bin width when using **stripplot**, which can be done with the **width()** option; a width of 1 is used in this example. In addition, the **msize()** option is used to reduce marker sizes to 60% of the original size. Note that even though marker size is reduced, there is still overlap because there are too many groups and observations.

Creating figure 5.7

```
. * stripplot command
. stripplot tempjan, over(division) stack width(1) msize(*0.6)
```

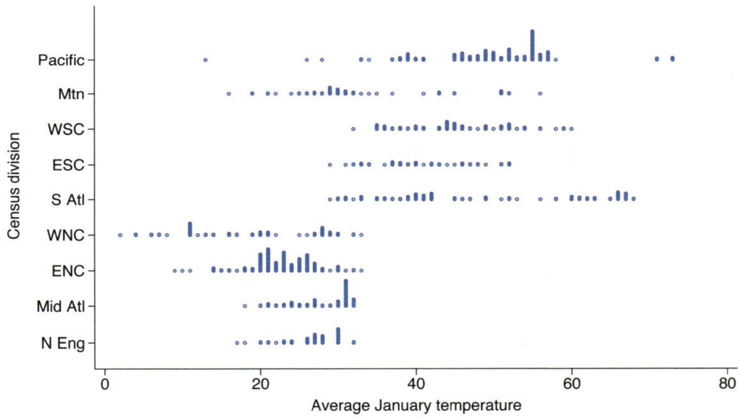

Figure 5.7. Multiple horizontal dot plots

Figure 5.7 is similar to figure 5.6 except that it has been rotated horizontally, making dot plot visualization more akin to a histogram. The horizontal aspect will likely be more user friendly because most people naturally examine distributions from left to right, not top to bottom.

5.5 Beam plots

One way to modify the basic horizontal dot plot shown in figure 5.7 is to use beams. Beam plots are standard dot plots positioned on beams extending from the minimum to maximum values for each element. The beams are balanced on a fulcrum, which, by default, is located at the mean of the data distribution. This visualization combines aspects of a box plot with those of a dot plot and provides users with a fast visual guide of the overall range of the data distribution and where the mean is located. The image of a beam plot is similar to a seesaw that contains weights distributed so that the seesaw is in perfect balance. This provides a powerful and intuitive visualization of where local data clusters are located in a distribution.

The community-contributed command **beamplot** (Cox 2005a) creates beam plots. It uses the same syntax as **dotplot** and **stripplot**, and it can be specified either with one or multiple y variables or with the **over()** option. A basic example with two y variables is shown below in figure 5.8.

Creating figure 5.8

```
. * beamplot
. beamplot tempjan tempjuly
```

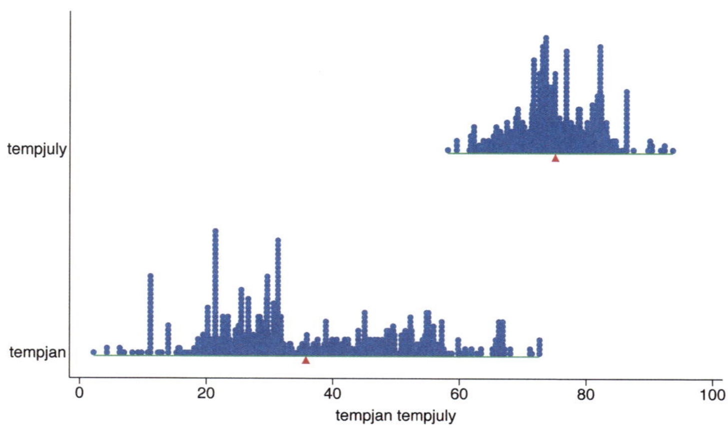

Figure 5.8. Beam plots

The beam plots in figure 5.8 are simple variations of dot plots but help to convey additional statistical information to end users. The **beamplot** command is useful if an understanding of means (or other statistical summaries that are accessed via the **summary()** option) is required to aid visual analysis of one or multiple distributions.

Like in other dot plot commands, the `over()` option can quickly generate many beam plots over a categorical variable. One should use the `dots()` option to specify further custom marker suboptions as shown below in figure 5.9. This allows users to fine-tune their beam plots to their liking. In the example below, markers are made smaller, are made hollow, have their line widths reduced, and have their lines colored black.

Creating figure 5.9

```
. * beamplot with options
. beamplot tempjan, over(division)                              ///
> dots(msize(*0.4) msymbol(Oh) mlwidth(*0.4) mlcolor(black))
```

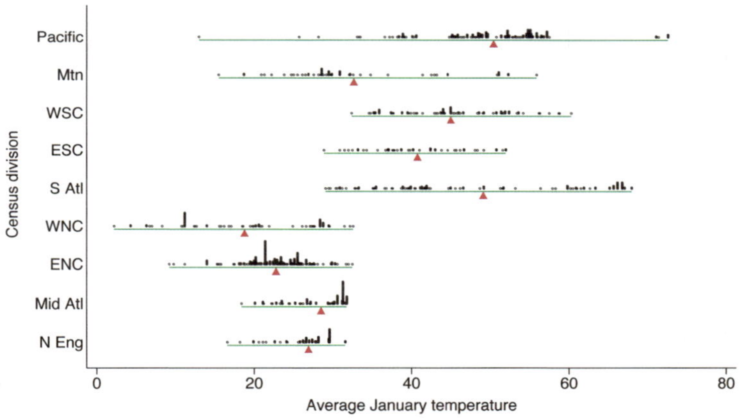

Figure 5.9. Multiple custom beam plots

Figure 5.9 shows how the use of the `dots()` and `over()` options results in a series of beam plots that allows users to quickly identify distributions across various groups and to identify mean values of all of these groups. This makes distributional comparisons across many groups fast at both the global and the local levels, and basic summary statistics are provided. Consider using `beamplot` when additional descriptive statistical information is required.

5.6 Dot plots with box plots

The concept of beam plots can be taken one step further. Instead of the simple beam that shows the range and mean, dot plots can be augmented with box plots that provide a five-number summary (minimum, first quartile, median, third quartile, and maximum) of the distribution. This can be done with the `stripplot` command and its `box` option, which adds box plots to each dot plot group. The `boffset()` option allows users to shift box plots left (negative numbers) or right (positive numbers). Figure 5.10 shows an example with a left shift. The `iqr()` option adds spikes to the boxes that reach as far as the maximum or minimum value within a specified interquartile range from the upper or lower quartile. The default is `iqr(1.5)`. Figure 5.10 also uses the option `vertical` to rotate the multiple dot plots into a vertical representation because this makes it easier to compare the many box plots. The `msize()` option is used to reduce marker sizes.

Creating figure 5.10

```
. * stripplot over groups and with box plots
. stripplot tempjan, over(division) stack width(1) vertical box ///
> boffset(-0.3) iqr(1.5) msize(*0.5)
```

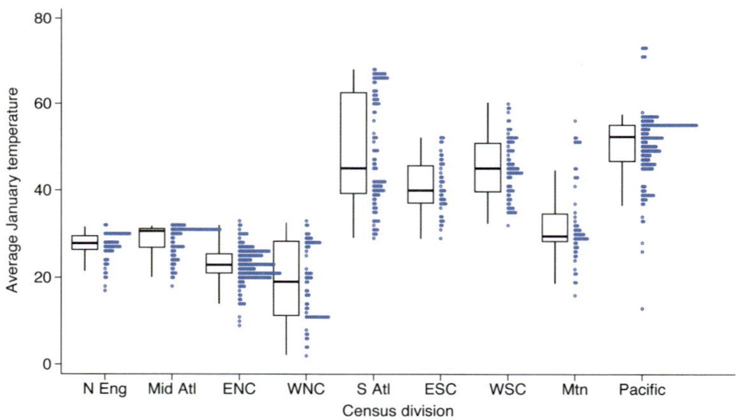

Figure 5.10. Multiple dot plots with box plots

Figure 5.10 extends the concept of beam plots and provides a lot of statistical information on a single graph. Multiple distributions are presented, individual data points can be identified, and box plots provide statistical summary information of each distribution. Users should be mindful, however, that such graphs can potentially overload readers with information and may be of little use to laypeople.

5.7 Cumulative dot plots

One unique feature of `stripplot` is the `cumulate` option, which indicates that data points should be plotted on a cumulative frequency scale. By default, `cumulate` plots only individual data markers. This can make reading the plot difficult because cumulative distribution plots require the area under the graph to be evaluated in conjunction with the gradients of lines. Therefore, users should connect the data markers with the `connect(L)` option, resulting in the plot demonstrated in figure 5.11. Note that figure 5.11 does not use the `width()` option, which means that data markers represent their true (cumulative) values. The `width()` option can be used if desired but will result in a markedly less detailed plot.

Creating figure 5.11

```
. * Cumulative dot plots
. stripplot tempjan, over(division) msize(*0.2) cumulate connect(L)
```

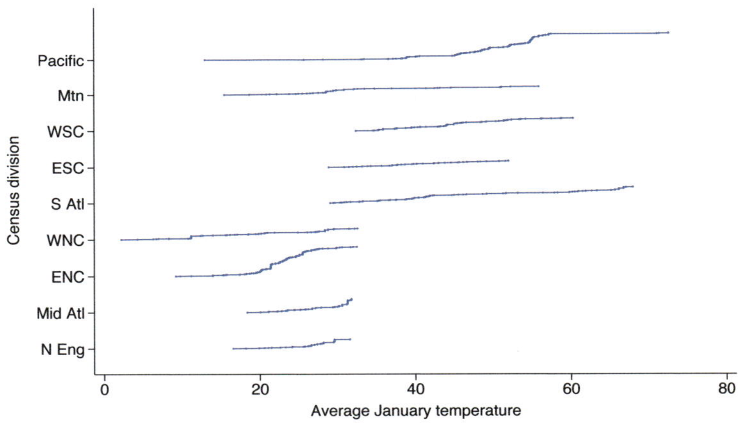

Figure 5.11. Multiple cumulative dot plots

Figure 5.11 demonstrates advanced usage of dot plots. It presents multiple cumulative frequencies of January temperatures across nine different divisional groupings. By evaluating the gradient of the relevant lines, it is possible to identify where data massing occurs on each distribution. For example, there is a large bunching of temperatures around 20–30 degrees Fahrenheit for the `ENC` region because the cumulative frequency increases rapidly at this point. At the same time, this plot allows users to identify which groups contain more data, because the final height of the cumulative plot is directly related to the number of observations in each group; the `Mtn` and `ESC` areas have few observations.

The `stripplot` command has many other advanced options. Users should examine its help file (`help stripplot`) if they wish to create advanced dot plots.

6 Stem-and-leaf plots

Advantages

1. Provide a basic distributional overview
2. Show every data point and data value
3. Are great for small samples
4. Retain all numeric information from raw data

Disadvantages

1. Are harder to understand than dot plots
2. Fare poorly with medium-to-large samples
3. Have more complex binning than other distributional plots
4. Are not frequently used, which may reduce familiarity

Required community-contributed package

To install the community-contributed command that is required to follow along with this chapter completely, type

```
. net install https://www.stata-journal.com/software/sj7-3/gr0028
```

6.1 Introduction

Stem-and-leaf plots, also called stem plots, are related to dot plots in that they represent every data point of an underlying variable. Rather than representing data with a simple marker (for example, a dot), stem-and-leaf plots convert markers into numbers representing the underlying numerical values. This creates a "double-numerical value" signal where markers represent both traditional markers (which need to be converted into values via a reading of the axes) and data values (which can be read instantly). This redundancy in information also means that stem-and-leaf plots are not traditionally plotted on graphs but rather as vertical tabular constructs. These tables are typically divided by a vertical bar; the characters to the left of the bar are the stem, while the numeric digits to the right are the leaves. Together, they mimic a tree structure that provides a visualization of the data distribution.

Stem-and-leaf plots excel at plotting distributions of small datasets (low hundreds), and their ability to retain numerical values in the visualization allows users to analyze each data point individually. This makes them a relatively unique plottype that allows users to examine variables in small datasets in detail. Stem-and-leaf plots can also be plotted on a graph for added visual effect. This can be a useful technique when such graphs are presented to laypeople because the graphical elements of the x and y axes will be familiar to most.

Stem-and-leaf plots are not used often because the disadvantages of using them are significant. They cannot handle substantial observation sizes, so most plots will be limited to 100 or 200 observations. Moreover, their representation is a complicated function of binning because both stems and leaves can be binned differently, creating additional complexity for users.

6.2 A basic stem-and-leaf plot

Stem-and-leaf plots show how observations of one variable are distributed across the data range. This chapter uses the automobile dataset (`auto.dta`), which contains information about car prices and car characteristics from 1978. The key to using this dataset to create a stem-and-leaf plot is its relatively small sample size of only 74 observations. Stem-and-leaf plots are specialized plots that work best with smaller datasets in the tens or hundreds range. A summary of the dataset is shown below.

```
. * Load data
. sysuse auto
(1978 automobile data)
. summarize
    Variable |        Obs        Mean    Std. dev.       Min        Max
-------------+---------------------------------------------------------
        make |          0
       price |         74    6165.257    2949.496       3291      15906
         mpg |         74     21.2973    5.785503         12         41
       rep78 |         69    3.405797    .9899323          1          5
    headroom |         74    2.993243    .8459948        1.5          5
-------------+---------------------------------------------------------
       trunk |         74    13.75676    4.277404          5         23
      weight |         74    3019.459    777.1936       1760       4840
      length |         74    187.9324    22.26634        142        233
        turn |         74    39.64865    4.399354         31         51
displacement |         74    197.2973    91.83722         79        425
-------------+---------------------------------------------------------
  gear_ratio |         74    3.014865    .4562871       2.19       3.89
     foreign |         74    .2972973    .4601885          0          1
```

The variable `mpg` measures car fuel consumption and ranges from 12 to 41 miles per gallon. A stem-and-leaf plot can provide a visualization of its distributional properties.

A basic stem-and-leaf plot is created with the `stem` command. This command allows only one variable to be specified and produces its results in the Results window of Stata as text; it will *not* create a standard graph. Users wishing to export the produced tabular plot can highlight the text and use a copy function. An alternative is to log the relevant output. An example of a basic stem-and-leaf plot for the variable `mpg` is shown below in figure 6.1.

Creating figure 6.1

```
. * Stem-and-leaf plot as text plot
. stem mpg
```

```
. * Stem-and-leaf plot as text plot
. stem mpg
Stem-and-leaf plot for mpg (Mileage (mpg))
  1t | 22
  1f | 44444455
  1s | 66667777
  1. | 88888888899999999
  2* | 00011111
  2t | 22222333
  2f | 444455555
  2s | 666
  2. | 8889
  3* | 001
  3t |
  3f | 455
  3s |
  3. |
  4* | 1
```

Figure 6.1. A basic stem-and-leaf plot

Figure 6.1 presents a basic stem-and-leaf plot in a tabular representation. The characters to the left of the vertical bar are the stem, while the numeric digits to the right are the leaves. When the first digit of the stem is placed next to a digit of its associated leaf, it re-creates an observation in the data. So the first stem, `1t`, along with its leaf, `22`, represent two observations of 12 MPG; likewise, the stem `2.` and its leaf, `8889`, represent three observations of 28 MPG and one observation of 29 MPG. The nonnumerical suffix of the stem is simply an identifier for the relevant leaf range: in this example, `*` identifies the leaf range 0–1, `t` identifies 2–3, `f` identifies 4–5, `s` identifies 6–7, and `.` identifies 8–9.

A useful feature of the stem plot is that it provides an overview of the data distribution while also displaying all numerical values. In figure 6.1, it is easy to identify that the values 18 and 19 occur the most, representing the "peak" of the `mpg` distribution.

6.2.1 Custom bins

Stem-and-leaf plots, like other plots that use binning, rely on binning options to modify their visual presentation. The options `lines()`, `width()`, and `digits()` can be used to modify the number of stems per interval, the stem width, and the number of digits per leaf, respectively. The `lines()` and `width()` options are mutually exclusive; both should generally be combined with `digits()`. Figure 6.2 presents an example that uses single lines as stems and single digits as leaves, causing the data to be banded in ranges of 10.

Creating figure 6.2

```
. * Stem-and-leaf plot as text plot with custom binning
. stem mpg, lines(1) digits(1)
```

```
. * Stem-and-leaf plot as text plot with custom binning
. stem mpg, lines(1) digits(1)
Stem-and-leaf plot for mpg (Mileage (mpg))
  1* | 22444444555666677778888888899999999
  2* | 0001111122222333444455555666888 9
  3* | 001455
  4* | 1
```

Figure 6.2. Stem-and-leaf plot with custom bins

Figure 6.2 shows how binning in stem-and-leaf plots can affect the final visualization of the data distribution. In this case, the larger bin width results in a stem-and-leaf plot that shows that most MPG values are located early in the data range. Unlike in density plots or histograms, the underlying numerical information is retained and depicted, which is a comparative advantage of stem-and-leaf plots. Thus, it is still possible to identify from figure 6.2 that the values 18 and 19 occur most, even though the "binning" has changed the visualization compared with figure 6.1.

6.3 A graphical stem-and-leaf plot

The official stem-and-leaf plot command relies on Stata's Results window to produce a text-based visualization of how a variable is distributed. While this is handy and relatively fast to produce when exploring data, those who would like a more graphical version of a stem-and-leaf plot can use the community-contributed command `stemplot` (Cox 2007c). This command creates a stem-and-leaf plot using Stata graphics, which, once created, can then be edited and modified via the Graph Editor. This command is useful if the plot must be exported or distributed more formally because copying and pasting tables from the Results window may not always be the most optimal way to transfer knowledge. The `stemplot` command is demonstrated in figure 6.3.

Creating figure 6.3

```
. * Graph stem-and-leaf plot
. stemplot mpg
```

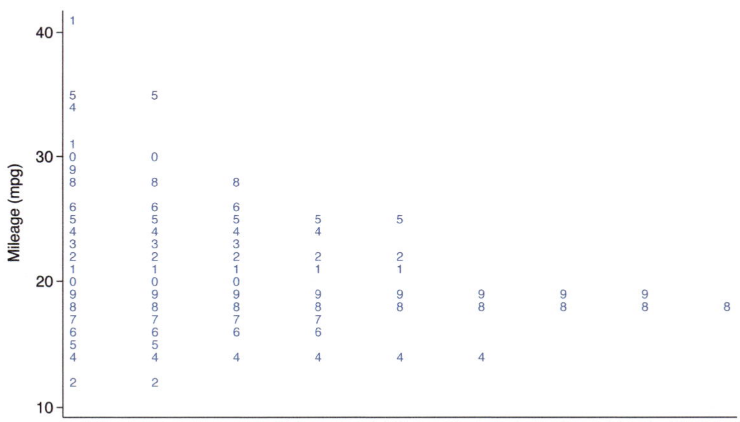

Figure 6.3. Graph-based stem-and-leaf plot

Figure 6.3 presents a graphical version of the stem-and-leaf plot. The y axis has become the stem and identifies numerical ranges, while individual markers represent leaves (data points). In this default version, only the last digit of the data value is shown. This makes the identification of data points more difficult compared with the `stem` command because users now need to associate which digit belongs to which numerical range. While numeric identification has become more challenging in the default version of this plot, the overall distributional properties of the data remain relatively easy to identify and mimic a basic histogram.

To adjust the number of digits, one can use the `digits()` option, which indicates how many digits should be shown as leaves (the default is `digits(1)`). Adjusting this option can make graph interpretation easier. Figure 6.4 demonstrates an example where two digits are used, resulting in the full numerical values being displayed for these data.

Creating figure 6.4

```
. * Graph stem-and-leaf plot with two digits
. stemplot mpg, digits(2)
```

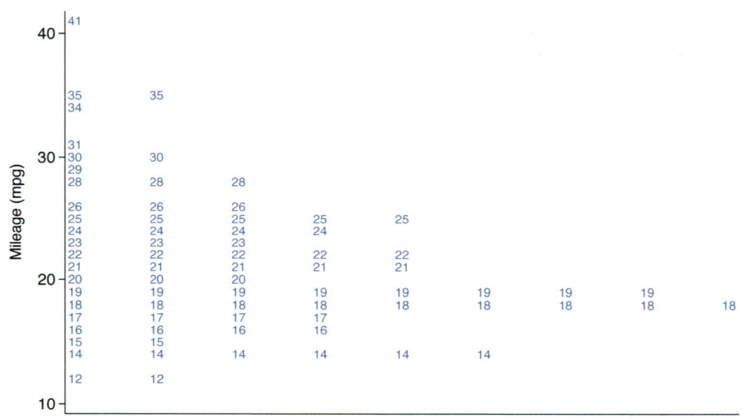

Figure 6.4. Graph-based stem-and-leaf plot with two digits

Figure 6.4 presents a more user-friendly graphical stem-and-leaf plot containing the full values of each data point. It is easy to identify, for example, the six individual observations of 14 MPG. Of course, more extensive numeric ranges could cause problems; for example, it would be impractical to show five digits for each observation unless the observation count were very low.

6.4 A horizontal stem-and-leaf plot

`stemplot` can rotate plots to be horizontal with the `horizontal` option. The visual narrative of this plot is more akin to a dot plot than a traditional stem-and-leaf plot because data values are now stacked on top of each other. Unlike in dot plots, the numerical information of each data value is retained in the stem-and-leaf plot. A horizontal stem-and-leaf plot can be useful if end users require the detail of each value but are unfamiliar with traditional vertical stem-and-leaf plots. A horizontal example is shown in figure 6.5, and the `mlabcolor()` option [which is part of *marker_label_options* (see `help marker_label_options`)] is used to color the leaves (numerical markers) black.

Creating figure 6.5

```
. * Graph stem-and-leaf plot horizontal plot
. stemplot mpg, digits(2) horizontal mlabcolor(black)
```

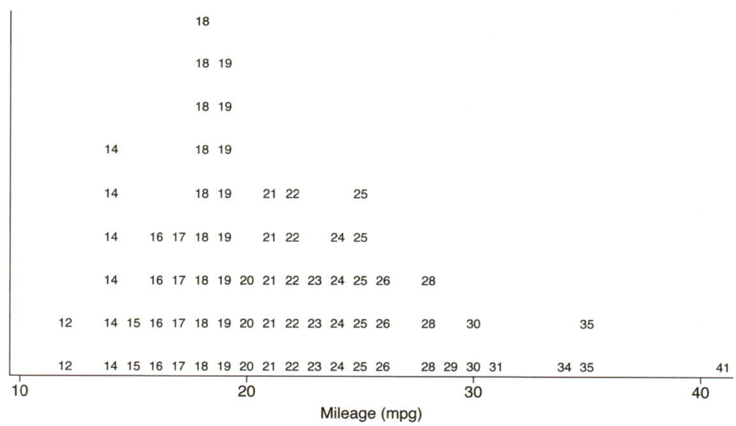

Figure 6.5. Horizontal stem-and-leaf plot

6.5 Mirrored stem-and-leaf plots

The `stemplot` command can also use the `by()` option, which does not exist for Stata's official `stem` command. This option allows users to draw multiple stem-and-leaf plots that are shown separately for each group of a categorical variable. Figure 6.6 below demonstrates this option with the dichotomous variable `foreign`. When the `by()` variable consists of only two groups, the `back` option can be used to create mirrored stem-and-leaf plots, which is also demonstrated in figure 6.6.

Creating figure 6.6

```
. * Graph stem-and-leaf plot over two variables
. stemplot mpg, digits(2) by(foreign) back mlabcolor(black)
```

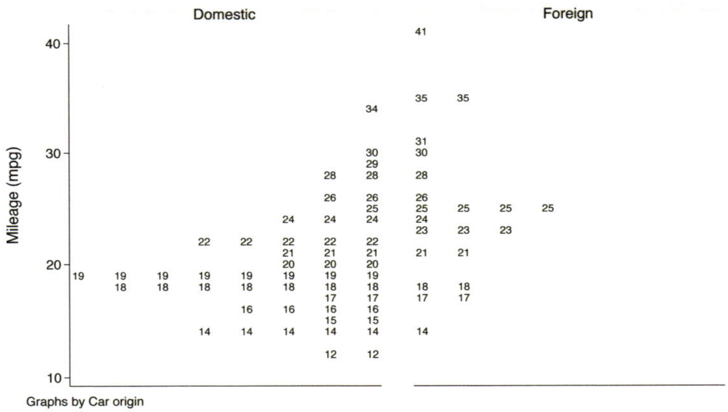

Figure 6.6. Multiple stem-and-leaf plots

Figure 6.6 presents two mirrored stem-and-leaf plots, aiding the comparison of data points across different MPG ranges. For example, it is easy to identify that domestic cars have lower MPG values than foreign cars. The most frequently occurring MPG value for domestic cars is 19, while the most frequent value for foreign cars is 25.

Stem-and-leaf plots are specialized plottypes that are appropriate when there are few observations. They retain numerical information by displaying all data values, making them perfect for "full disclosure" graphs where everything is plotted. Stem-and-leaf plots should generally be avoided when dealing with large or complex datasets.

7 Distributional diagnostic plots

Advantages

1. Provide visual comparisons for a range of specific distributional properties
2. Provide fast visual tests via reference lines
3. Are easier to understand than statistical tests
4. Show all data points

Disadvantages

1. Transform data such that inferring the actual distribution and interpreting data points are more complex
2. Cannot test complex distributional properties
3. Require a good understanding of the reference distribution

Required community-contributed package

To install the community-contributed command that is required to follow along with this chapter completely, type

```
. ssc install skewplot
```

7.1 Introduction

A common requirement of modern data analytics is to compare observed empirical distributions with known theoretical distributions. For example, in regression analysis, the construction of confidence intervals or the hypothesis testing of coefficients often assumes that the residual error term is normally distributed. Evaluating residuals from a regression can be numerically or visually done. Often, the visual approach is faster and more intuitive. Most users will understand whether a data distribution is appropriately bell shaped and thus fits the assumption of a normally distributed error term in a regression model. However, such visual evaluation can still be difficult with standard histograms or density plots because users need to "imagine" what the theoretical distribution looks like on top of the actual data. Such challenges are made more complex when the theoretical distribution is nonstandard.

Symmetry, skew, and quantile plots are common plottypes that visually perform distributional checks. These plots transform the data in a particular way and plot the transformed values on a graph with a reference line for a visual comparison. This provides an immediate visual cue of whether data follow a particular distribution: When data are on top of the reference line, the distributional properties are met, and the data conform with the theoretical distribution. When the transformed data are away from the reference line, they are not distributed according to the theoretical distribution. A disadvantage of such graphs is that the transformation makes data interpretation more complex, so it is challenging to infer how the entire data distribution looks.

Distributional diagnostic plots illustrate how observations of one continuous variable match a specific theoretical distributional property. `citytemp.dta` is used in this chapter. It contains the variable `tempjan`, which measures temperature data in degrees Fahrenheit across US cities. The variable contains 954 observations and is continuously distributed with primarily unique values. A summary is displayed below.

```
. * Load data
. sysuse citytemp
(City temperature data)

. summarize

    Variable |        Obs        Mean    Std. dev.       Min        Max
-------------+---------------------------------------------------------
    division |        956    5.135983    2.694249          1          9
      region |        956    2.623431    1.057724          1          4
      heatdd |        953    4425.533    2199.605          0      10816
      cooldd |        953    1240.413    937.6679          0       4389
     tempjan |        954    35.74895    14.18813        2.2       72.6
-------------+---------------------------------------------------------
    tempjuly |        954    75.05377    5.495036       58.1       93.6
```

7.2 Symmetry plots

Symmetry plots examine whether empirical data distributions are symmetric or not. A symmetric distribution is a distribution that can "fold" perfectly in half. The fold point is at the median, and all data points to the left of the median should overlap with all data points to the right of the median in a perfect fold. That is why these plots commonly transform the data into the distance above (y axis) and below (x axis) the median. In a symmetry plot, each data point's distance below the median is compared with the corresponding data point's distance above the median. For example, if the data are sorted, the distance below the median for the first observation is compared with the distance above the median for the last observation, the second observation is compared with the second-to-last observation, and so on. Data values that are symmetric will then align along a 45-degree line, indicating perfect symmetry. A symmetry plot can be created using the `symplot` command, as demonstrated below in figure 7.1.

Creating figure 7.1

```
. * Symmetry plot
. symplot tempjan
```

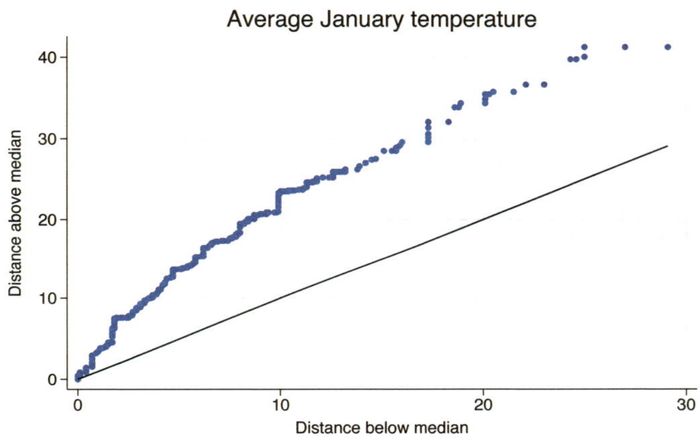

Figure 7.1. Symmetry plot

Figure 7.1 presents a basic symmetry plot of average January city temperatures. The full data range for the variable `tempjan` is [2.2, 72.6]. If this variable were perfectly symmetric, distances from individual data values to the median would range between approximately 0 and 35. In a symmetry plot, data points on the reference line are symmetric because their distances above and below the median are identical. In this case, there is evidence that the temperature distribution is not symmetric. Data points diverge away from the symmetry line almost from the start. The fact that most data points are located above the 45-degree line indicates a right skew in the data, whereas data points being below the line would suggest a left skew.

7.3 Skew plots

If the concepts of skewness and symmetry are important considerations in the analysis of a variable, then the community-contributed **skewplot** command (Cox 1999a) should be considered. It offers several additional features over the official command **symplot**. A key feature of **skewplot** is its ability to analyze multiple variables in one plot. The **skewplot** command generates a plot of the mid-summary (also called the mid-range) against the variable's spread (also called the range). Hence, it is also known as a mid-versus-spread plot. The transformation of these plots results in visualization that is compared with the horizontal reference lines instead of a 45-degree line. This makes comparison easier on a graph and allows multiple variables to be plotted at once. An example graph is demonstrated in figure 7.2.

Creating figure 7.2

```
. * Skew plot
. skewplot tempjan
```

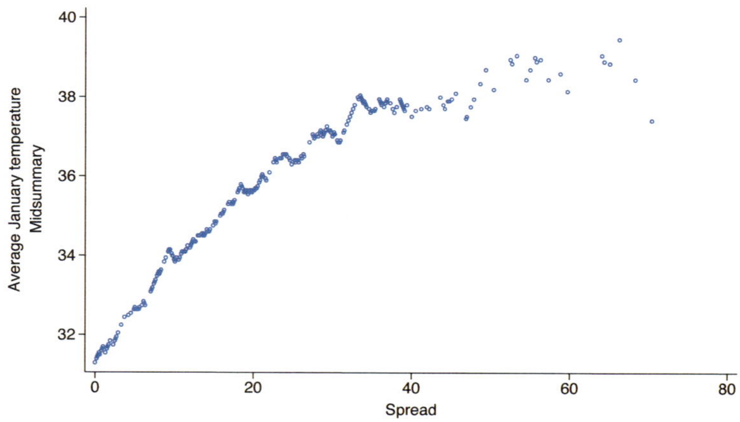

Figure 7.2. Skew plot

Figure 7.2 presents a basic skew plot. Note the absence of a 45-degree reference line. A perfectly symmetrical distribution would yield a perfectly horizontal line; thus, the standard horizontal reference lines (which can easily be adjusted) serve as the primary visual test. Departure from a horizontal data series indicates skewness in the data. An upward departure, as shown in figure 7.2, suggests right skew, while a downward departure from the horizontal reflects left skew. In this case, there is evidence that average January temperatures are not symmetrically distributed and that the distribution has a right skew.

An advantage of **skewplot** is its ability to create visualizations of multiple variables. This can be achieved by adding more y variables after the command. This concept is demonstrated in figure 7.3.

Creating figure 7.3

```
. * Multiple skew plots
. skewplot tempjan tempjuly
```

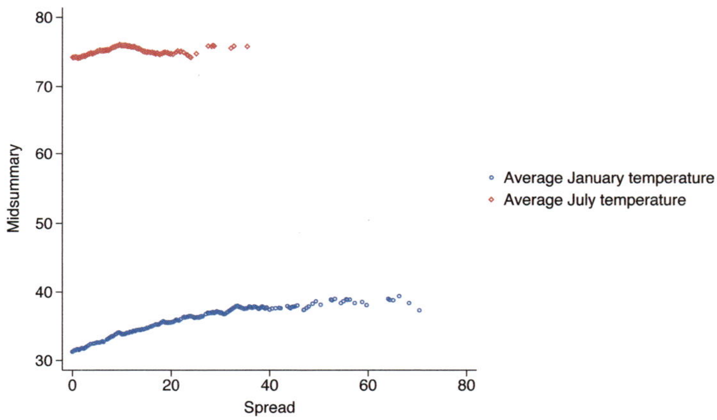

Figure 7.3. Multiple skew plots

Figure 7.3 visually compares two data distributions against symmetry and skew. It would appear that average January temperatures are somewhat right-skewed compared with average July temperatures, where values seem more horizontally distributed, suggesting more symmetry in that distribution.

7.4 Quantile-uniform plots

Quantile-uniform plots continue the fashion of transforming data values and plotting these against a reference line. Specifically, the transformed data in these plots evaluate whether the underlying data are uniformly distributed. A uniform (also known as a rectangular) distribution has a horizontal probability density function, indicating that each value has the same probability of occurring. Like symmetry plots, a quantile-uniform plot's advantage is that it highlights areas of the distribution where uniformity breaks down. The command to create quantile-uniform plots is **quantile**, and only one variable can be specified. Figure 7.4 shows an example.

Creating figure 7.4

```
. * Quantile-uniform plot
. quantile tempjan
```

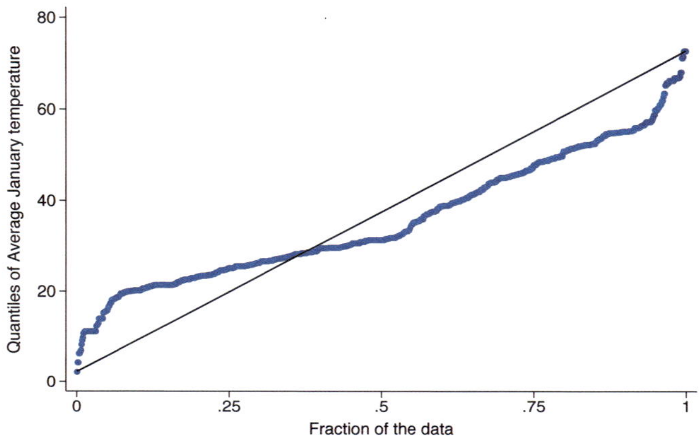

Figure 7.4. Quantile-uniform plot

In figure 7.4, each value of the variable is plotted against the proportion of the data with values less than that fraction. If temperature were uniformly distributed, all the data points would be plotted on the reference line. In this case, reading the x axis from left to right, early data points are plotted above the reference line, while later data points are plotted below the reference line. This suggests the presence of more low-temperature values than expected and fewer high-temperature values than expected. Overall, the majority of the data points deviate from the 45-degree line, suggesting these data are not uniformly distributed.

Working out the underlying distribution from such a plot is complex because it requires an understanding of the tested distribution (uniform in this case) and how data deviations away from the tested distributions could be depicted. Users should stick to the primary purpose of these plots, which in this case is to test whether a variable is uniformly distributed. Here the visual test suggests it is not.

7.5 Quantile-normal plots

To test whether a variable is normally distributed, one can use the **qnorm** command, which evaluates whether data are normally (bell-shaped) distributed. Many statistical processes make assumptions about normality, and **qnorm** provides a quick visual identification of whether a continuous variable is normally distributed. As with other commands in this chapter, a reference line is key to visual analysis. An example is shown in figure 7.5.

Creating figure 7.5

```
. * Quantile-normal plot
. qnorm tempjan
```

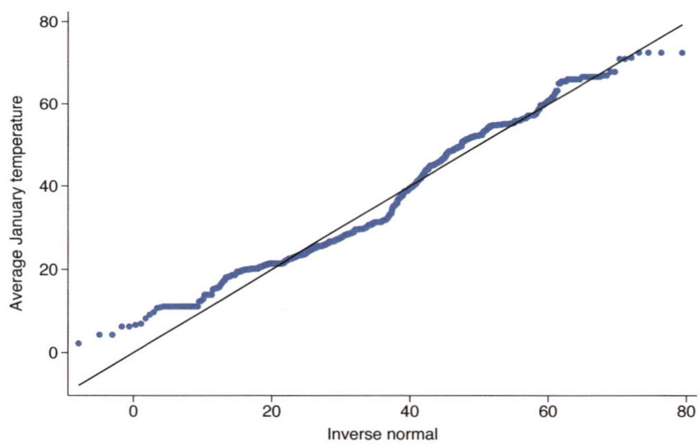

Figure 7.5. Quantile-normal plot

Figure 7.5 presents a quantile-normal plot. The reference line indicates where data would lie if they were normally distributed. In this case, there is a hint that, globally, the data are normally distributed, although points in the early and later parts of the January temperature range lie above the reference line while points in the middle lie below the reference line. This suggests two data peaks clustered to the left and right of the middle of the distribution. Overall, this result, while initially suggestive, is not evidence of a normally distributed variable.

7.6 Quantile χ^2 plots

The command qchi allows users to compare variables against the χ^2 distribution. Each data point is appropriately transformed, and values that lie on the reference line are χ^2 distributed. An example is shown in figure 7.6.

Creating figure 7.6

```
. * Quantile chi-squared plot
. qchi tempjan
```

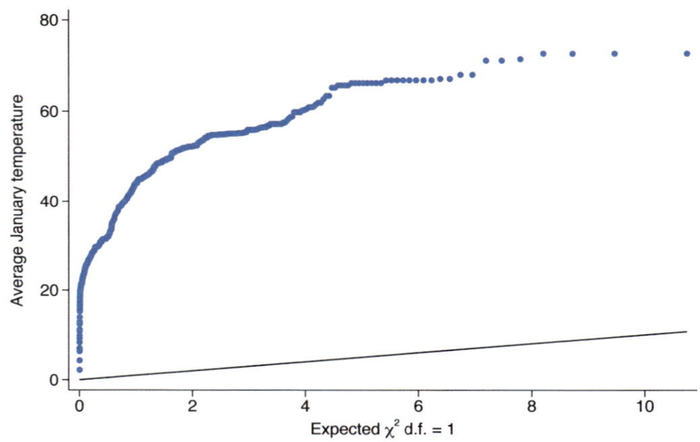

Figure 7.6. Quantile χ^2 plot

Figure 7.6 clearly shows that average January temperatures are not χ^2 distributed with 1 degree of freedom. None of the points are on or near the reference line. However, χ^2 distributions are subject to a degrees-of-freedom parameter, and different degrees of freedom result in different theoretical distributions. Degrees of freedom can be adjusted with the `df()` option; the default is `df(1)`. Figure 7.7 below demonstrates a quantile χ^2 plot with 40 degrees of freedom.

Creating figure 7.7

```
. * Quantile chi-squared plot with custom degrees of freedom
. qchi tempjan, df(40)
```

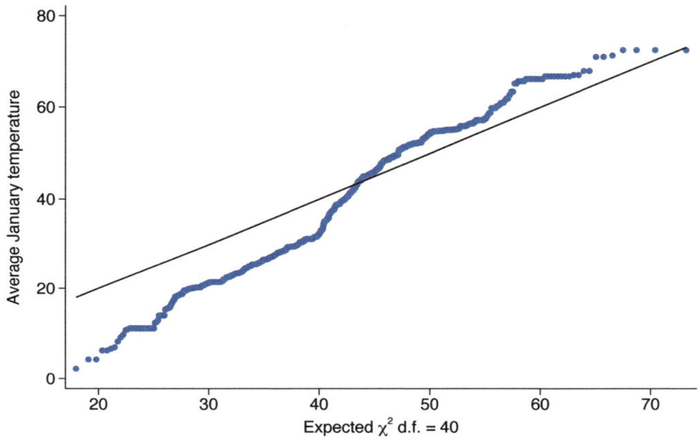

Figure 7.7. Quantile χ^2 plot with 40 degrees of freedom

A χ^2 distribution with 40 degrees of freedom fits the data better than the previous distribution with 1 degree of freedom. Nonetheless, even when the distribution has 40 degrees of freedom, it is easy to see that the data points are not located on the reference line, which means that the data remain non-χ^2 distributed.

7.7 Quantile–quantile plots

Quantile–quantile (QQ) plots are scatterplots of two variables and their quantiles plotted against one another. The idea is that if both sets of quantiles came from the same distribution, the data points would form a line that follows a 45-degree gradient. In other words, QQ plots evaluate whether two variables are similarly distributed, not whether they follow a particular theoretical distribution. Note that if both data distributions are linearly related but not identical, the plot will show a straight line but not necessarily on the reference line $y = x$. Data distributions that are shifted but parallel will have the same distribution but be mean shifted; data distributions that are shifted and differently sloped will be mean shifted and differ in other parameters (such as the variance).

QQ plots are great for comparing two continuous distributions with each other. Importantly, the distributions do not need to be compared with a theoretical reference distribution. QQ plots are created using the `qqplot` command, which needs two variables to be specified. An example is shown in figure 7.8.

Creating figure 7.8

```
. * Quantile-quantile plot
. qqplot tempjan tempjuly
```

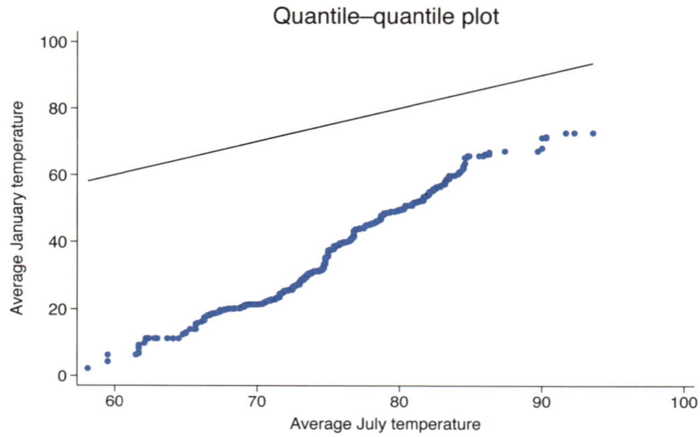

Figure 7.8. QQ plot

Figure 7.8 demonstrates a QQ plot of the quantiles of average January and July temperatures in US cities. A priori, it would be reasonable to expect January temperatures to be differently distributed than July temperatures because January temperatures are, on average, colder than July temperatures. However, one may expect the temperature variance to be similar. There will be warmer and colder cities in January. The same will happen in July, implying that the two distributions may be linearly related. The QQ plot shows little evidence of this. It is clear that January and July temperatures are not identically distributed; the data values are not on the reference line. In addition, the July temperature distribution is not simply a right-shifted version of January temperatures; the data plot is not parallel to the reference line.

One should use distributional diagnostic plots as quick visual checks of specific distributional hypotheses. The commands outlined in this chapter are fast, simple to use, and do not offer many options. If users wish to be more explorative with the data, then standard kernel density plots or histograms are better ways of understanding the distributional aspects of the data and variables.

8 Rootograms

Advantages

1. Excel at comparing empirical distributions with theoretical distributions
2. Can simplify the interpretation of distributional shapes via hanging the results from the theoretical distribution
3. Emphasize tails of the distribution

Disadvantages

1. May confuse general users with square root transformation
2. Are subject to binning choices
3. Are limited to frequency rather than density
4. Are not frequently used, which may reduce familiarity

Required community-contributed package

To install the community-contributed command that is required to follow along with this chapter completely, type

```
. ssc install hangroot
```

8.1 Introduction

Rootograms are graphical ways to compare empirical (observed) distributions with theoretical distributions. Often, the distributional analysis of a continuous variable is limited to graphing the empirical distribution and then leaving it to end users to visually judge whether the observed distribution mimics some known theoretical distribution. A common theoretical distribution often tested against observed data is the normal (or Gaussian) distribution with its well-known bell shape. However, many empirical distributions have long tails and small observation counts at the edge of distributions, making visual identification challenging.

Rootograms and hanging rootograms attempt to overcome this problem by plotting the square root transformation of frequency count on the y axis, hence the name "rootogram". This emphasizes small outlying values of empirical distributions, making it easier to compare actual distributions with theoretical distributions. In the case of hanging rootograms, the comparison is simplified by suspending (or "hanging") the results from the theoretical distribution. This allows differences to be observed against a simple horizontal axis, which is faster and less error-prone than examining discrepancies against a more complex function. This eases visual identification.

Importantly, for both rootograms and hanging rootograms, the vertical axis is scaled to the square root of the frequencies, which helps highlight and identify differences in the distribution tails, often a critical but difficult component when identifying distributions visually. In practice, one should use hanging rootograms because the visualization is more distinct and layperson-friendly. In particular, when complex theoretical distributions are tested, hanging rootograms avoid users needing prior knowledge of complex theoretical shapes.

Like many histogram-type graphs, rootograms are subject to binning choices. These can greatly alter the visual image, but the significance is somewhat less for rootograms because the comparative theoretical distribution would not change. It is rare for binning choices to modify the empirical distribution to such an extent that it matches the theoretical distribution under some circumstances and not under others. Nonetheless, users should always vary binning choices to explore the potential visual effects.

8.2 A basic rootogram

Rootograms illustrate how observations of a continuous variable compare with theoretical distribution. `citytemp.dta` is used in this chapter. It contains the variable `tempjan`, which measures temperature data in degrees Fahrenheit across various US cities. Basic summary statistics are shown below.

```
. * Load data
. sysuse citytemp
(City temperature data)

. summarize

    Variable |        Obs        Mean    Std. dev.       Min        Max
-------------+---------------------------------------------------------
    division |        956    5.135983    2.694249          1          9
      region |        956    2.623431    1.057724          1          4
      heatdd |        953    4425.533    2199.605          0      10816
      cooldd |        953    1240.413    937.6679          0       4389
     tempjan |        954    35.74895    14.18813        2.2       72.6
-------------+---------------------------------------------------------
    tempjuly |        954    75.05377    5.495036       58.1       93.6
```

A basic rootogram can be produced using the `spikeplot` command in Stata with the `root` option. With the `root` option, the `spikeplot` command creates a standard spike plot (a dense histogram) but transforms the vertical scale to show the square root of frequency. Figure 8.1 below exemplifies a basic use case.

Creating figure 8.1

```
. * Rootogram
. spikeplot tempjan, root
```

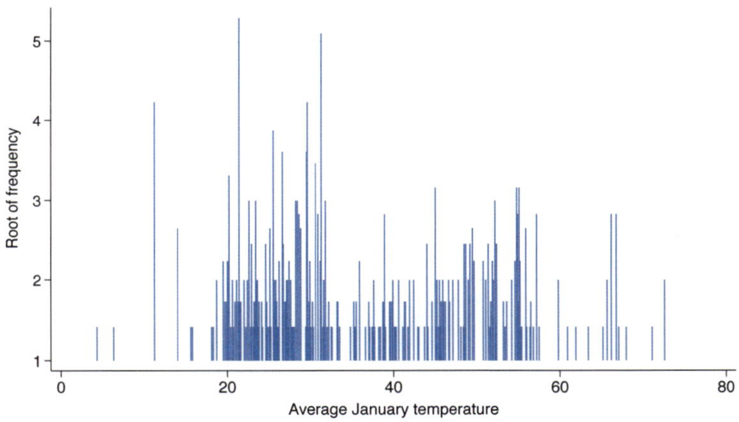

Figure 8.1. Basic rootogram

The rootogram presented in figure 8.1 is identical at first glance to a basic spike plot, a histogram with many bins. However, because of the square-root transformation, observations in the tails are emphasized in this plot. Unfortunately, the `root` option does not superimpose a theoretical distribution over the empirical distribution, making it difficult to gauge whether the observed data match any known theoretical distributions. Although the `addplot()` option can be used to overlay custom `twoway function` plots, which, in turn, can call upon the plethora of statistical functions within Stata (see `help density_functions`), this requires a lot of extra coding work. Users will first have to manually estimate the parameters of an appropriate theoretical distribution from the underlying empirical data and then apply those parameters in the `addplot()` option.

8.3 Hanging rootograms

An easier option is to use the community-contributed command **hangroot** (Buis 2007), which comes with a wide range of theoretical functions. **hangroot** produces hanging rootograms where data values are "hung" from a theoretical distribution. By default, the normal distribution is used. Only one variable can be specified with this command. An example is demonstrated in figure 8.2 below.

Creating figure 8.2

```
. * Hanging rootogram
. hangroot tempjan
```

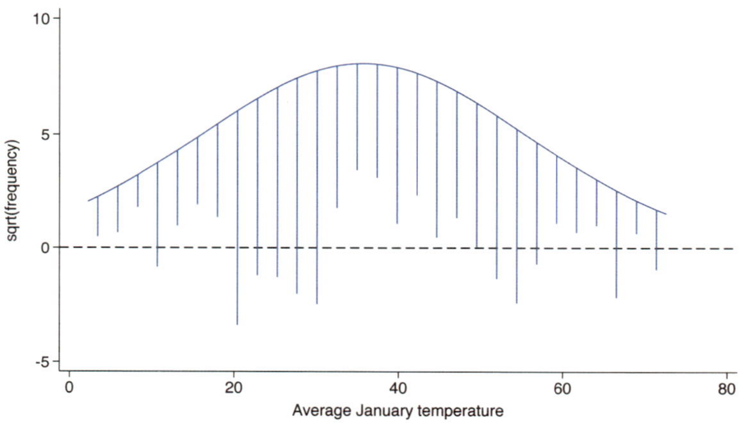

Figure 8.2. Hanging rootogram

Figure 8.2 demonstrates a hanging rootogram of average January temperatures. Individual frequency bars are hanging from the theoretical distribution toward a dashed horizontal line at 0. If the hanging bars all end exactly at the dashed line, then the data match that particular distribution (a normal distribution in the default case.)

Bin choices can change the shape of the empirical distribution. The `bin()` option or the `width()` option sets the number of bins or the width of the bins, respectively, to a numeric value. Only one of these two options can be used. Figure 8.3 demonstrates an example using the `width()` option to specify a smaller bin width.

Creating figure 8.3

```
. * Hanging rootogram with more detail
. hangroot tempjan, width(1)
```

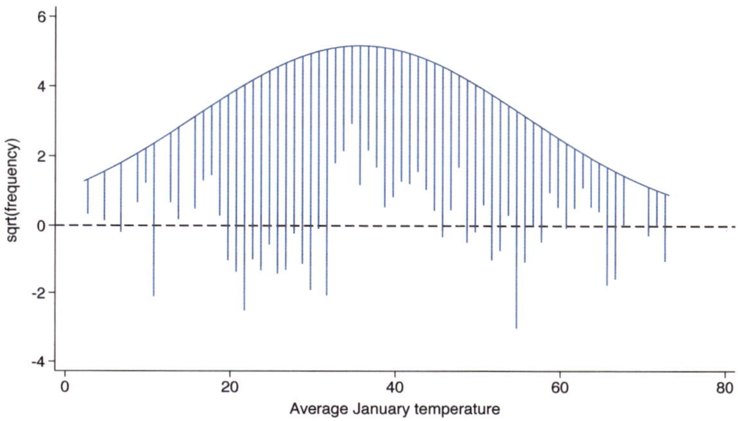

Figure 8.3. Hanging rootogram with custom bin width

Changing the bin width did not materially affect the visualization of the empirical distribution: average January temperatures continue not to match the normal distribution.

8.3.1 Different theoretical distributions

The most important option of `hangroot` is `dist()`, which allows users to define the theoretical distribution that the data should be tested against. A wide range of theoretical distributions are available, including discrete distributions, making this option also useful for noncontinuous variables. A list of possible distributions is given in table 8.1 below.

Table 8.1. `hangroot` distribution options

Theoretical distribution	Estimation command	`dist(`*name*`)`
Normal/Gaussian	`regress`	`normal` or `gaussian`
Lognormal	`lognfit`	`lognormal`
Logistic		`logistic`
Weibull	`weibullfit`	`weibull`*
χ^2		`chi2`
Gamma	`gammafit`	`gamma`
Gumbel	`gumbelfit`	`gumbel`
Inverse gamma	`invgammafit`	`invgamma`
Wald/inverse Gaussian	`invgaussfit`	`wald`
Beta	`betafit`	`beta`
Pareto	`paretofit`	`pareto`
Fisk/log-logistic	`fiskfit`	`fisk`*
Dagum	`dagumfit`	`dagum`*
Singh–Maddala	`smfit`	`sm`*
Generalized beta II	`gb2fit`	`gb2`*
Generalized extreme value	`gevfit`	`gev`*
Exponential		`exponential`
Laplace		`laplace`
Uniform		`uniform`
Geometric		`geometric`
Poisson	`poisson`	`poisson`
Zero-inflated Poisson	`zip`	`zip`*
Negative binomial I	`nbreg`	`nb1`*
Negative binomial II	`nbreg` and `gnbreg`	`nb2`*
Zero-inflated negative binomial	`zinb`	`zinb`*

In table 8.1 above, for all `dist()` options with asterisks, users must self-estimate the relevant distributional properties using a Stata command in the table (some of these may be community-contributed commands). `hangroot` will then use the postestimation output from the relevant estimation command to populate the distributional parameters of the theoretical distribution. In most use cases, the `dist()` option can be used in the standalone syntax without any prior or manual estimation of parameters. An example using the `lognormal`, `logistic`, `uniform`, and `exponential` distributions is shown below in figure 8.4. Each graph is individually saved by using the `name()` option and then merged using the `graph combine` command. In addition, the bar width is set to 0.2 for finer detail.

Creating figure 8.4

```
. * Hanging rootograms testing different distributions
. hangroot tempjan, width(0.2) dist(lognormal)  ///
> title(Lognormal) name(g1, replace)
. hangroot tempjan, width(0.2) dist(logistic)   ///
> title(Logistic) name(g2, replace)
. hangroot tempjan, width(0.2) dist(uniform)    ///
> title(Uniform) name(g3, replace)
. hangroot tempjan, width(0.2) dist(exponential) ///
> title(Exponential) name(g4, replace)
. graph combine g1 g2 g3 g4
```

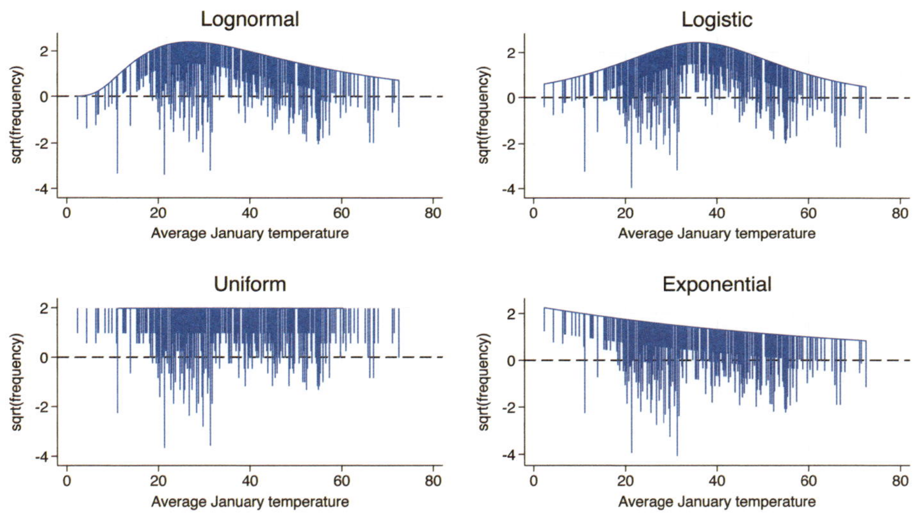

Figure 8.4. Hanging rootograms tested against multiple theoretical distributions

Figure 8.4 demonstrates the full power of `hangroot`. A wide array of theoretical distributions can be quickly visually compared against a continuous variable. It is easy to identify from figure 8.4 that the average January temperatures do not match any of the four theoretical distributions.

Users may wish to self-define theoretical distribution parameters, for example, to test the same theoretical distribution but with changing parameters. This can be achieved with the `par()` option. Depending on the distribution, one or multiple parameters can be defined. Figure 8.5 below demonstrates custom parameters for the χ^2 distribution.

Creating figure 8.5

```
. * Hanging rootograms testing different chi-squared distributions
. hangroot tempjan, width(0.2) dist(chi2) par(1)  ///
> title({&chi}{sup:2} d.f. = 1) name(g1, replace)
. hangroot tempjan, width(0.2) dist(chi2) par(5)  ///
> title({&chi}{sup:2} d.f. = 5) name(g2, replace)
. hangroot tempjan, width(0.2) dist(chi2) par(10) ///
> title({&chi}{sup:2} d.f. = 10) name(g3, replace)
. hangroot tempjan, width(0.2) dist(chi2) par(50) ///
> title({&chi}{sup:2} d.f. = 50) name(g4, replace)
. graph combine g1 g2 g3 g4
```

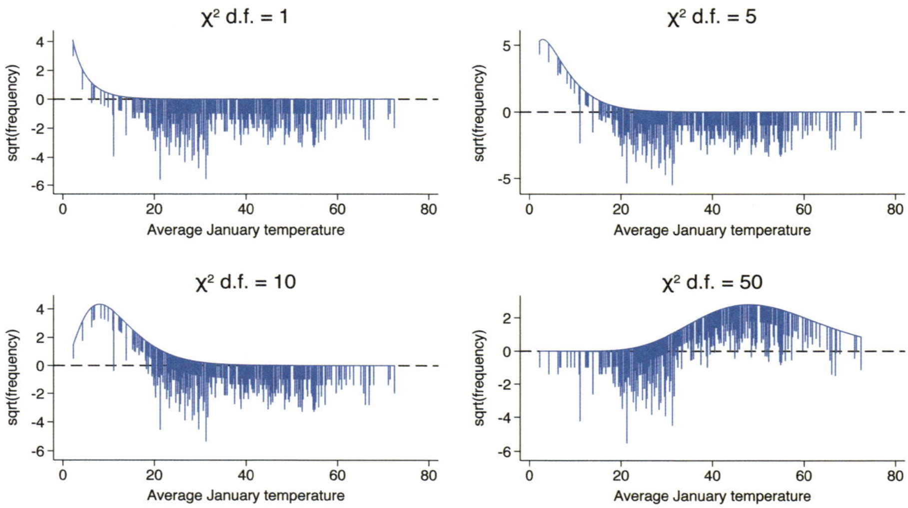

Figure 8.5. Hanging rootograms tested against multiple χ^2 distributions

It is clear from figure 8.5 that the underlying data are not χ^2 distributed across a range of degrees-of-freedom parameters.

8.3.2 Confidence intervals

Users may wish to add confidence intervals to their hanging rootograms to better identify whether there is evidence that bars are different from the horizontal reference line. This can be done with the `ci` option, which adds 95% confidence intervals to the bottom of the bars. Figure 8.6 demonstrates this option.

Creating figure 8.6

```
. * Hanging rootogram with confidence intervals
. hangroot tempjan, ci
```

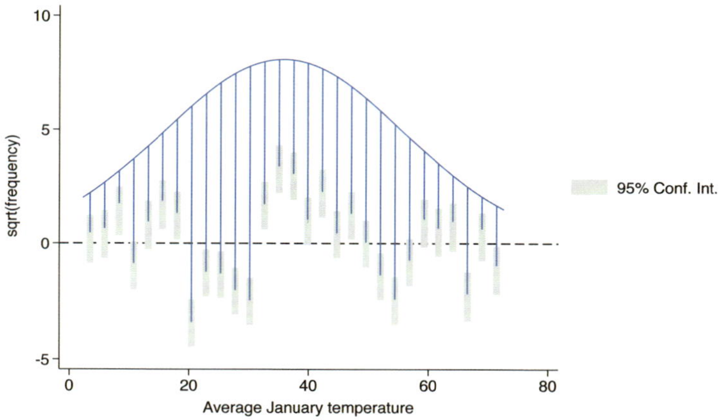

Figure 8.6. Hanging rootogram with 95% confidence intervals

Figure 8.6 demonstrates 95% confidence intervals around each bar, allowing users to better gauge statistical significance. All confidence intervals should overlap with the horizontal dashed line for an empirical distribution to match a theoretical distribution. If any bars do not, then it would be evident that the data do not conform to the theoretical distribution. In this case, it is evident that most of the confidence interval bars are either entirely above or entirely below the reference line, suggesting that average January city temperatures are not normally distributed.

9 Univariate bar charts

Advantages

1. Are easy to understand
2. Excel for variables with many categories
3. Can take a variety of visual appearances
4. Can easily handle multiple groups of discrete variables

Disadvantages

1. Can lead to false impressions if bars are not properly spaced or scaled
2. Are less useful for variables with few categories
3. Do not handle extreme number ranges well
4. Are poor at presenting exact statistics

Required community-contributed packages

No additional installations are required to follow along with this chapter completely.

9.1 Introduction

Bar charts (bar graphs) are by far the most common type of data visualization largely because they are tremendously flexible. They can be used for both continuous and discrete variables, although this chapter will focus on demonstrating bar charts for discrete variables.

These graphs visualize both discrete and categorical data using rectangular bars. They can have either a vertical or a horizontal orientation. To an extent, bar charts are similar to histograms except that bin sizes are discrete and encapsulate every category. The sizes of the bars represent statistical values. This means that there is no need to think about bin widths because each value of the variable is now its own category.

Bar charts excel at depicting discrete categories by providing easy visual cues via bar lengths. Often, this means that they lose statistical accuracy because it is not always easy to follow bar lengths to the axis and identify exact statistics. Bars also make more sense when variables have many discrete categories, especially over multiple groups. For example, graphing only 2 categories leads to a visual plot that takes a lot of space but conveys relatively little statistical information, while graphing 10 or more groups leads to a lot of statistical information that the eye can quickly compare and contextualize.

Users should be wary of axis scaling and axis labeling when using bar charts. It is easy to accidentally misrepresent bars using a wrong scale, too few labels, or variables with extreme value differences in their categories. However, when used responsibly, bar charts are one of the few visualization techniques that can significantly impact data analysis across all industries, ages, and professions.

9.2 A basic frequency bar chart

Bar charts can help illustrate how observations of a categorical variable are distributed across different categories. The 1988 US National Longitudinal Survey of Young Women dataset (`nlsw88.dta`) is used in this chapter. It contains data about a group of women in their 30s and 40s from 1988, with the prime focus of the data on their labor market characteristics. The dataset contains 2,246 observations; multiple categorical variables, including `race` and `occupation`; and dichotomous variables, such as `south` and `union`. Basic summary statistics are displayed below.

```
. * Load data
. sysuse nlsw88
(NLSW, 1988 extract)
. summarize
    Variable |        Obs        Mean    Std. dev.       Min        Max
-------------+---------------------------------------------------------
      idcode |      2,246    2612.654    1480.864          1       5159
         age |      2,246    39.15316    3.060002         34         46
        race |      2,246    1.282725    .4754413          1          3
     married |      2,246    .6420303    .4795099          0          1
never_marr~d |      2,246    .1041852    .3055687          0          1
-------------+---------------------------------------------------------
       grade |      2,244    13.09893    2.521246          0         18
    collgrad |      2,246    .2368655    .4252538          0          1
       south |      2,246    .4194123    .4935728          0          1
        smsa |      2,246    .7039181    .4566292          0          1
      c_city |      2,246    .2916296    .4546139          0          1
-------------+---------------------------------------------------------
    industry |      2,232    8.189516    3.010875          1         12
  occupation |      2,237    4.642825    3.408897          1         13
       union |      1,878    .2454739    .4304825          0          1
        wage |      2,246    7.766949    5.755523   1.004952   40.74659
       hours |      2,242    37.21811    10.50914          1         80
-------------+---------------------------------------------------------
     ttl_exp |      2,246    12.53498    4.610208   .1153846   28.88461
      tenure |      2,231     5.97785    5.510331          0   25.91667
```

The basic bar chart is created with the `graph bar` command. The general syntax of this command is `graph bar` *yvars*, *options*. There are two important considerations when using `graph bar` code in Stata: 1) a list of y variables with specific statistics can be specified, and 2) several options are available to break down statistics by one or more discrete variables. The set of statistics can be altered by each y variable, and this code will take the form (*statistics*) [*varlist*]. Common statistics are `count` and `percent`, and more can be found in the help file (see `help graph bar`). When examining categorical data, the option most used is the `over()` option, which specifies a categorical variable over which the statistics are created. The `over()` option can be repeated up to three times.

For this chapter, the interest is primarily in graphing single-variable discrete data distributions. This results in a code setup where no y variables are specified in the core `graph bar` syntax, though the required statistics continue to be specified here. All variables of interest are generally specified in one or multiple `over()` options. To create a basic frequency bar chart of the categorical variable `race` from `nlsw88.dta`, the following command is used:

Creating figure 9.1

```
. * Simple bar chart
. graph bar (count), over(race)
```

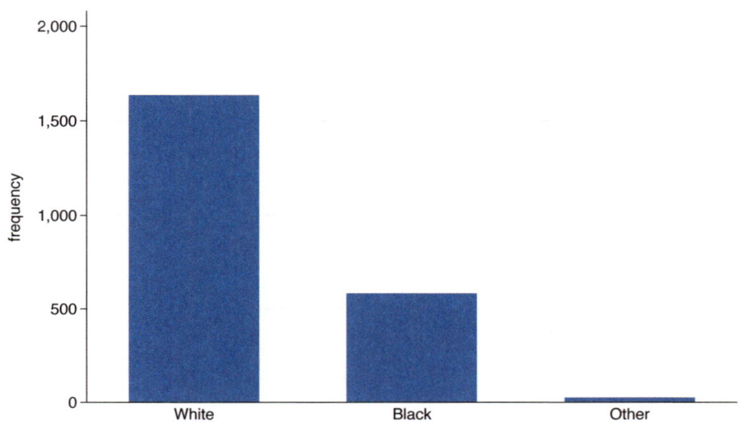

Figure 9.1. A basic bar chart

Figure 9.1 is a visualization of how many observations are in each particular category of the `race` variable. End users can quickly identify that the majority of observations are White and that the next biggest category is Black.

Suppose users wish to have bars in different colors. In that case, they can use the `asyvars` option, which will differentiate the categories of the first variable that is specified in the `over()` option. This option will remove the x-axis category labeling and introduce a legend identifying the relevant colors. If users wish to retain individual color control, then the `bar()` option can be used to specify custom colors for each bar. Figure 9.2 presents an example with the `asyvars` option.

Creating figure 9.2

```
. * Simple bar chart with asyvars
. graph bar (count), over(race) asyvars
```

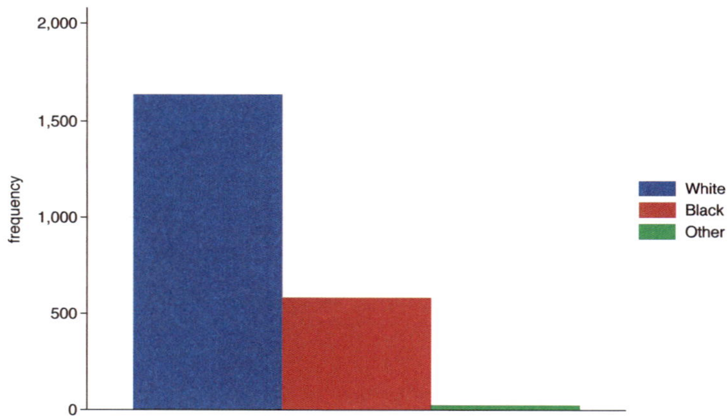

Figure 9.2. A basic bar chart with `asyvars`

9.3 Percentage bar charts

To create percentage bar charts of discrete variables, users need to specify `(percent)` in the `graph bar` command. Figure 9.3 presents a basic percentage bar chart.

Creating figure 9.3

```
. * Percentage bar chart
. graph bar (percent), over(race) ylabel(0(20)100)
```

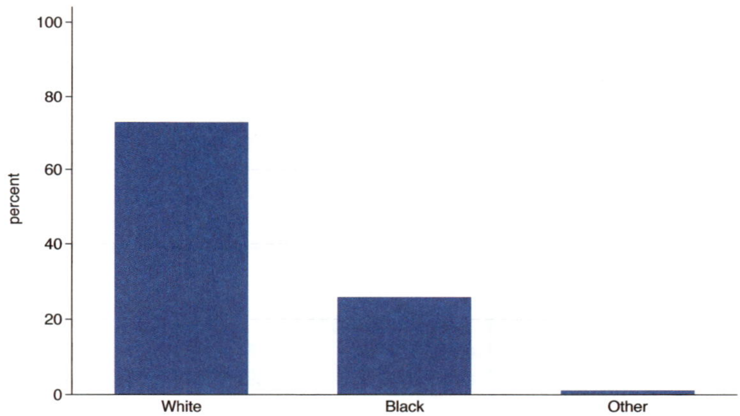

Figure 9.3. A basic percentage bar chart

Figure 9.3 is identical to figure 9.1 except that the y axis is rescaled to a percentage. Users should pay careful attention to the y-axis labels when using percentage bar charts. These labels can be modified with the `ylabel()` option. An example of labels going from 0 to 100 in increments of 20 is demonstrated in figure 9.3.

9.4 Horizontal bar charts

Bar charts excel at presenting many categories; however, when many categories are used, labeling on the x axis may become cluttered. The solution is often to angle the text or even introduce vertical text, but this can hinder the rapid identification of categories. Another, perhaps better, solution is to use horizontal bar charts, which rotate the basic bar chart so that category labeling occurs on the y axis. This, in turn, results in long text being displayed in a regular left-to-right fashion, speeding up visual acquisition of the individual categories.

To create horizontal bar charts, the user can use the `graph hbar` command, which is identical to `graph bar` and retains all of its options. An example using the `occupation` variable, which contains 13 discrete categories and long label names, is shown in figure 9.4.

Creating figure 9.4

```
. * Horizontal percentage bar chart
. graph hbar (percent), over(occupation)
```

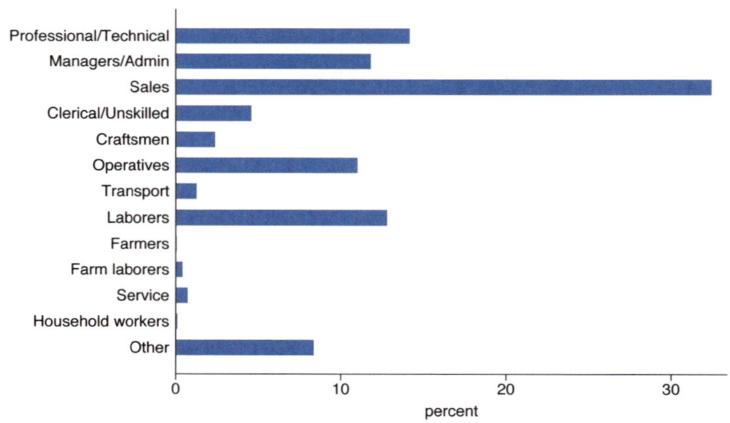

Figure 9.4. A horizontal bar chart

Figure 9.4 shows how horizontal bars complement long label text. The decision to use horizontal or vertical bar charts is often a matter of preference, but when there are long labels over many categories, horizontal bar charts have a distinct advantage.

9.5 Sorted bar charts

The `over()` option is key to using bar charts with discrete variables. It contains various suboptions (see `help graph bar##over_subopts`) that can be used to modify the visualization of the bars for each discrete category. `sort()` and `descending` are important suboptions that put bars in height order and reverse height order, respectively. `sort()` can be specified without a variable or, less commonly, with a variable to define the prescribed order based on the specified variable's value. If no variable is specified in the `sort()` option, the order will be determined by the requested statistical values for the relevant `over()` variable. Figure 9.5 demonstrates an example with the `graph hbar` command, where data are sorted on the variable `occupation` (the `over()` variable) in descending order. Because percentages were requested as the statistical measure, bars are sorted by percentage values for the categories of `occupation`.

Creating figure 9.5

```
. * A sorted bar chart
. graph hbar (percent), over(occupation, sort(1) descending)
```

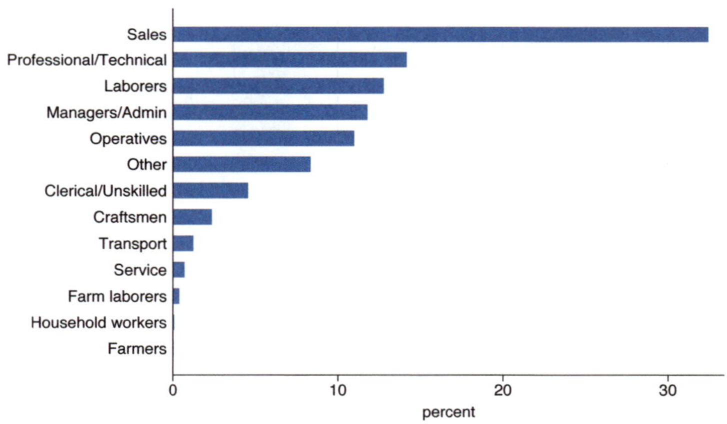

Figure 9.5. Sorted bar chart

From figure 9.5, it is simple to identify that the sales occupations dominate these labor market data. Sales is by far the biggest category, followed by professional occupations and laborers.

9.6 Multiple bar charts

Multiple bar charts of discrete variables can be created via repeated use of the over() option or via the by() option. The over() option can be repeated up to three times and draws multiple subgroups on a single graph. The by() option is generally less flexible and creates separate graphs for each by() category. An example of the by() option is shown in figure 9.6.

Creating figure 9.6

```
. * Multiple bar charts
. graph bar (count), over(race) by(south)
```

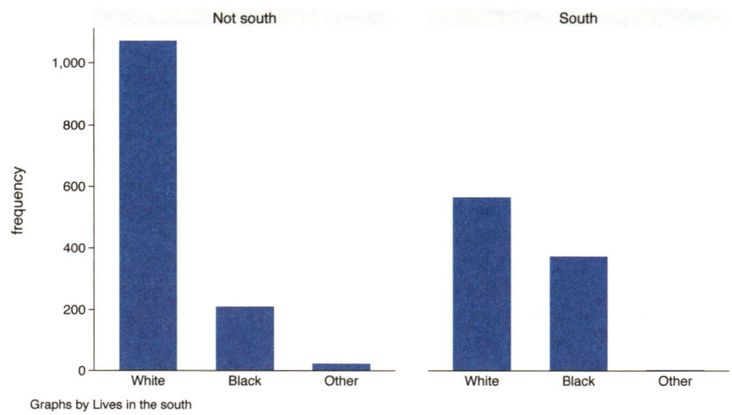

Figure 9.6. Multiple bar charts with by()

Figure 9.6 presents multiple bar charts using one over() option and the by() option. Notice that the visualization consists of two merged charts rather than one. One graph visualization can be created with the use of two over() options, as demonstrated in figure 9.7.

Creating figure 9.7

```
. * Multiple bar charts with two over() options
. graph bar (count), over(race) over(south)
```

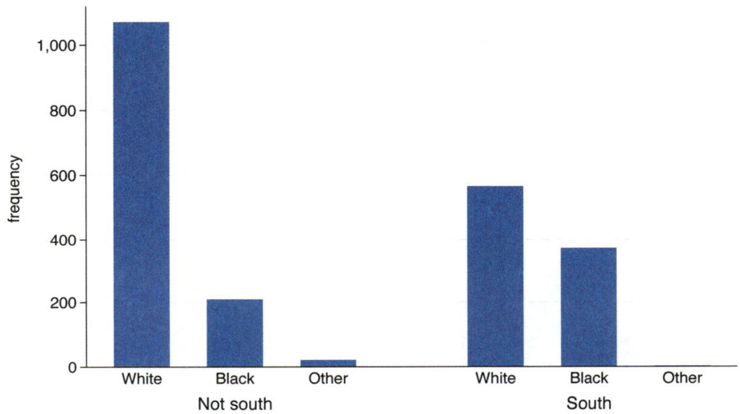

Figure 9.7. Multiple bar charts with two `over()` options

Figure 9.7 presents a similar visualization to figure 9.6. An advantage of using the `over()` option is that it can be repeated three times. Figure 9.8 demonstrates using three `over()` options.

Creating figure 9.8

```
. * Multiple bar charts with three over() options
. graph bar (count), over(race) over(south) over(union)
```

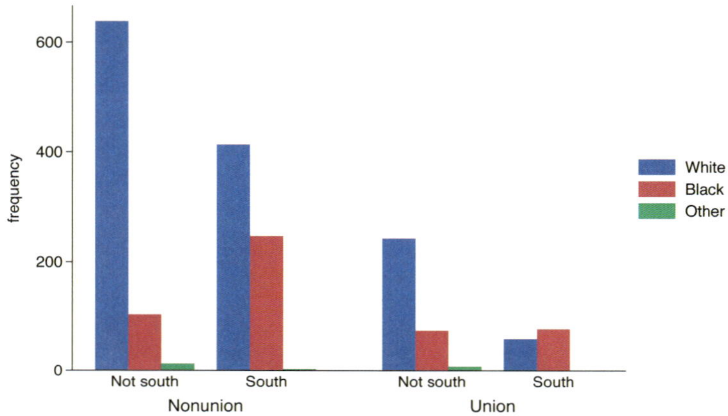

Figure 9.8. Multiple bar charts with three `over()` options

9.7 Dot charts

For an alternative visualization style, users can switch to dot charts via the `graph dot` command. Dot charts are almost identical to bar charts in their underlying concept. They replace the bars with lines of (often 100) dots and then plot a marker on the line of dots to indicate where each discrete value statistic is located. The `graph dot` command follows the same syntax as `graph bar` and shares many of the same options. Users need only switch the word `bar` with `dot` in their code to generate a different visual style. An example is shown below in figure 9.9.

Creating figure 9.9

```
. * Dot chart
. graph dot (count), over(occupation)
```

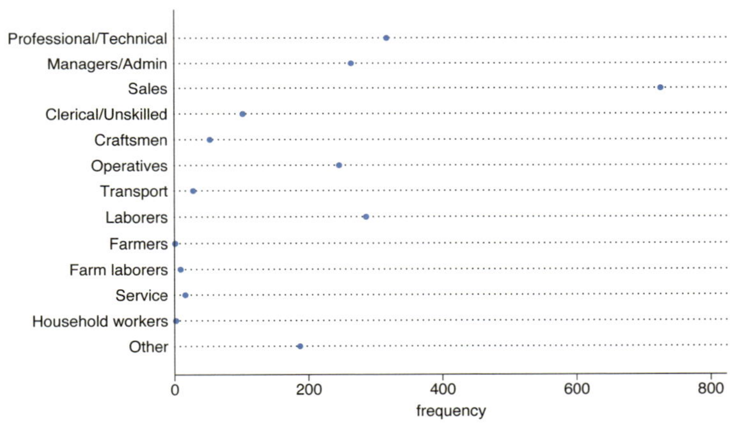

Figure 9.9. Dot chart

Dot charts excel at depicting many categories and often have better visualization characteristics when detailed categorical comparison is required. This is because users do not need to follow individual bars to the axis to read statistical information but can use the individual dots in each category as easy visual comparators. Consider using dot charts when an exact comparison is required.

10 Pie charts

Advantages

1. Feature uninterrupted circular planes of visualization that are easy to track
2. Are good for showing part-to-whole relationships
3. Are visually appealing and very user friendly
4. Can emphasize specific categories via exploding pies

Disadvantages

1. Have problems conveying exact statistics or comparing similar slices
2. Become less useful when there are many categories
3. Are not useful for comparing groups of discrete variables

Required community-contributed packages

No additional installations are required to follow along with this chapter completely.

10.1 Introduction

Pie charts are circular graphics that split data into slices, also called "pies". Each pie represents a statistical value wthat can be measured via its arc length, central angle, or area.

The primary advantage of this method is that the circular nature of the visualization has a particular aesthetic appeal. There are no sharp discontinuities, and the eye can scan overall graph information relatively easily. This also makes pie charts excellent for communicating basic statistical quantity information to generalists and nondata analysts. Arguably, they are even more accessible than bar charts. Moreover, the ability to explode a slice of the pie allows users to draw particular attention to one or multiple categories. This is not possible with bar charts because moving bars across the x or y plane would change the representation of statistical information.

A disadvantage of pie charts is that they are poor at revealing underlying values. Angle measurements are complex for end users, and the radial computation of sums to

360 is difficult. Moreover, pie charts rely on color to identify groups, which limits the useful number of discrete categories that can be displayed. Pie charts are also poor at illustrating multiple groups of discrete variables. Often, one pie chart must be created per group, leading to many different subgraphs that are not integrated and are visually disruptive.

Pie charts are generic graph types, and most professionals will stay away from them because the disadvantages often outweigh the advantages. However, if the primary motivation is to present a simple visual message using one discrete variable, then pie charts have some use in modern data visualization.

10.2 A basic pie chart

Pie charts can help illustrate how observations of a categorical variable are distributed over different numerical or string categories. This chapter uses `nlsw88.dta`, which contains 2,246 observations of a group of women in their 30s or 40s. The variables in the data relate to individual and labor market characteristics. The dataset contains several categorical variables, including `race` and `occupation`, and several dichotomous variables, such as `south` and `union`. Basic summary statistics are displayed below.

```
. * Load data
. sysuse nlsw88
(NLSW, 1988 extract)

. summarize

    Variable |        Obs        Mean    Std. dev.       Min        Max
-------------+---------------------------------------------------------
      idcode |      2,246    2612.654    1480.864          1       5159
         age |      2,246    39.15316    3.060002         34         46
        race |      2,246    1.282725    .4754413          1          3
     married |      2,246    .6420303    .4795099          0          1
never_marr~d |      2,246    .1041852    .3055687          0          1
-------------+---------------------------------------------------------
       grade |      2,244    13.09893    2.521246          0         18
    collgrad |      2,246    .2368655    .4252538          0          1
       south |      2,246    .4194123    .4935728          0          1
        smsa |      2,246    .7039181    .4566292          0          1
      c_city |      2,246    .2916296    .4546139          0          1
-------------+---------------------------------------------------------
    industry |      2,232    8.189516    3.010875          1         12
  occupation |      2,237    4.642825    3.408897          1         13
       union |      1,878    .2454739    .4304825          0          1
        wage |      2,246    7.766949    5.755523   1.004952   40.74659
       hours |      2,242    37.21811    10.50914          1         80
-------------+---------------------------------------------------------
     ttl_exp |      2,246    12.53498    4.610208   .1153846   28.88461
      tenure |      2,231     5.97785    5.510331          0   25.91667
```

The command **graph pie** creates pie charts in Stata. Note that the command can be run in three different configurations:

- The first mode is invoked if users specify more than two variables after **graph pie**. Then each pie will display the relevant slices, or proportions, as the totals of each variable. This is useful for continuous variables that measure similar items.
- The second mode is invoked when only one variable is specified after **graph pie**. Here an `over()` option with a discrete variable must be specified. The first variable's slices are then split across the relevant `over()` groups.
- The third mode is the most used. Here no variable is specified after **graph pie**, but the `over()` option is. This will produce a pie chart that displays slices, or proportions, as frequencies within the `over()` groups. Each pie slice will correspond to one discrete category of the specified `over()` variable. Figure 10.1 demonstrates this third option using the variable **race**.

Creating figure 10.1

```
. * Basic pie chart
. graph pie, over(race)
```

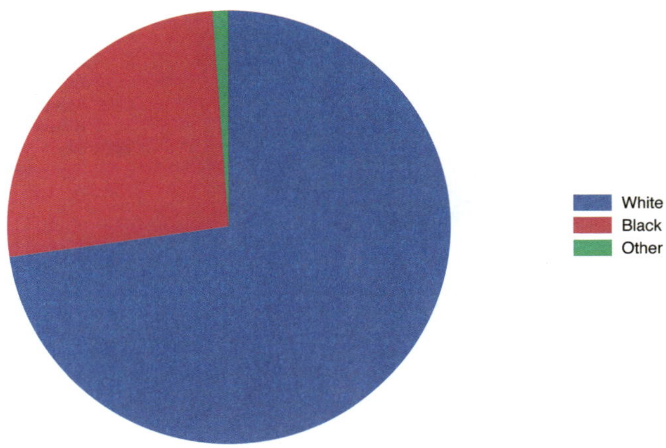

Figure 10.1. A basic pie chart

Figure 10.1 presents a basic pie chart. The visualization allows users to quickly identify that the White group is the largest group in the dataset, making up almost three-quarters of the sample.

10.2.1 Too many categories

Note that pie charts have natural limits to how many categories can be used. The key limiter is color differentiation. There are only so many colors that the human eye can quickly discern, and when too many categories are used, some colors will start to resemble each other. Moreover, a pie chart with many categories requires a legend with many color categories. This makes identification of individual categories more challenging. Figure 10.2 demonstrates these problems with the variable `occupation`, which contains 13 discrete values.

Creating figure 10.2

```
. * An overloaded pie chart
. graph pie, over(occupation)
```

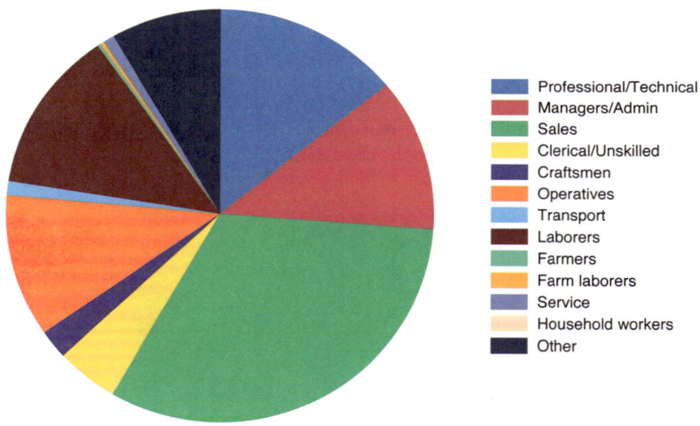

Figure 10.2. An overloaded pie chart

Figure 10.2 is a good example of a poor pie chart. Some categories have small angles that are hard to read, while others have similar colors that are difficult to identify in the legend. For such data, a bar chart is a much better choice.

10.2.2 Labeling pie charts

Stata does not label pie charts by default. In many cases, you may wish to do so because there often remains significant real estate on a pie chart that can be used to enhance the presented information. Often, you will want to add labels or numerical values to the pies to make them easier to read and interpret.

This can be done with the `plabel()` option. Users can either specify individual pies via a pie number or specify all pies using the `_all` suboption. They can then select from three mutually exclusive options that exist for labeling. The `name` option names each pie for easy identification, the `sum` option displays the sum of a y variable for each pie, and the `percent` option reveals the percentage in each slice of the pie. Figure 10.3 demonstrates this using the `_all` and `percent` suboptions.

Creating figure 10.3

```
. * Labeling pies
. graph pie, over(race) plabel(_all percent)
```

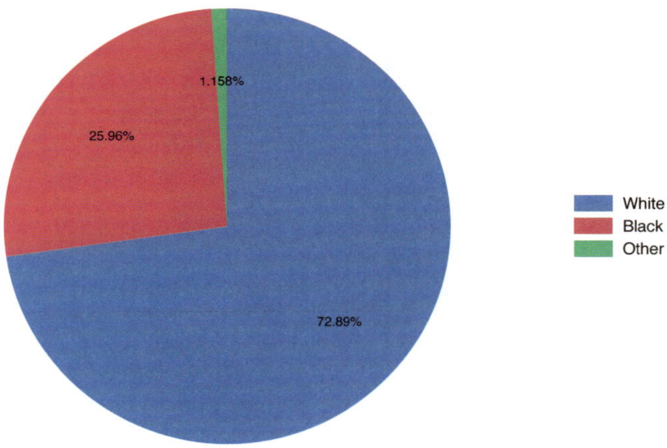

Figure 10.3. A labeled pie chart

Figure 10.3 shows how users can add basic labeling information to pies. The labeled text can be further modified via the textbox options (see `help textbox_options`) or via active graph management using the Graph Editor.

10.2.3 Exploding pie charts

One helpful aspect of a pie chart is its ability to display particular discrete categories via exploding pies. This provides a second alternative for categorical emphasis in addition to color. Exploding a pie slice is an essential concept because pie charts are visual and not meant to provide exact statistics.

To explode pies, one must use the `pie()` option, which allows users to adjust slice colors via the `color()` suboption and explode them via the `explode()` suboption. Users must specify which slice they want to explode and the "strength" of the explosion. Figure 10.4 demonstrates this concept.

Creating figure 10.4

```
. * Exploding pies
. graph pie, over(race) plabel(_all name) pie(2, explode(10))
```

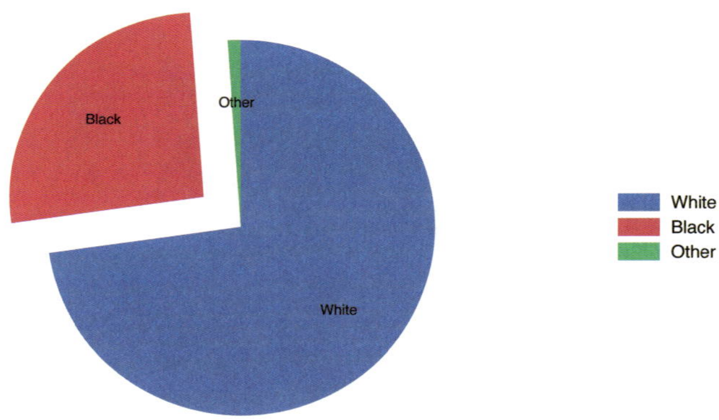

Figure 10.4. Pie chart with exploding slice

Figure 10.4 shows how exploding pies can be used to emphasize particular discrete categories. All pie slices can be separately exploded. However, keeping the explosions to one category where possible is generally advised because exploding multiple categories interferes with statistical inference. The slices create a visual discontinuity that makes it harder to compare individual pie slices.

10.2.4 Sorting pie charts

Pie slices can be ranked to make identifying large and small categories easier. This is done with the `sort` option. By default, pies are sorted in ascending order, although the `descending` option can be used to reverse the sort order. Figure 10.5 demonstrates sorting with the variable `occupation`, which contains many categories.

Creating figure 10.5

```
. * Sorting pies
. graph pie, over(occupation) sort
```

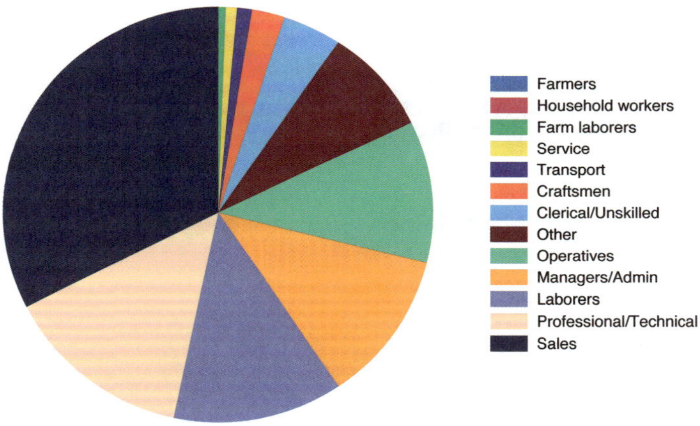

Figure 10.5. Sorted pie chart

Figure 10.5 sorts slices into ascending order, making identifying large and small discrete groups easier. Unfortunately, the `sort` option does not sort the associated legend, making inference with many groups difficult because users must track legends to random areas of the pie chart. One can use the `plabel()` option to add category names to each pie.

10.3 Multiple pie charts

Pie charts do not lend themselves well to displaying multiple discrete variables. Unlike with bar charts, users have only the `by()` option, which repeats the pie chart over groups of another categorical variable (or even multiple variables). The `by()` option can be added before or after the `over()` option and specifies over which discrete variable or variables multiple pie charts are drawn. Unlike **`graph bar`**, there are no repeated uses of the `over()` option.

Caution is advised because the `by()` option may result in many pie charts being drawn. The result is that data analysis requires end users to identify multiple angles and surface areas across multiple pie charts. This is visually challenging and results in further loss of statistical precision. Multiple pie charts work best if the original `over()` option or the `by()` option contains few categories. An example is shown below in figure 10.6 using the variables `union` (which contains 2 categories) and `occupation` (which contains 13 categories).

Creating figure 10.6

```
. * Multiple pie charts with by()
. graph pie, over(union) by(occupation)
```

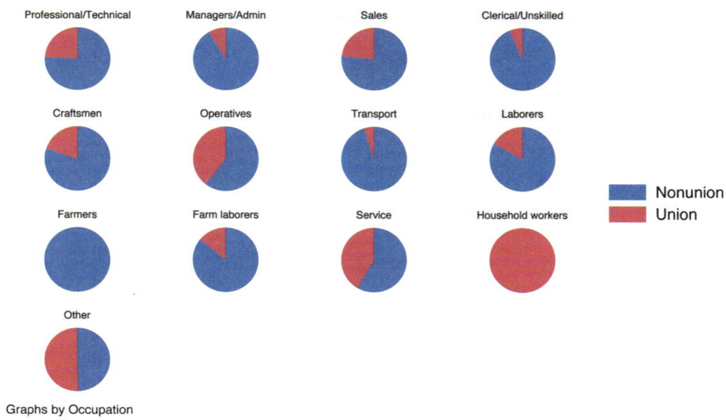

Figure 10.6. Multiple pie charts

Figure 10.6 demonstrates how multiple pie charts can be drawn quickly over various discrete groups. Because having multiple pie charts makes data evaluation harder rather than easier, one should use such graphs carefully. In most cases, one should use bar charts instead.

11 Radar charts

Advantages

1. Are good for comparing many discrete categories over different variables
2. Provide easy-to-follow circular visual narratives
3. Highlight strengths and weaknesses across categories

Disadvantages

1. Can be more complex to interpret than pie and bar charts
2. Carry statistical information only through spoke length rather than area and angle
3. Can make discrete categories difficult to compare if they are not next to each other
4. Can be difficult to follow if categories have extreme value differences
5. May need careful scaling to avoid misinterpretation

Required community-contributed package

To install the community-contributed command that is required to follow along with this chapter completely, type

```
. ssc install radar
```

11.1 Introduction

Radar charts, also called spider plots, are alternative ways to graph discrete data. They are not as intuitive as bar charts or pie charts but can plot many categorical values on a single graph, making them useful for the comparison of many discrete categories.

Radar charts are composed of several equally angled spokes called radii, with each spoke representing a distinct category. The data length of the spoke is proportional to what is being measured, usually frequencies or percentages. Finally, a line is drawn between values on adjacent spokes, giving the plot a star-like or spider-like visual appearance. Radar charts are often particularly good at comparing many discrete categories over a range of different variables.

Radar charts come with some disadvantages, however. First, only the spoke length matters. There is no statistical significance to area or angle measurements, which may confuse laypeople who are used to pie charts. Moreover, it can be difficult to compare categories that are far apart on the radar chart, especially when there are many categories. In addition, overall spoke length matters greatly in the visualization of the radar chart. Users should always manually vary the spoke length to examine what impact this has on the radar chart.

Currently, no official Stata command creates radar charts. The community-contributed command `radar` is required to do so. By default, this command treats each observation as a separate category in the radar chart. This means that sample sizes should generally be low (less than 100). For large datasets, the implication is that collapsing data into smaller datasets is required, which requires additional coding effort. A simple example of collapsing a large dataset into a smaller one is shown at the end of this chapter, with the resulting radar chart presented in figure 11.8.

11.2 A basic radar chart

Radar charts can help illustrate how observations of single or multiple discrete variables are distributed. This chapter uses `auto.dta`, which contains information about car prices and car characteristics from 1978. Importantly, this dataset is relatively small, with a sample size of only 74 observations. This matters because observations rather than variable categories will form the spoke categories. A summary of the dataset is shown below.

```
. * Load data
. sysuse auto
(1978 automobile data)

. summarize

    Variable |        Obs        Mean    Std. dev.       Min        Max
-------------+---------------------------------------------------------
        make |          0
       price |         74    6165.257    2949.496       3291      15906
         mpg |         74     21.2973    5.785503         12         41
       rep78 |         69    3.405797    .9899323          1          5
    headroom |         74    2.993243    .8459948        1.5          5
-------------+---------------------------------------------------------
       trunk |         74    13.75676    4.277404          5         23
      weight |         74    3019.459    777.1936       1760       4840
      length |         74    187.9324    22.26634        142        233
        turn |         74    39.64865    4.399354         31         51
displacement |         74    197.2973    91.83722         79        425
-------------+---------------------------------------------------------
  gear_ratio |         74    3.014865    .4562871       2.19       3.89
     foreign |         74    .2972973    .4601885          0          1
```

The community-contributed command **radar** (Mander 2007) allows Stata users to create radar charts. The command is relatively simple to use, although it generally requires many options to produce desired visual results. Users must always specify at least two variables. The first variable is the group identifier for the spokes and should be categorical. Note that the categories are individual observations in the dataset. This variable can be string or numeric. The second variable must be numeric and represents the statistic that is plotted on the spokes of the radar chart.

A basic example is shown below in figure 11.1. Only the first 12 observations are plotted using the `in` qualifier to reduce initial visual overload. The variable `make` identifies the car name of each observation and acts as a categorical variable. The variable `weight` identifies the corresponding weight for each car.

Creating figure 11.1

```
. * Radar chart
. radar make weight in 1/12
```

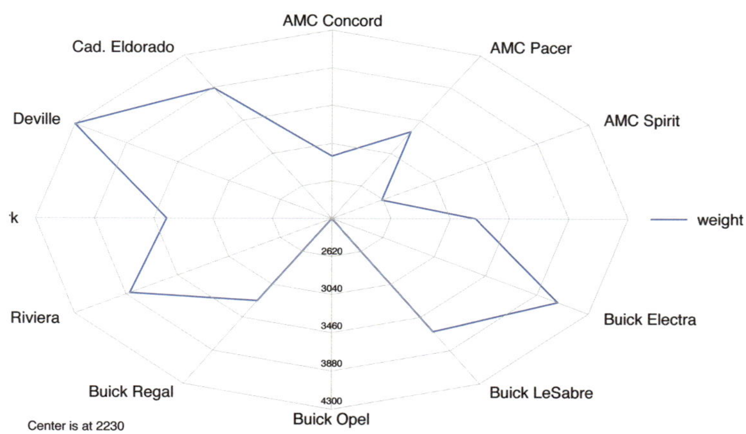

Figure 11.1. A basic radar chart

Figure 11.1 presents an initial radar chart of car `make` against car `weight`. Users should note two important points here: 1) The presented graph is not a univariate distributional plot but a plot of how a second variable varies by categories of a first variable. The complexity of code to achieve a basic distributional plot using **radar** necessitates that this is shown toward the end of this chapter. 2) By default, radar charts have aspect and color problems. The graph in figure 11.1 is too stretched and has text that disappears. Older versions of Stata (17 or earlier) will also encounter two different background colors.

To fix the aspect ratio of the default radar chart, users can invoke the option `aspectratio()` with an aspect ratio of 1. This will normalize the aspect of the graph and result in a more standardized picture. Users with Stata 18 or later are encouraged to specify `xlabel(, nogrid)` to remove the light-gray grid lines that come from the default graph scheme. Users with older versions of Stata (17 or earlier) can specify the `graphregion(color())` option to change the gray-blue inner graph color to white. Both options are exemplified in figure 11.2, although users will need only one of these options, depending on their version of Stata. The remainder of this chapter will use the `xlabel(, nogrid)` option.

Creating figure 11.2

```
. * Radar chart with aspect ratio
. radar make weight in 1/12, aspect(1) graphregion(color(white))  ///
> xlabel(, nogrid)
```

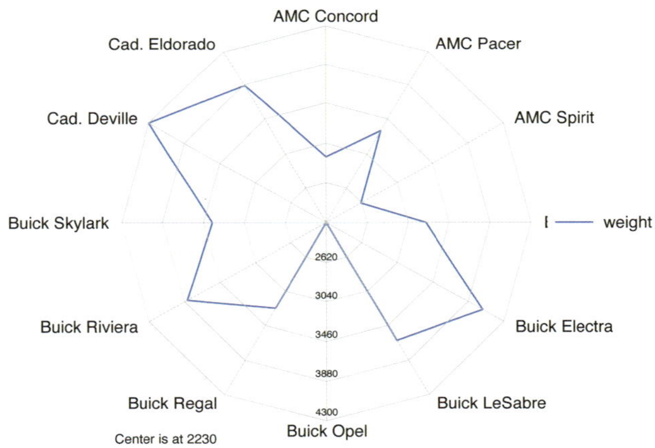

Figure 11.2. Radar chart with adjusted aspect ratio

Figure 11.2 presents a basic, usable radar chart. It is easy to identify that the Cadillac Deville has the highest weight of the cars in question. The Buick Opel has the lowest weight.

11.2.1 Custom spoke labels

A problem with the chart in figure 11.2 is that, by default, the **radar** command centers the radar chart at the minimum value of the presented data. In addition, it sets the maximum spoke length at the maximum value of the data. This means that, by default, all radar charts will have at least one spoke on the edge and one spoke at the center of the chart. This eases rapid maximum and minimum identification, but it may not be visually appealing in other circumstances. Users will often want to set the center of a radar chart to 0. This can be achieved with the **rlabel()** option, where users define a range of numbers that they want as ticks on the spokes. An example is shown in figure 11.3. In this example, the data are restricted to all `foreign` observations via the `if` qualifier.

Creating figure 11.3

```
. * Radar chart with custom spoke lengths
. radar make weight if foreign ==1, aspect(1)    ///
> rlabel(0 500 1000 1500 2000 2500 3000 3500) xlabel(, nogrid)
```

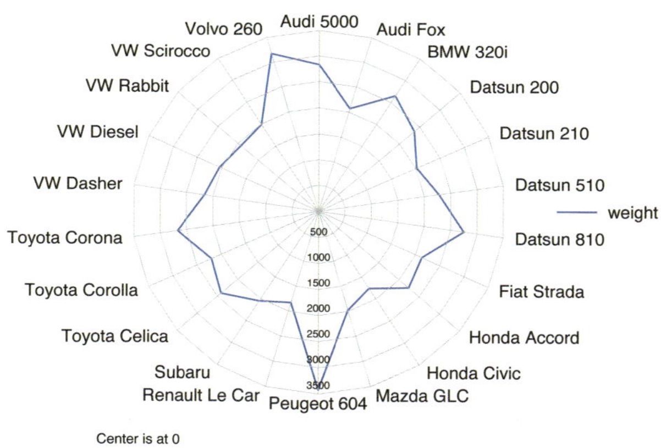

Figure 11.3. Radar chart with custom spoke lengths

Figure 11.3 provides a more visually pleasing radar chart where the connected data line does not touch the inner or outer spoke bounds. It also displays less visual variance.

11.3 Multiple-radar charts

A significant advantage of radar charts is that they can display multiple variables easily. Multiple-radar charts use overlay rather than repeat plots, which allows users to quickly acquire and compare numerous statistics over a range of categorical groups. To achieve this effect with the **radar** command, users simply add extra variables after the second variable. Each additional variable adds a further radar chart as an overlay. Users may need to be cautious of graph legends. By default, legends are placed on the right-hand side of Stata graphs; however, radar charts already include horizontal labels in a graph, often reducing the space available for legend text. In this case, there would be too much overlap, so the `legend()` option is used below with the `position()` suboption to place the legend at the 6 o'clock position. The suboption `row()` places the three labels in one row.

A multiple-radar chart with multiple variables is demonstrated in figure 11.4.

Creating figure 11.4

```
. * Radar chart with multiple variables
. radar make turn mpg trunk if foreign == 1, rlabel(0 10 20 30 40 50)  ///
> aspect(1) legend(position(6) row(1)) xlabel(, nogrid)
```

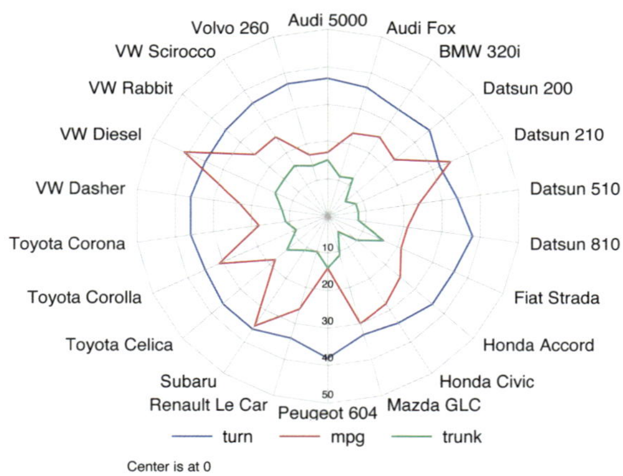

Figure 11.4. Multiple-radar chart with multiple variables

Because the spoke measurements are identical for all three variables in figure 11.4, it is easy to identify that the variation in miles per gallon is more extreme than the variation in turn radius. In addition, the VW Diesel has a higher miles-per-gallon value. A lot of statistical information is present on this graph; this is a key advantage of radar charts.

When the category count becomes large, it may be worth helping users focus on specific categories and data points to identify statistics. A useful option is `connected`, which adds marker points to each spoke, thereby better signifying the statistical value of each category and variable. The option `radial()` causes a variable to be plotted as spikes that emerge from the center and travel along the spokes. `radial()` can be useful to emphasize one particular variable in a multiple-radar chart. Figure 11.5 presents an example using both options.

Creating figure 11.5

```
. * Radar chart with multiple variables and radial representation
. radar make turn mpg if foreign == 1, rlabel(0 10 20 30 40 50) ///
> aspect(1) connected radial(trunk) legend(position(6) row(1))  ///
> xlabel(, nogrid)
```

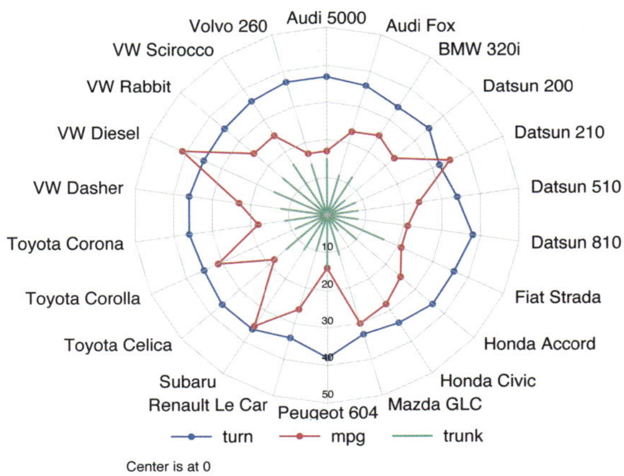

Figure 11.5. Multiple-radar chart with multiple variables and radial representation

Figure 11.5 demonstrates how marker points and spikes can lead to a different visual effect that allows more statistical detail to be inferred by end users, especially when the plot contains many categories.

11.4 Radar charts with too few or too many categories

Users should limit radar charts to a reasonable set of categories. Radar charts with too few or too many categories become difficult to interpret. Two examples are shown below. The first example, in figure 11.6, presents a radar chart for three categories.

Creating figure 11.6

```
. * Radar chart with too few categories
. radar make weight in 1/3, aspect(1) rlabel(0 1000 2000 3000 4000 5000) ///
> labsize(*0.5) xlabel(, nogrid)
```

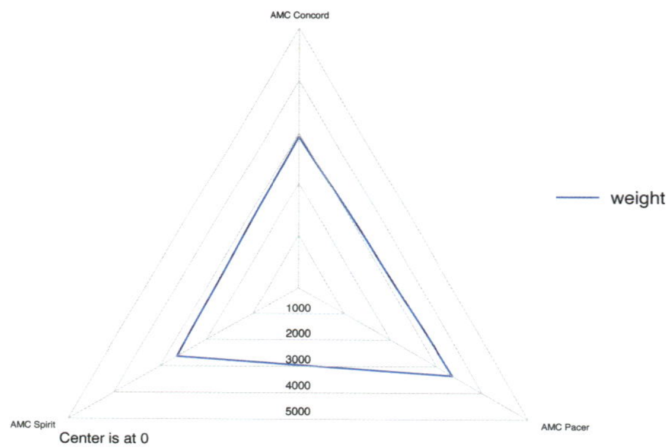

Figure 11.6. Too few categories in a radar chart

This is a poor radar chart. The eye must travel far across this picture to identify statistics and perform comparative actions. Figure 11.7 below presents an opposite use case with 74 categories, which is too many.

Creating figure 11.7

```
. * Radar chart with too many categories
. radar make weight, aspect(1) rlabel(0 1000 2000 3000 4000 5000) ///
> labsize(*0.5) legend(position(6) row(1)) xlabel(, nogrid)
```

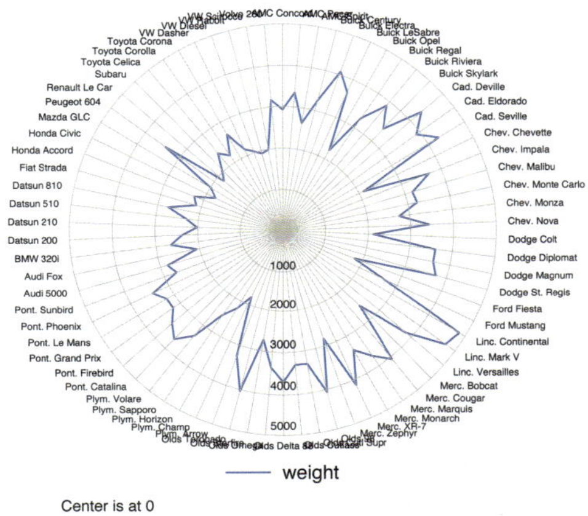

Figure 11.7. Too many categories in a radar chart

Interestingly, the chart remains relatively well presented; with some careful examination, individual statistics and comparisons could be identified. However, this number of observations is likely toward the upper reasonable limit for radar charts. Point-to-point comparison is made more complex, and such graphs should be limited to making assessments of the general categorical distribution of the data rather than individual category-to-category comparisons.

11.5 Frequency radar chart

The `radar` command plots observations as categories, with a second variable providing the spoke statistics. If users wish to plot the univariate frequency distribution of discrete variables, they must resort to more complex code. In this case, they must first use the `collapse` command to create an appropriate new dataset. An example with the discrete variable `rep78` is shown below.

```
. tabulate rep78

     Repair |
record 1978 |      Freq.     Percent        Cum.
------------+-----------------------------------
          1 |          2        2.90        2.90
          2 |          8       11.59       14.49
          3 |         30       43.48       57.97
          4 |         18       26.09       84.06
          5 |         11       15.94      100.00
------------+-----------------------------------
      Total |         69      100.00
```

The variable rep78 has five distinct categories. To visualize the frequency distribution of this variable using the radar command, users must collapse the data. Before the data are collapsed, a new variable, obs, is created that takes the value 1 for each observation. This variable is then summed in the collapse command by each value of rep78, leaving behind a small new dataset with five observations containing the sum of frequencies of each value of the categories in rep78. The radar command can then be used to plot this information. The result is shown in figure 11.8.

```
. * Collapse data
. generate obs = 1
. collapse (sum) obs, by(rep78)
```

Creating figure 11.8

```
. * Radar chart of collapsed data
. radar rep78 obs, rlabel(0 10 20 30 40) aspect(1) xlabel(, nogrid)
```

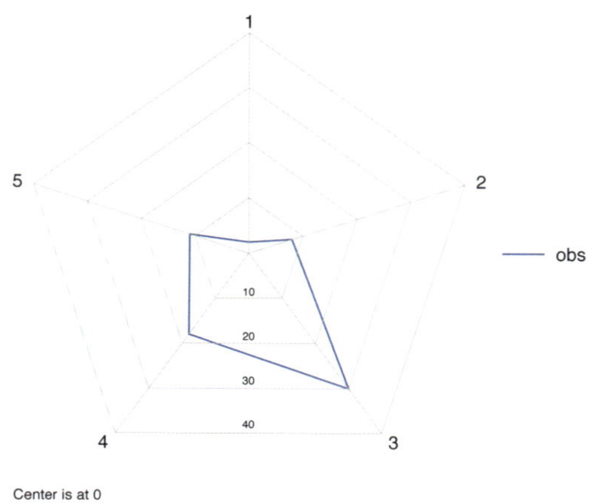

Figure 11.8. Frequency radar chart

Figure 11.8 shows how to use the radar command to create distributional plots similar to bar charts and pie charts for categorical data.

Part II

Graphs for bivariate data

Bivariate analysis is one of the primary forms of testing basic relationships between two variables. The variables are jointly examined, and the association between them is hypothesized and tested. Numerically, this often results in some form of regression or correlation analysis where one or more parameters indicate the strength of the association. Depending on the variables and context, a causal interpretation is sometimes imparted on the relationship (for example, a change in x causes a change in y). Alternatively, a correlated interpretation is used when both variables randomly covary. One significant disadvantage of numerical associations is that they usually reduce complex variables with many values into one parameter or a handful of parameters.

Graphical analysis of how two variables are related can often overcome this limitation and, in turn, allow users more freedom in interpreting how a relationship varies across a joint distribution. It can highlight local parts of the joint density, identify outliers, and explore the appropriateness of various parametric and nonparametric fitted plots. The graphical representation of two variables is usually one of the most powerful visualizations one can produce because the visual dimensions are perfectly suited to plotting two variables along the Cartesian plane.

Bivariate analysis is generally the second step after users carefully explore individual variables. Usually, it will be a precursor to more serious and advanced model building, especially parametric models that require assumptions about functional form. The bivariate visualization of data often helps users establish initial functional-form conditions that are then fed into more complex statistical models. In addition, it shows areas of high data density, helping users understand what part of the joint density might be driving their results. It is rare for modern data analysts to proceed straight to multivariate modeling without first carefully examining bivariate relationships for coding, functional-form, density, and outlier problems. Especially in regression modeling, where a dependent variable (y) is often assumed to be driven by many independent variables (x), it is common to visually check the bivariate relationship between y and the many x variables.

12 Scatterplots

Advantages

1. Display raw data
2. Are great for small datasets
3. Do not impose parametric assumptions
4. Allow decisions on the nature of the data relationship
5. Have many customization options for markers

Disadvantages

1. May provide different perceptions of data relationships to different readers
2. Fare poorly with discrete variables
3. Do not work very well for large datasets (overplotting)
4. Can be influenced by scaling and axis limits

Required community-contributed packages

To install the community-contributed commands that are required to follow along with this chapter completely, type

```
. ssc install fabplot
. ssc install sepscatter
. net install http://digital.cgdev.org/doc/stata/MO/Misc/superscatter
. ssc install binscatter
```

12.1 Introduction

Scatterplots are the go-to visualizations for anyone seeking to explore how two continuous variables might be related. They are the mainstays of bivariate data analysis, and no graphical visualization is simpler or more intuitive than a basic scatterplot. Essentially, scatterplots plot data values from two variables on a 2D plane where each plane (axis) represents a variable. Individual data markers (often visualized as dots, but any type of symbology is possible) representing pairs of x–y values (coordinates) for each unique observation.

Identification of a correlational relationship is the most common use of scatterplots, and assuming a small- or medium-sized dataset, the resulting scatterplot will often clearly show the form of the relationship between x and y. Relationships between variables can be described in many ways—for example, as positive or negative, strong or weak, or linear or nonlinear. Other uses of scatterplots include identifying outliers in data, examining differences in variation in y (or x) among parts of the x (or y) distribution, or simply understanding the scale of both variables.

A core advantage of scatterplots is their nonparametric nature: they plot the raw data as Cartesian coordinates on a graph. Because readers see the actual data, they can make up their own minds on the nature of the data relationship. This, in turn, is also their biggest disadvantage; if anyone can make up their mind, then some viewers might interpret the data relationship in opposing or different ways. Moreover, there is only so much information the eye can process. Once observations reach 1,000 or more, it can be difficult to keep track of patterns, especially if there are many overlapping observations. Overplotted scatterplots are of little use to analysts.

12.2 A basic scatterplot

Scatterplots are designed to answer basic relationship questions of underlying individual data. Users will want to examine how two continuous variables are related. This chapter uses `auto.dta`, which contains information about car prices and characteristics from 1978. The small sample size of only 74 observations means that scatterplots from these data are unlikely to suffer from overplotting. A summary of the dataset is shown below.

```
. * Load data
. sysuse auto
(1978 automobile data)

. summarize

    Variable |        Obs        Mean    Std. dev.       Min        Max
-------------+---------------------------------------------------------
        make |          0
       price |         74    6165.257    2949.496       3291      15906
         mpg |         74     21.2973    5.785503         12         41
       rep78 |         69    3.405797    .9899323          1          5
    headroom |         74    2.993243    .8459948        1.5          5
-------------+---------------------------------------------------------
       trunk |         74    13.75676    4.277404          5         23
      weight |         74    3019.459    777.1936       1760       4840
      length |         74    187.9324    22.26634        142        233
        turn |         74    39.64865    4.399354         31         51
displacement |         74    197.2973    91.83722         79        425
-------------+---------------------------------------------------------
  gear_ratio |         74    3.014865    .4562871       2.19       3.89
     foreign |         74    .2972973    .4601885          0          1
```

In Stata, the command to create scatterplots is the `scatter` command. `scatter` is also a plottype within the `twoway` family of plots, and it is common to overlay multiple scatterplots on top of each other with the `twoway` command. It is worth getting into the habit of using the `twoway` command with the `scatter` plottype as much as possible because scatterplots are often combined with additional plottypes. This will be demonstrated later in the chapter. To create a basic scatterplot in Stata, specify a y and an x variable after the `scatter` command.

Creating figure 12.1

```
. * Basic scatterplot
. twoway (scatter price mpg)
```

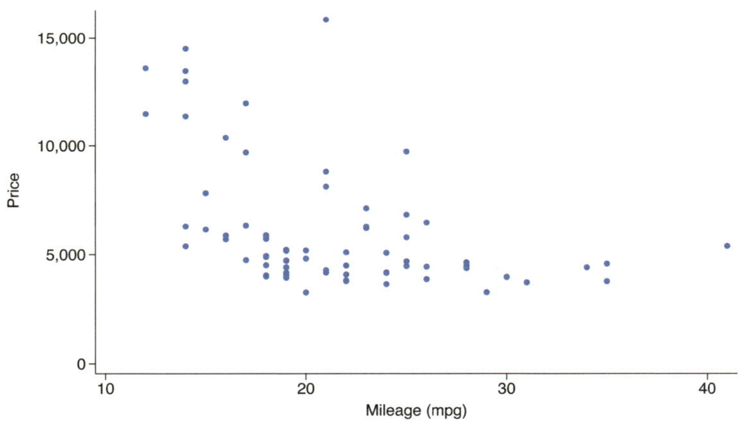

Figure 12.1. A basic scatterplot

Figure 12.1 presents a basic scatterplot of car price versus car mileage, and the results suggest that the relationship between price and mileage is negative. In other words, as car price increases, car mileage decreases and vice versa. The visualization is powerful, is simple to understand, and perfectly portrays how the variables are related. The raw data are plotted, and no parametric assumptions are required in creating this visualization.

One disadvantage is that the nonparametric nature of the plot can lead to different opinions. While many will see a negative relationship, some may see a negative relationship that then becomes linear. In other words, the relationship may well be interpreted as piecewise linear, quadratic, or even fractional. The most appropriate choice of relationship interpretation can be hard to justify and remains subject to individual preferences.

12.2.1 Important marker options

A key feature of any scatterplot is the versatility of the markers. Markers are not required to be small blue dots; they can take any shape, size, or color. This can be an important feature for marketing, presentation, and other visual purposes. Stata offers a wide range of marker and marker label options that can be used to customize how individual markers look. The most important options are `msymbol()`, `mcolor()`, `msize()`, and `mlabel()`; these are exemplified below in figure 12.2. These four options change the symbol of the marker, change the color of the marker, change the size of the marker, and add labels next to the markers, respectively. The default symbol is a solid circle, but other shapes can be requested.

The example below defines solid square shapes via the `msymbol(S)` option. The color of the symbol is changed by specifying `mcolor(edkblue)` as the *colorstyle*, which corresponds to *The Economist* magazine's dark-blue color. The color is made somewhat transparent by specifying `%70` in the `mcolor()` option after the color is specified. The `msize()` option changes the size of markers, and although there are various choices for *markersizestyle* available, an easy way to modify sizes is to specify `msize(*3)`, which multiplies marker sizes by 3. Numbers above 1 lead to larger markers, and numbers between 0 and 1 lead to smaller markers. Finally, the `mlabel()` option adds labels next to the data markers. These labels can be text or numeric and help to identify the individual observations. In this case, the variable `make` is specified as containing marker label information. Many additional options exist, and users should examine the help file of `scatter` (`help scatter`) to learn more about the many options available to modify scatterplots visually.

Creating figure 12.2

```
. * Square, colored, and sized markers
. twoway (scatter price mpg in 1/15, msymbol(S) mcolor(edkblue%70) ///
> msize(*3) mlabel(make))
```

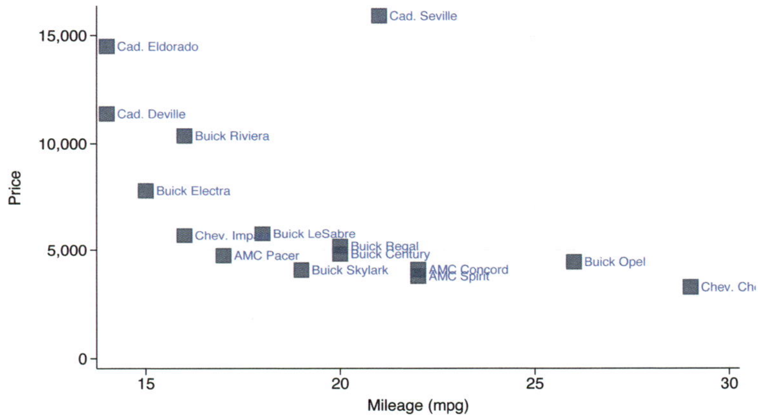

Figure 12.2. Scatterplot with custom markers

12.3 Scatterplots with multiple y and x variables

A useful feature of the `scatter` command is that it can plot multiple y variables against one x variable, all within the same command syntax. This avoids using lengthier `twoway` code, where each scatterplot is specified independently and subsequently overlaid. For example, it is simple to examine how the variables `trunk`, `turn`, and `mpg` relate to the variable `price`. Insert all the relevant y variables after `scatter`, and conclude with the relevant x variable. When using the multiple-variable syntax of `scatter`, you can specify multiple `scatter`-level options, each corresponding to a combination of y and x variables. This makes it faster to specify multiple marker colors, marker symbols, or other options than to significantly change the graph outlook. In figure 12.3 below, the `msymbol()` and `mcolor()` options are used to create square (`S`), triangular (`T`), and diamond (`D`) marker symbols in three different grayscale colors (`gs1`, `g6`, and `g12`).

Creating figure 12.3

```
. * Multiple y variables against one x variable
. twoway (scatter trunk turn mpg price, msymbol(S T D) mcolor(gs1 gs6 gs12))
```

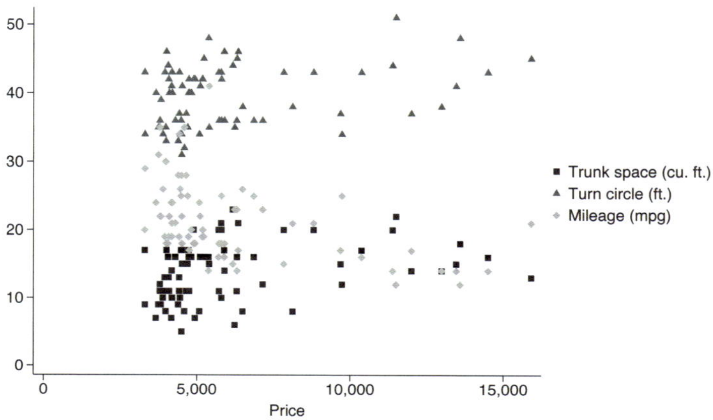

Figure 12.3. Scatterplot with multiple y variables

Some care must be taken when using the built-in ability of `scatter` to plot many y variables. `scatter` does not automatically place multiple y variables on different y axes; this results in one shared axis for all y variables. This shared axis can quickly turn unpleasant when the specified y variables have different scales, as shown below in figure 12.4. Here the variables `weight`, `trunk`, `turn`, and `mpg` are all treated as y variables to be scattered against the x variable `price`. As seen below, three of the four y variables are clustered toward the bottom of the graph because the fourth y variable, `weight`, has values in the thousands and not in the tens.

Creating figure 12.4

```
. * Beware of scales when using multiple y variables
. twoway (scatter weight trunk turn mpg price)
```

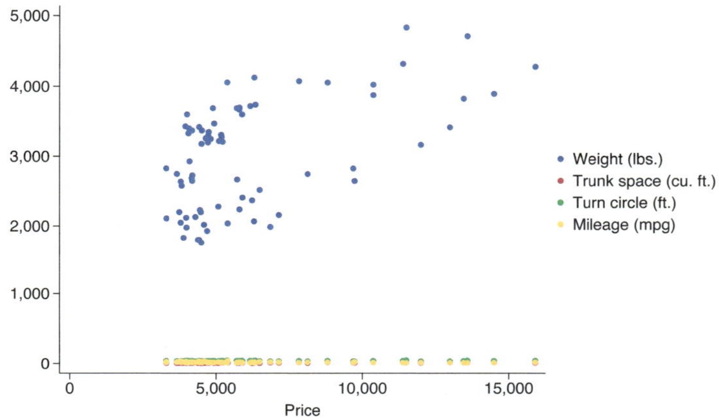

Figure 12.4. Scatterplot with multiple y variables and extreme scaling

The solution to this problem is to specify multiple y axes via the axis-choice options, which allow users to specify different x and y axes for different variables. Users must invoke the `twoway` command and create multiple individual scatterplots that are then overlaid. Each overlay is then individually adjusted via the axis options.

In figure 12.5 below, the `yaxis()` option is used to create four different y axes, one for each y variable. The initial instinct to specify `yaxis(1 2 3 4)` as an option within the `scatter` command, such as in the code for figure 12.3, does not produce the correct graph. Instead, this approach would create four different y axes of the original scale with values in the thousands. The correct way is to separate each graph into independent scatterplots and overlay these via the `twoway` command. In each `scatter` subgraph, the `yaxis()` option is used to place each y variable on an independent y axis. By default, y axes are stacked on the left side of the graph when there are three or more. Similarly, x axes are stacked on the bottom when there are three or more. To avoid overstacking, the `alt` suboption in the `yscale()` option is used to move some y axes to the right side in the final plot. When only two axes are specified, Stata will automatically place these on the left or right and top or bottom of the graph.

Creating figure 12.5

```
. * Adjusting scales with multiple axes
. twoway (scatter weight price, xaxis(1))                    ///
> (scatter trunk price, yaxis(2) yscale(axis(2) alt)) ///
> (scatter  turn price, yaxis(3))                            ///
> (scatter mpg price, yaxis(4) yscale(axis(4) alt))
```

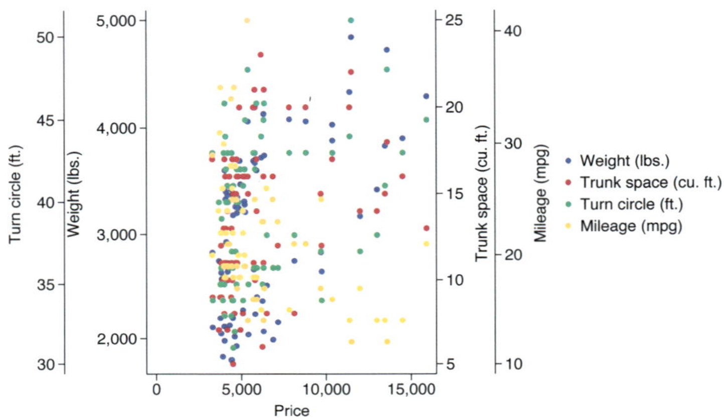

Figure 12.5. Scatterplot with multiple y axes

Figure 12.5 is an example of four scatterplots overlaid on top of each other using the `yaxis()` and `yscale()` options to modify the axis behavior and overcome differential scale problems in the y variables. Note that figure 12.5 is not a particularly useful plot for data analysis. In most cases, users are advised to plot such data on different graphs rather than resort to multiple y axes on a single graph. The same can be done for the x axis, and a slightly simpler plot is demonstrated below in figure 12.6 where two different x axes are used.

Creating figure 12.6

```
. * Advanced syntax, two axes
. twoway (scatter price mpg, xaxis(1) yaxis(1 2))   ///
> (scatter price turn, xaxis(2))
```

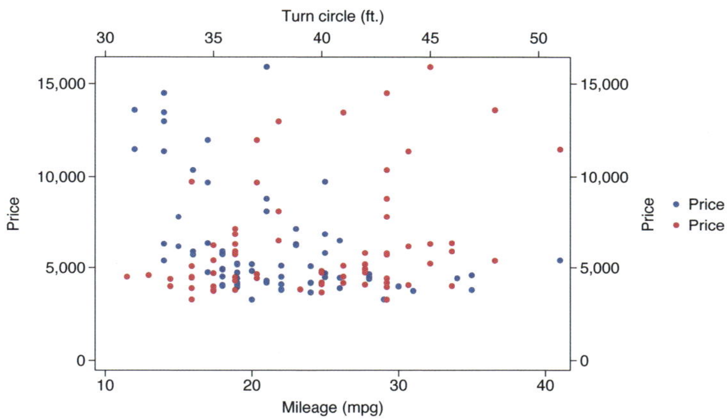

Figure 12.6. Scatterplot with multiple x axes

12.4 Scatterplots over categorical groups

Another common use of scatterplots is overlaying multiple scatterplots by categorical groups. The previous examples demonstrated how multiple continuous variables can be overlaid in `scatter`, but arguably, an even more common use of scatterplots is the overlay of the same variables but by different categorical groups. In official Stata, this can often be done in three primary ways: 1) using the `if` qualifier, 2) using the `by()` option, or 3) using the `colorvar()` option. Figure 12.7 demonstrates an example using the `if` qualifier, where five x individual scatterplots, separated by categories of the variable `rep78`, are overlaid.

Creating figure 12.7

```
. * Using the if qualifier
. twoway (scatter price mpg if rep78 == 1)  ///
> (scatter price mpg if rep78 == 2)         ///
> (scatter price mpg if rep78 == 3)         ///
> (scatter price mpg if rep78 == 4)         ///
> (scatter price mpg if rep78 == 5)
```

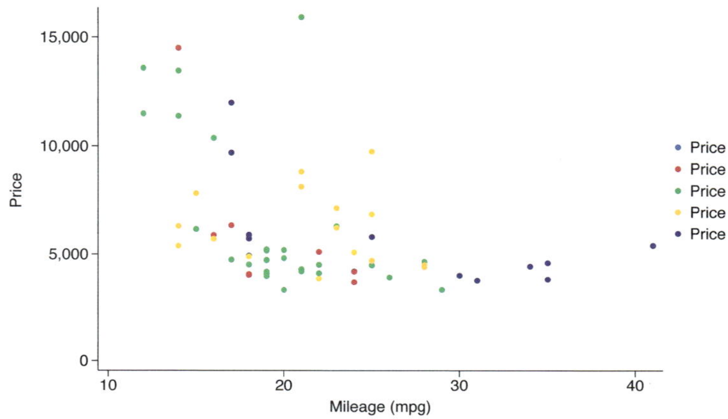

Figure 12.7. Scatterplot with multiple groups using `if`

Figure 12.7 demonstrates how `scatter` can be used to scatter multiple categories of a single variable on a graph. The `if` qualifier is perhaps the most intuitive and common way to achieve multiple scatterplots over a group. As can be seen by the visualization, however, the default labeling is poor and color choices may not be optimal. In addition, it requires rather lengthy coding.

The second method is invoking the `colorvar()` option, which colors the points in the scatterplot to reflect different groups. Simply specify the relevant group variable in `colorvar()`, and add the `colordiscrete` and `coloruseplegend` options. The option `colordiscrete` treats the `colorvar()` variable as discrete instead of continuous, and the option `coloruseplegend` ensures that the legend is also discrete. An example is shown in figure 12.8. Note that the qualifier `if !missing(rep78)` is added to the `scatter` command to ensure that no missing values of the variable `rep78` are plotted.

Creating figure 12.8

```
. * Using the colorvar() option
. scatter price mpg if !missing(rep78), colorvar(rep78) ///
> colordiscrete coloruseplegend
```

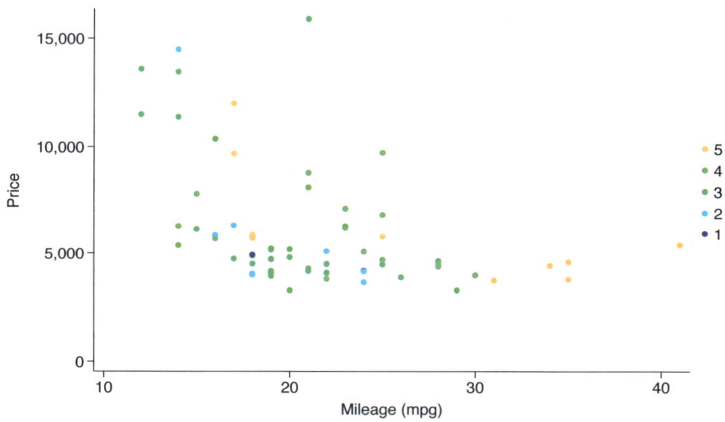

Figure 12.8. Scatterplot with multiple groups using the `colorvar()` option

Figure 12.8 is very similar in style to figure 12.7 but has better default legend labeling. It is the preferred method to create such plots because it avoids lengthier code but requires users to have Stata 18 or higher. A third option is the `by()` option. Rather than overlaying each subgraph, this option produces multiple side-by-side scatterplots. This is exemplified in figure 12.9, where the suboption `compact` is also used to ensure a minimal distance between the individual graphs.

Creating figure 12.9

```
. * Using the by() option
. twoway (scatter price mpg, by(rep78, compact))
```

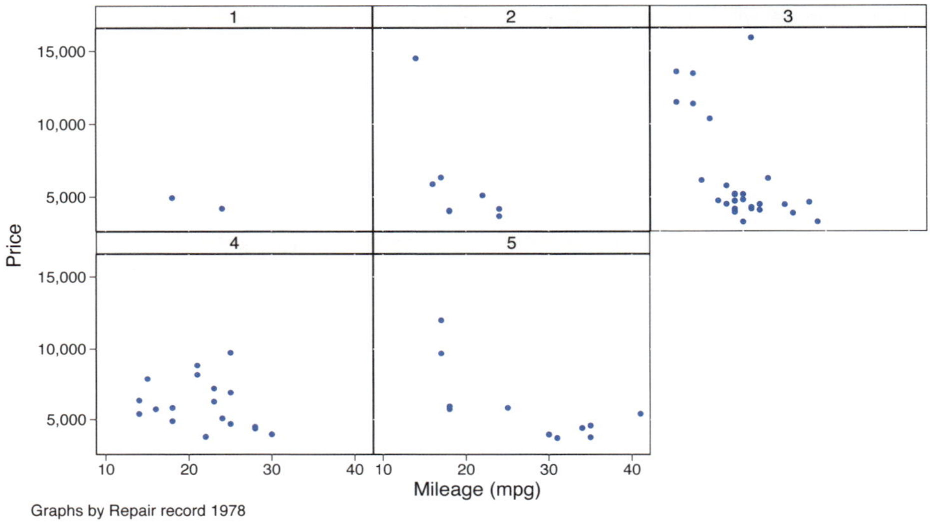

Figure 12.9. Scatterplot with multiple groups using the `by()` option

From a coding perspective, figure 12.9 presents an easier method to identify the different relationships between `price` and `mpg` among the five different `rep78` categories compared with figure 12.7. At the same time, multiple side-by-side graphs make it harder to compare the relationships across categories because the eye must travel across the graphs and potentially lose focus with respect to the axes' ranges. A partial way to overcome this is to specify the `total` suboption within the `by()` option, which will create a final graph with all categories merged.

A neater method is to use the community-contributed command `fabplot` (Cox 2018). `fabplot` requires a Stata command to be specified before the variables. This could be any of the `twoway` graph commands, and its use is not limited only to `scatter`, although only a scatterplot is demonstrated here. Essentially, `fabplot` is similar to a standard `scatter` command with the option `by()` specified, but it sets each panel within the context of all the data. In other words, the total relationship between y and x is used as a backdrop in each subpanel. The advantage of this is that it becomes much easier to contextualize each categorical relationship within all the data space. An example is shown in figure 12.10.

Creating figure 12.10

```
. * fabplot
. fabplot scatter mpg weight, by(rep78)
```

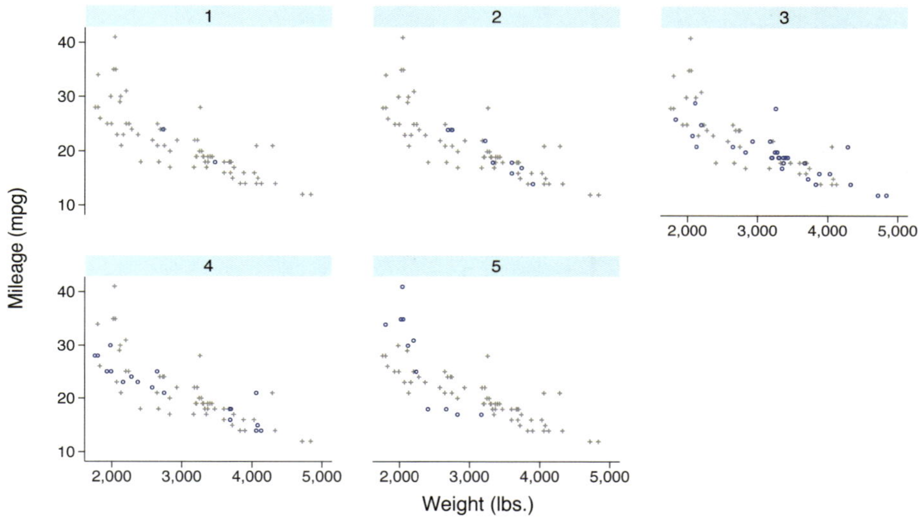

Figure 12.10. Scatterplot with multiple groups using `fabplot`

Figure 12.10 demonstrates how useful the `fabplot` command can be in showing how each subpanel's data fit within the entire dataset. Note that `fabplot` can be used with many other Stata graph commands; users should explore the help file of `fabplot` for more information (see `help fabplot`).

`fabplot` creates multiple side-by-side scatterplot panels, which may not always be desired. Such scatterplots create more visual clutter via repetition and require more careful examination by casual users. The competing choice is usually a variation of figure 12.7, which overlaid multiple groups. To avoid overplotting disasters, markers need careful adjustment (for example, size, shape, or color) to identify individual categories on a single scatterplot. An easy way to accomplish this without specifying a plethora of *marker_options* is to use the community-contributed `sepscatter` command (Cox 2014b). `sepscatter` works just like a standard scatterplot (see the examples shown in figures 12.7 and 12.8), but in addition to creating different marker colors, it also creates different marker symbols by a categorical variable specified in the `separate()` option. An example is shown in figure 12.10.

Creating figure 12.11

```
. * sepscatter
. sepscatter mpg weight, separate(rep78)
```

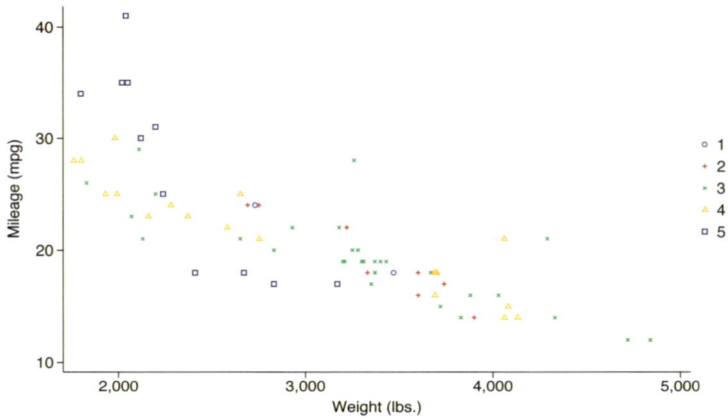

Figure 12.11. Scatterplot with multiple groups using `sepscatter`

12.5 Scatterplots with marginal distributions

Scatterplots are basic relationship plots, and when comparing relationships, it may be helpful for users to have some understanding of the underlying individual data distributions that make up the bivariate scatterplot. This can help users better understand where data are clustered or located across the x and y ranges. Marginal distributions (for example, the individual univariate distributions of the bivariate relationship) can be added to both the x and the y axes to create a better sense of data density.

While it is perfectly feasible to create and merge many graphs, the easiest way to perform such an operation is with the community-contributed command `superscatter` (Over 2020). `superscatter` is a versatile command offering many customization options for creating unique scatterplots, and its help file (`help superscatter`) offers more than just marginal distributions. Its default operational mode is to add marginal distribution plots to both the x and the y axes. Users simply specify y and x variables after the primary command. An example is shown below in figure 12.12.

Creating figure 12.12

```
. * Marginal distributions
. superscatter price mpg, ylabel(5000(5000)15000)
```

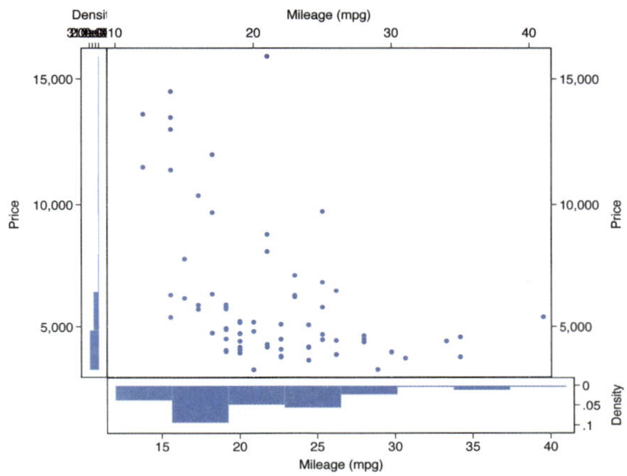

Figure 12.12. Scatterplot with marginal distributions

Figure 12.12 presents a standard scatterplot of `price` versus `mpg`, and a histogram with the relevant univariate distribution complements each axis. It is easy to see that most cars (data points) have a low price and mileage. Users can specify the `kdensity` option to redraw the graph with kernel density plots. In this case, the histogram y axis does not perfectly fit the scatterplot y axis. This requires either 1) tinkering with axis-choice options `ylabel()` and `xlabel()` or 2) invoking the `kdensity` option to redraw the graph with kernel density plots.

An example with the `kdensity` option is shown below in figure 12.13, which also uses the `detail`, `tabulate()`, and `means` options to display additional statistics, tabulate counts in each quadrant, and quarter the scatterplot by means, respectively.

Creating figure 12.13

```
. * Marginal distributions with options
. superscatter price mpg, detail kdensity tabulate(count) means
```

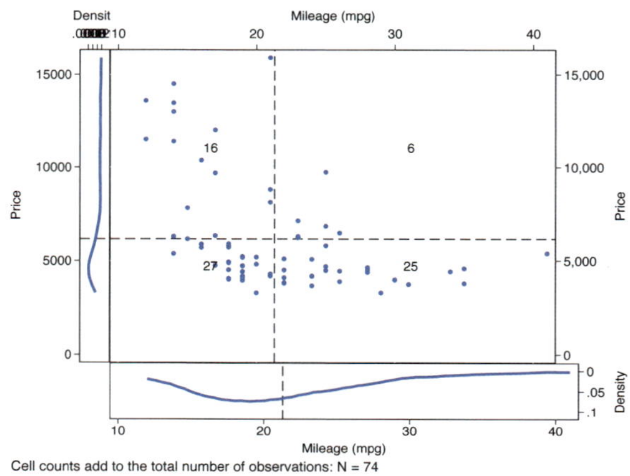

Figure 12.13. Scatterplot with marginal distributions and means

The resulting scatterplot contains an incredible amount of detail, which allows users to carefully examine the univariate distributions and the bivariate relationships between `price` and `mpg`. Many more options exist as part of `superscatter`.

12.6 Binned scatterplots

Scatterplots tend to work poorly when faced with extremely large datasets. Overplotting is the most common reason to move beyond traditional scatterplots and toward other specialized plottypes. However, scatterplots can still be used even when data samples are large. Several examples of this using the National Health and Nutrition Examination Survey dataset (`nhanes2.dta`) are shown below.

`nhanes2.dta` is a medical dataset that contains information on various health characteristics of a sample of over 10,000 individuals. A basic analysis might involve examining the relationship between the variables `weight` and `height`.

```
. * Binned scatterplots
. webuse nhanes2
. summarize weight height
```

Variable	Obs	Mean	Std. dev.	Min	Max
weight	10,351	71.89752	15.35642	30.84	175.88
height	10,351	167.6509	9.655916	135.5	200

Both `weight` and `height` are continuously measured, but a traditional scatterplot would quickly become overloaded with over 10,000 markers. The first approach should always use smaller and transparent marker symbols to see whether a traditional scatterplot can still be constructed. In figure 12.14 below, the `mcolor()` option is used to adjust the markers.

Creating figure 12.14

```
. * The problem
. twoway (scatter weight height, mcolor(%50))
```

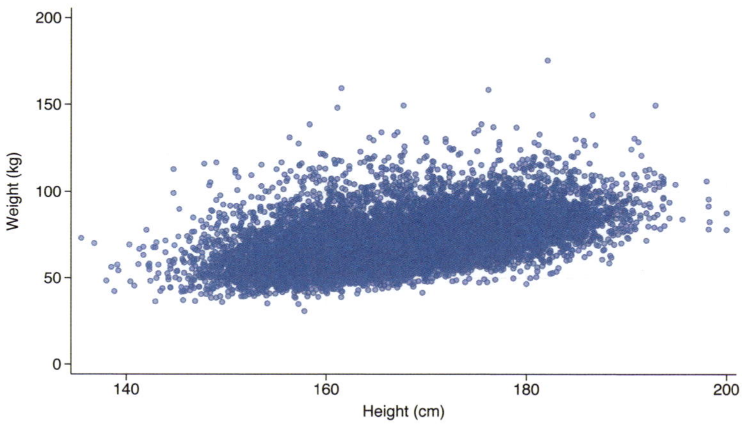

Figure 12.14. An overplotted scatterplot

Figure 12.14 shows the limits of a traditional scatterplot. Even with smaller and transparent markers, only the outer circumference of the bivariate relationship is visible. It is unknown whether the upward-sloping relationship between `weight` and `height` is maintained inside the blue blob.

As an alternative option, the community-contributed `binscatter` command (Stepner 2013) is available. The `binscatter` command divides the x-axis variable into equal-sized bins, calculates the average values for both the x axis and the y axis within each bin, and then produces a scatterplot of these values. This results in a scatterplot that visualizes the conditional expectation (also called conditional mean) function. In other words, it is a binned scatterplot. By default, `binscatter` also plots a line of best fit, but this can be removed (or adjusted) with the `linetype()` option. Moreover, because of the automatic adjustment of the y axis due to binning, the `ylabel()` option often must be used to ensure comparability with a previous overplotted scatterplot (figure 12.14).

Creating figure 12.15

```
. * Using binscatter to reduce number of markers
. binscatter weight height, linetype(none) ylabel(0(50)200)
```

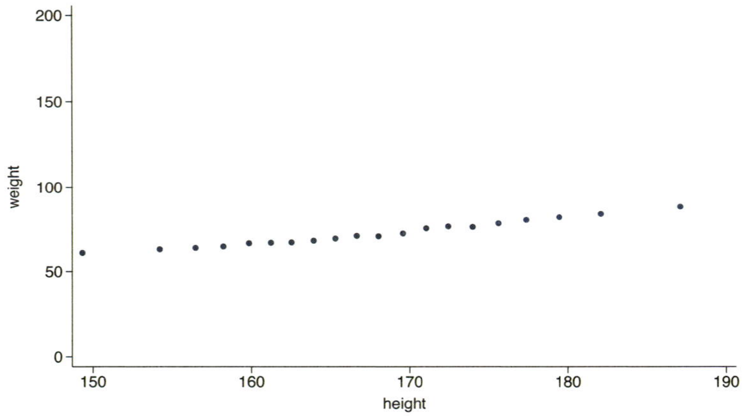

Figure 12.15. Binned scatterplot

Figure 12.15 presents a basic binned scatterplot. Instead of an overplotted scatterplot with too many markers, the result is a much-reduced scatterplot that displays the relationship between `weight` and `height`.

Many options are available as part of `binscatter`. Lines of best fit can be added via the `linetype()` option, and the number of bins can be adjusted with the `nquantiles()` option. The `medians` option plots within-bin medians instead of means, and the `by()` option creates a separate overlaid scatterplot for each value of the categorical variable specified in the option. Each individual binned scatterplot can be further adjusted using standard `scatter` options such as `mcolor()`. An example is shown in figure 12.16.

Creating figure 12.16

```
. * binscatter with line fit, median, and over group
. binscatter weight height, linetype(lfit) nquantiles(100) ///
> ylabel(0(50)200) medians by(sex) mcolor(%50 %50)
```

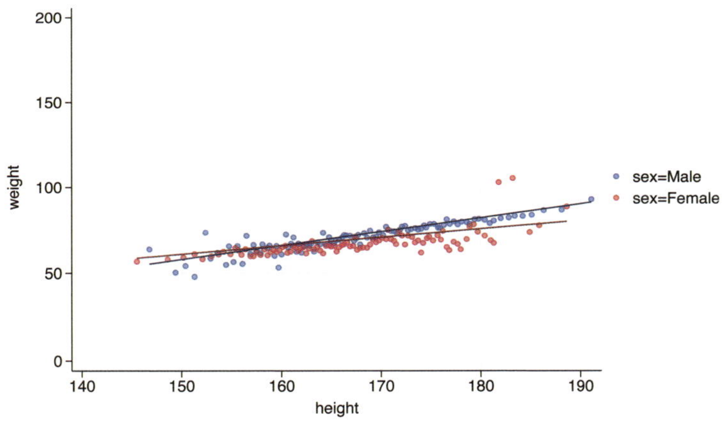

Figure 12.16. Multiple binned scatterplots

Binned scatterplots are great visualization options if users want to retain the traditional visual approach of a scatterplot, but a standard scatterplot would be overplotted.

13 Heat plots

Advantages

1. Are great for examining bivariate distributions in large datasets
2. Deal well with overplotting
3. Provide strong, easy-to-understand narratives via colors

Disadvantages

1. Are not useful for small datasets
2. Use color shading, which may not provide accurate statistical information
3. Can be greatly affected by bin sizes and color choices
4. Use rectangular bins, which emphasize vertical and horizontal correlations
5. May be less interpretable for viewers with color vision deficiencies

Required community-contributed packages

To install the community-contributed commands that are required to follow along with this chapter completely, type

```
. ssc install heatplot
. ssc install palettes
. ssc install colrspace
```

13.1 Introduction

Heat plots, also called heat maps, are graphs that use color gradients to represent some aspect of the data being displayed, often bivariate data density. Two (sometimes three) color combinations are used to depict areas of data on a graph that contain few or many data points. Strong- or light-shaded colors are often associated with maximum or minimum values, while mixed or gradient colors are associated with numerical values between such minimums or maximums. Heat plots have many practical applications, including displaying topography, mouse or eyeball tracking on websites, highlighting the movement of players in sports games, or simply correlating variables. In all cases, heat plots generally display bivariate density distributions of underlying data placed on the x and y planes. Some heat plots are used for trivariate data analysis and will represent a z variable; this is explored in a later chapter. In this chapter, the focus will be on the bivariate cases.

A common aspect of all heat plots is colorization and their binning. Colors play an essential role in heat plots, and some thought should be given to what color scheme is most suited to the analysis in question. Are red–green colors the right choice, or is a yellow–blue combination more appropriate? Which colors should represent maximum and minimum values? These are important considerations for all heat plots. Binning also plays a key role in the creation of heat plots, which are sometimes called two-way histograms because their mechanisms rely on binning data markers into squares (or other tessellation patterns). Like histograms, bin size and shape choices can significantly affect the final graph. Careful thought should be given to whether larger or smaller bins are required for a particular analysis or visualization.

13.2 A basic heat plot

A typical application of heat plots is to reveal bivariate densities when basic scatterplots become overloaded with observations (overplotting). While reducing marker sizes is an effective strategy (this can often be achieved with the `msize()` option), this solution reaches a natural limit once sample sizes cross a certain threshold, when many data are concentrated in one particular graph region, or when editing requirements limit marker or symbology size. An example of an overplotted scatterplot is shown below in figure 13.1, which seeks to visually identify the association between person weight and person height using `nhanes2.dta`. This dataset contains about 10,000 individual observations and 58 variables. Only the variables `weight` and `height` are summarized below.

```
. webuse nhanes2
. summarize weight height
    Variable |        Obs        Mean    Std. dev.       Min        Max
-------------+---------------------------------------------------------
      weight |     10,351    71.89752    15.35642      30.84     175.88
      height |     10,351    167.6509    9.655916      135.5        200
```

Creating figure 13.1

```
. * The basic problem
. twoway (scatter weight height)
```

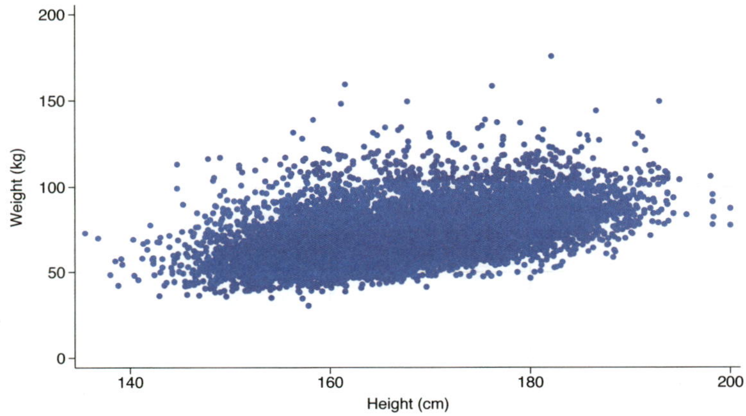

Figure 13.1. An overplotted scatterplot

Figure 13.1 is a classic example of when scatterplots begin to lose information efficiency. A rough outline of the relationship and data concentration between person weight and height is shown, but any inference around a relationship is determined primarily via outliers and edge cases, not the central data mass. The positive (and apparently linear) relationship depicted by this scatterplot may not hold within the "blue blob" of this graph. This "blue blob" is where most data values lie and, therefore, where inference will have the most importance. A different visualization technique is required.

A standard heat plot is akin to a bivariate histogram that bins x and y values into rectangles. But rather than displaying a 3D plot where an imaginary z axis measures observation count, it displays a 2D plot with rectangles, where colors mimic the third z axis that measures observation count. Each rectangle is assigned a color value dependent on the number of observations in the rectangle. The exact color values will often be a linear gradient of two primary colors representing maximum and minimum bin values. This leads to a simple and colorful plottype that avoids pseudo-3D representation but can effectively communicate bivariate densities.

Stata's official `twoway contour` command includes a heat plot option, but the community-contributed command `heatplot` (Jann 2019) offers more options and flexibility. It works in three different modes: it can create heat plots from 1) variables, 2) a standard Stata matrix, and 3) a Mata matrix. The first option is the most commonly used, requiring the command `heatplot` to include y and x variables. The command has many options that allow for customization of heat plots. A basic heat plot of the scatterplot in figure 13.1 is demonstrated below in figure 13.2.

Creating figure 13.2

```
. * Basic heat plot bivariate distribution
. heatplot weight height
```

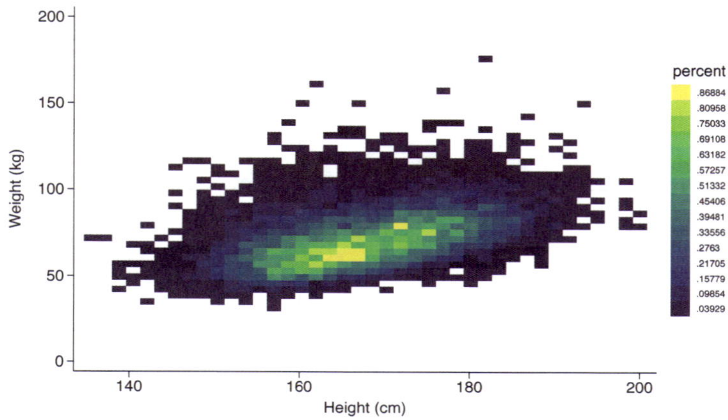

Figure 13.2. A basic heat plot

The general outline of figure 13.2 is the same as the scatterplot in figure 13.1. However, the user is now presented with a series of colorized rectangles that denote what proportion (percent) of data sits inside each bin. It is now easy to identify that there is a massing of observations in the center of this bivariate distribution. Central sections with yellow colors have many observations, while purple areas of the graph contain few observations. The exact amount can be identified by the legend. For example, one bright yellow rectangle contains around 0.87% of the data in that rectangle only. From this, it appears that the relationship between person weight and person height remains positively sloped within this central data mass. So the overall relationship between `weight` and `height` is positive, which can be confirmed via a heat plot but not via the outlying data points from an overloaded scatterplot.

13.2.1 Heat plot with frequency statistics

`heatplot` has many options that allow users to customize their heat plots. For example, the default legend display of percentages may not be to everyone's liking. The `statistic()` option allows users to choose from many summary statistics (in fact, all of those supported by the `collapse` command) that can be used as legend labels instead. To change the percentage statistics to frequency statistics, users can specify the `count` statistic in the `statistic()` option. This will result in a heat plot with colors based on the number of observations in each bin. Figure 13.3 demonstrates this example and also uses the `keylabels()` option to format the legend output to 1 decimal place for a better visual presentation. Users should note that the legend reports the midpoint of each bin color, and there will be some variation between bins with the same color. To identify the exact range of each bin, users can also specify the suboption intervals within the `keylabels()` option, although this is not shown here.

Creating figure 13.3

```
. * A heat plot bivariate distribution using count statistics
. heatplot weight height, statistic(count) keylabels(, format(%9.1f))
```

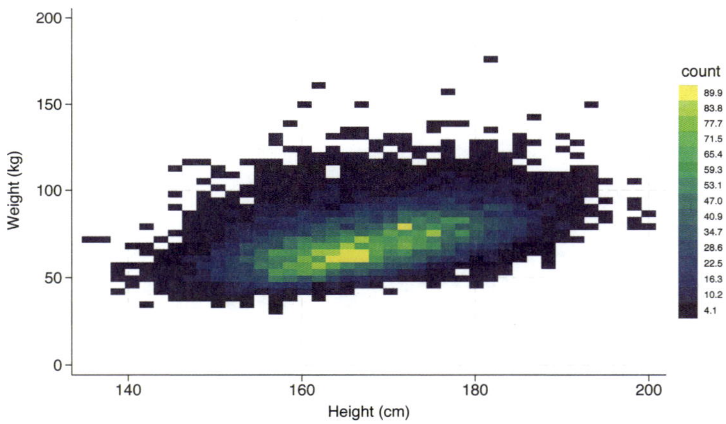

Figure 13.3. Heat plot with frequency statistics

13.2.2 Custom colors

An important aspect of heat plots is color presentation. Color choice and color cutoffs can greatly affect how a heat plot is presented, and there are several options available to users to tune their colors. By default, `heatplot` uses the `viridis` color palette, exhibiting a color gradient ranging from dark purple to bright yellow with green as the middle color. The `levels()` option allows users to set the number of color bins into which the observed frequencies are banded. High values will lead to smoother color transitions and less sharp cutoffs, while lower values will lead to more distinct color transitions between the x and y bins. For those with very specific requirements, the `cuts()` option allows custom thresholds to categorize the bin ranges. Finally, the `colors()` option can be used to specify custom colors that differ from the default presentation. Users should have the community-contributed command `colorpalette` installed (part of the `palette` package [Jann 2017]), which will allow them to choose from the available list of sequential color palettes (see `help colorpalette`). Figure 13.4 presents an example with only 4 distinct color levels, while figure 13.5 uses 50 different color levels and also uses the `oranges` color palette.

Creating figure 13.4

```
. * Heat plot with four levels
. heatplot weight height, levels(4)
```

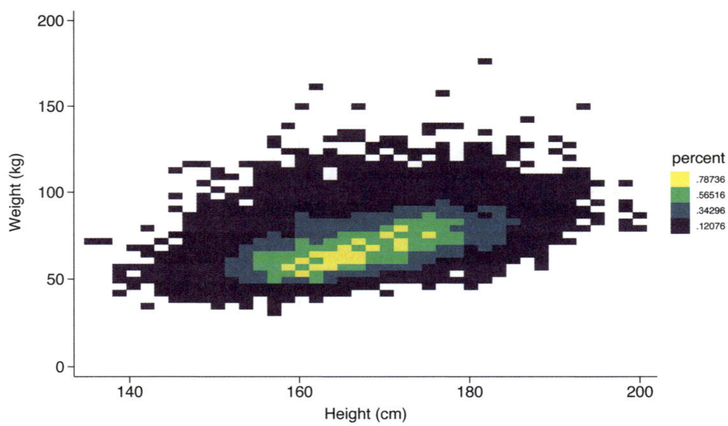

Figure 13.4. Heat plot with four colors

Creating figure 13.5

```
. * Heat plot with 50 levels and colors
. heatplot weight height, levels(50) colors(oranges)
```

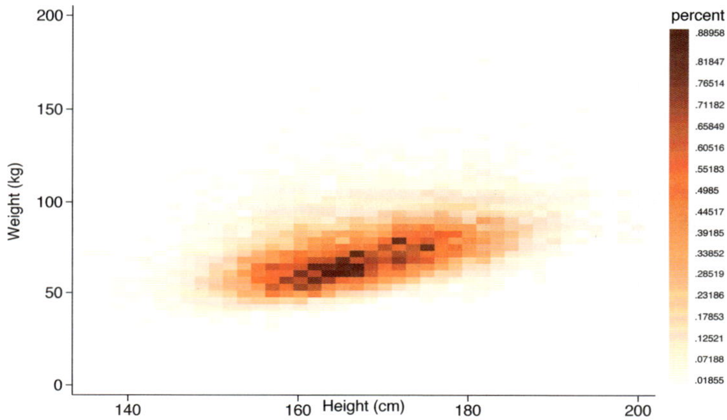

Figure 13.5. Heat plot with 50 colors and `oranges` color palette

Note that figures 13.4 and 13.5 present only different *colors*. The number of bins in each graph remains the same. In figure 13.4, the limitation of only four colors appears to lead to a less detailed visualization, while in figure 13.5, more detail is presented. But this does not imply that limited color ranges should be discounted and that users should always run with the option `levels(50)`. An advantage of figure 13.4 is that the demarcation between different density regions is much clearer than in figure 13.5. The central data mass is easier to identify and, therefore, to contextualize within the surrounding data space.

13.2.3 Custom bins

Heat plots are sometimes called bivariate histograms because the visualization is similar to two histograms interlaced on the x and y axes. This means that bandwidth (or bin size widths) changes are possible, which can significantly affect the visualization of the heat plot. More or less granular visualizations are the result. The `heatplot` command has multiple options on this front; the most important are the `bins()`, `xbwidth()`, and `ybwidth()` options. The `bins()` option allows users to specify the same number of bins to be used for both the x and the y variables. The `xbwidth()` and `ybwidth()` options allow users to specify custom x- and y-bin widths and give users the ability to create rectangular bins. This is particularly useful when there are significant scale differences between the two variables. Figures 13.6 and 13.7 demonstrate these concepts.

Creating figure 13.6

```
. * Heat plot with different bin widths
. heatplot weight height, bins(10)
```

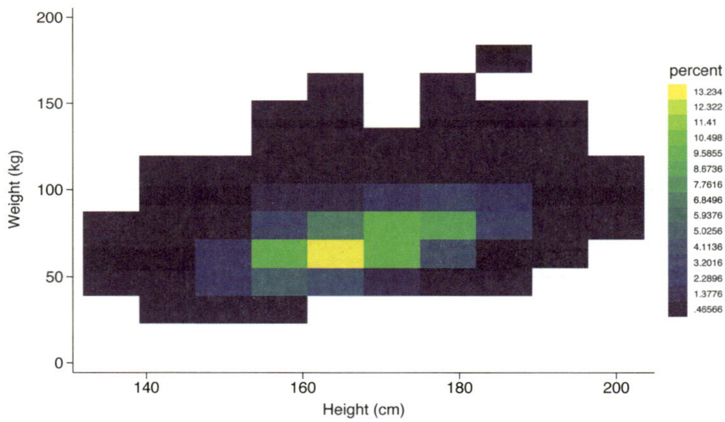

Figure 13.6. Heat plot with 10 bins

Creating figure 13.7

```
. * Heat plot with different bin widths
. heatplot weight height, ybwidth(1) xbwidth(5) hexagon
```

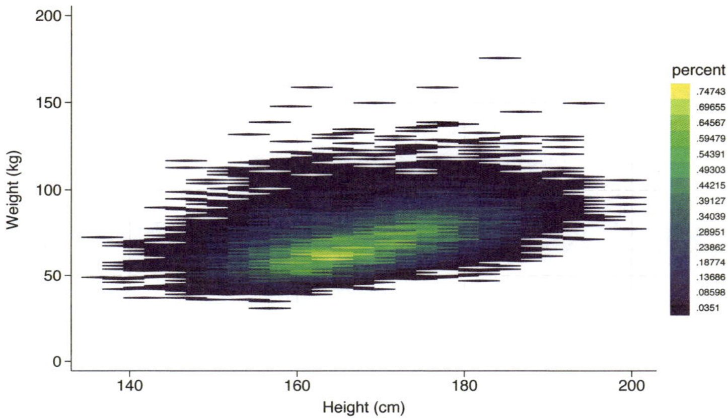

Figure 13.7. Heat plot with custom rectangular bin widths

Figures 13.6 and 13.7 demonstrate how heat plots vary when different bin size options are specified. Users can create more or less granular heat plots and take their variables' scalings into account to ensure an appropriate representation. Adjusting bin widths can be particularly useful when data naturally band into groups. For example, there are 7 days in a week and 12 months in a year, and such variables can easily be grouped into appropriate bin widths.

13.2.4 Proportional bin sizes

`size` is another neat option that helps to impart a strong visual sense of where data are located. It is not a bandwidth option per se but a pure visual option that scales bins according to their sample size. The bin color calculation remains a function of the default (or custom) specified bin sizes, but the pictorial representation scales each bin according to its underlying observation count. This visualization essentially repeats the colorization effect of a heat plot but this time via scaled bins, which makes it even easier for users to identify areas of light or dense data concentration because the information is effectively presented twice. This option might be useful when a monochrome or grayscale color scheme is used because of publication limitations or when accessibility issues need be considered, for example, because of color vision deficiencies. An example is shown in figure 13.8.

Creating figure 13.8

```
. * Heat plot with proportional bins
. heatplot weight height, size
```

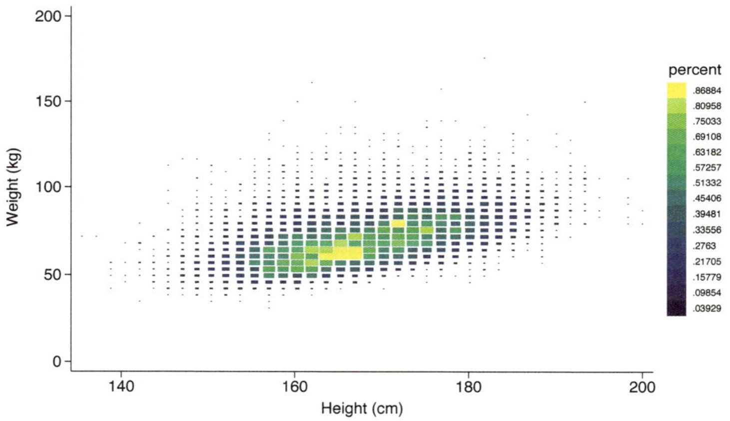

Figure 13.8. Heat plot with proportional bin sizes

13.2.5 Scatter heat plots

Scatterplots are the most common methods to illustrate how bivariate data are distributed and related. It might make sense to mimic a basic scatterplot design in certain situations while retaining the data binning and colorization methodology of a heat plot. This could be useful when presenting heat plots to laypeople or when a particular pictorial symbology is required. However, the resulting plot firmly remains a heat plot; any resemblance to a scatterplot is in looks only. Data markers are not true data markers but simply heat bins transformed into "nontouching squares".

Data markers can be transformed into *pseudo*-scatterplots using the `scatter()` option. Bins are recast as regular circles, creating the illusion of a scatterplot with heat plot elements. The `scatter()` option also takes any *symbolstyle*, thereby allowing users access to many more marker options. Figure 13.9 demonstrates four different examples that include the default markers (circle) and three custom markers (squares, triangles, and hollow circles). All graphs are combined using the `graph combine` command.

Creating figure 13.9

```
. * Heat plot with different bin widths
. heatplot weight height, scatter title(Circle)                        ///
> name(circle, replace)
. heatplot weight height, scatter(square) title(Square)                ///
> name(square, replace)
. heatplot weight height, scatter(triangle) title(Triangle)            ///
> name(triangle, replace)
. heatplot weight height, scatter(circle_hollow) title(Hollow circle)  ///
> name(chollow, replace)
. graph combine circle square triangle chollow
```

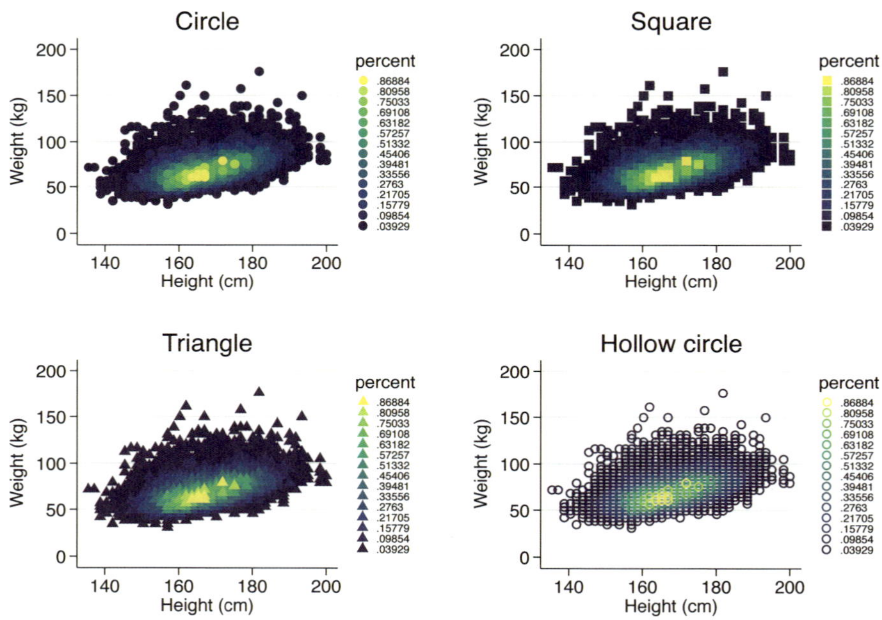

Figure 13.9. Four types of scatter heat plots

Figure 13.9 demonstrates how heat plots can be configured to resemble scatterplots while maintaining the aspects of the underlying heat plots. There is no conceptual difference between the plots in figure 13.9 and the plot in figure 13.2. Only the visual style is different.

13.3 Hex-heat plots

A valuable extension of scatter heat plots is hex plots. Hex-heat plots use hexagonal binning as opposed to regular square binning. From a visual perspective, hexagons look more like circles but can be joined at their sides without leaving gaps. From a statistical perspective, hexagons have improved efficiency in covering elliptical shapes on a 2D plane. Because most real-world bivariate data relationships will exhibit some trend, hex plots are more appealing options than regular heat plots based on squares. Lastly, traditional square bins bias the analysis into the horizontal and vertical, while hexagonal bins allow for a better balance between the horizontal, vertical, and diagonal.

There are two main ways to access hex plots. The `hexagon` option of the `heatplot` command creates a hex-heat plot. Alternatively, when users install the `heatplot` command, they also gain access to the `hexplot` command. Figure 13.10 provides a basic example using the `hexplot` command, and figure 13.11 provides an example using the `hexagon` option.

Creating figure 13.10

```
. * Hex-heat plot
. hexplot weight height
```

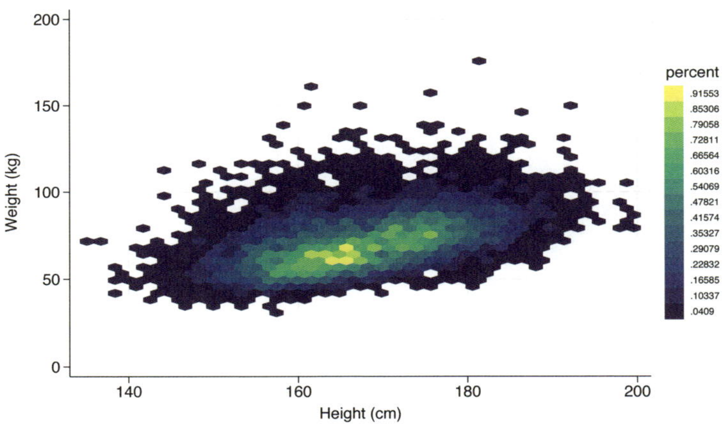

Figure 13.10. A basic hex-heat plot

13.3.1 Custom bins

Like all other heat plots, hex-heat plots can be customized to have larger or smaller hexagonal bins. The previously mentioned options `bins()`, `xbwidth()`, and `ybwidth()` all work and can be used to specify larger, smaller, or custom hexagonal bins. An example of larger hex bins is shown in figure 13.11 below.

Creating figure 13.11

```
. * Hex-heat plot with custom bins
. heatplot weight height, bins(10) hexagon
```

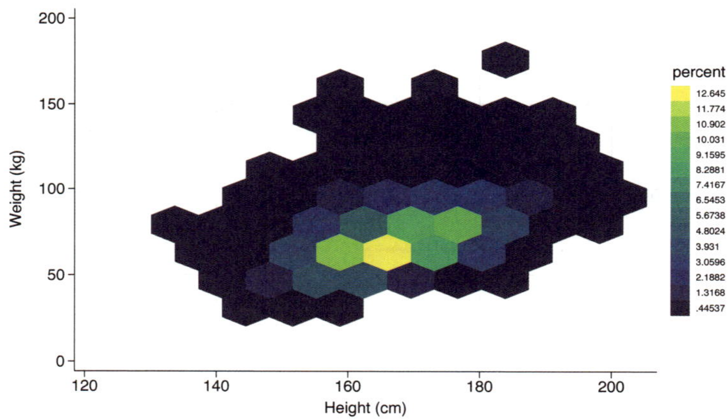

Figure 13.11. Hex-heat plot with custom bins

13.4 Sunflower plots

Stata, though it includes only limited capabilities for heat plots, does include a built-in command for creating sunflower plots. Sunflower plots (Dupont and Plummer 2002) are similar to heat plots in that they try to reduce overplotting by binning data into "sunflowers". Their visualization in Stata relies less on color gradients and more on symbols. Round markers, hexagonal bins, a two-tone color scheme, and "petals" are used to create a visualization of a bivariate density distribution. The idea is that dark and light sunflowers and petal symbols represent low-, medium-, and high-density data areas. While their initial presentations may appear more complex than those of heat plots (especially to laypeople), they combine some of the best elements of the previously shown graphs. Some individual data markers are displayed. Hexagonal bins are used to increase efficiency. Simple light and dark colors are used to clearly differentiate low- or high-density areas. Finally, a petal symbol is used to act as an additional, noncolor-specific indicator of density.

Sunflower plots are created with the official Stata command `sunflower`. This command will create a sunflower plot that consists of individual markers that represent singular data points, primarily in areas where there is little density. Where data density increases, yellow hexagonal bins are created, which are called light sunflowers. In each light sunflower, a line segment is shown radiating out from the center. Each line represents one observation and is called a petal. Light sunflowers with more petals represent more data within that particular bin. Once data density increases, orange sunflowers are created. These are called dark sunflowers; their petals carry higher weights than light petals. This creates a multitier heat map that provides a visualization of data density by combining symbols and colors. Figure 13.12 shows a basic sunflower plot.

Creating figure 13.12

```
. * A basic sunflower plot
. sunflower weight height
```

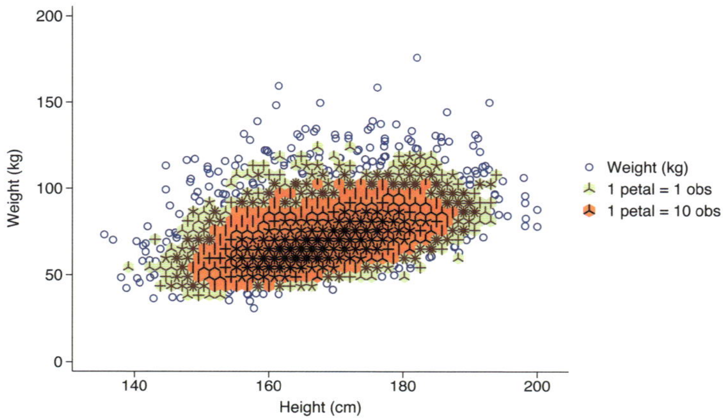

Figure 13.12. A basic sunflower plot

The sunflower plot in figure 13.12 looks similar to a standard heat plot. A positive relationship between weight and height is identified via the upward slope of the central data mass. Underlying cell sizes can be inferred via the legend. Individual blue circles represent single data points, while light (yellow) sunflowers represent one observation per petal in each hexagonal sunflower. For example, a light sunflower with 3 petals will contain 3 observations in that hexagonal field. Dark (orange) sunflowers contain 10 observations per petal. It is easy to see that the central data mass, which is colored almost black, has significantly more observations than the outer edges of this plot.

13.4.1 Custom bins

Sunflower plots come with many options that allow for custom bin and petal configurations. Like the previous **heatplot** command, various options allow users to change the aspect ratio and width of the hexagonal bins. The color styles of both bins and petals can also be changed via multiple options. Important basic options to be aware of are `dark()`, `light()`, and `petalweight()`. Both the `dark()` and `light()` options change the cutoff point for when individual markers are turned into hexagonal bins. The defaults are `dark(13)` and `light(3)`. Higher values will lead to more individual data markers being shown, while lower values will lead to more hexagonal bins. The `petalweight()` option changes the weight of the petals in the dark sunflowers only. Low values will lead to many petals being drawn, while high values will lead to few. Light sunflower petals will always represent only one observation. An example of a sunflower plot with custom bins and custom petal weights is shown in figure 13.13.

Creating figure 13.13

```
. * Sunflower plot with custom bins
. sunflower weight height, petalweight(2) dark(10) light(2)
```

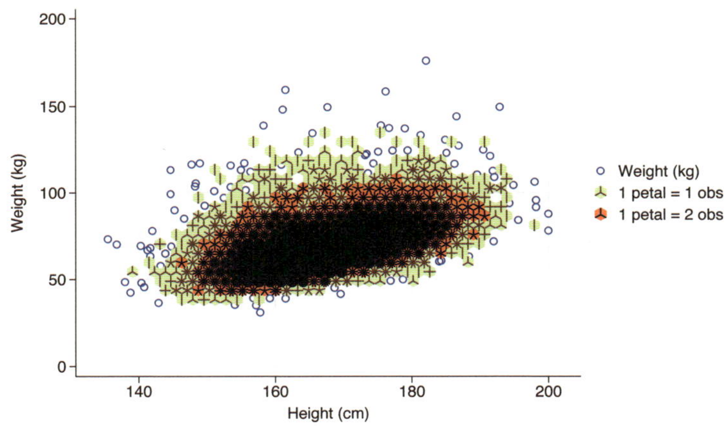

Figure 13.13. Sunflower plot with custom bins

14 Line plots

Advantages

1. Provide nonparametric overview of data relationship
2. Excel at working with flow variables, such as time
3. Clearly depict outliers and trends
4. Are useful for trending data
5. Facilitate comparison between groups via multiple line plots

Disadvantages

1. Require data ordering
2. Less clearly depict exact values
3. Can result in differing interpretations among different users because of their nonparametric natures
4. Are easier to misinterpret as causal because of their connected and sorted natures
5. Do not work as well with nontrending or high-variation data
6. Are prone to overplotting if too many lines are plotted
7. Can be challenging to interpret with missing data points

Required community-contributed package

To install the community-contributed command that is required to follow along with this chapter completely, type

```
. ssc install sparkline
```

14.1 Introduction

Line plots, which graph data values in the Cartesian $[y, x]$ space, are natural extensions to scatterplots. Unlike scatterplots, line plots require that one variable is ordered (typically the x variable), and data markers are subsequently joined via line segments (typically straight lines). Often, the actual data markers are suppressed, leaving only a series of connecting lines that are excellent for picking out global trends in bivariate data relationships but provide less detail on exact marker values.

Line plots are often used when the x variable is time; these variables have a natural ordering, and investigators often wish to examine how a y variable evolves. The advantage of a line plot over a best fit plot is that it remains purely nonparametric. It simply connects the data points, and it is left to end users to identify a relationship from the data. This makes line plots excellent graphs for exploratory purposes.

Line plots must be approached with some caution, however. Some aspects of line plots—the ordered values and the connecting lines—make them more susceptible to causal interpretations, which are misleading. Line plots, like scatterplots, are correlates, and users should not mislead audiences into causal interpretations unless there are clear reasons to do so. In addition, line plots are not helpful when there is significant variation from one data marker to the next. This can create confusing zigzag patterns that are more clearly understandable in other plot styles such as bar charts. Finally, line plots also perform poorly when there are many lines, especially lines that overlap a lot. The eye quickly becomes confused and has problems discerning which line belongs to which group.

14.2 A basic line plot

Line plots help users answer basic questions about how two variables are related. This chapter uses the US life expectancy dataset (`uslifeexp.dta`), which contains time-series information on various population groups from 1900 to 1999. A summary of the dataset is shown below.

```
. * Load data
. sysuse uslifeexp
(US life expectancy, 1900-1999)
. summarize
```

Variable	Obs	Mean	Std. dev.	Min	Max
year	100	1949.5	29.01149	1900	1999
le	100	64.829	9.158628	39.1	76.7
le_male	100	62.302	8.436369	36.6	73.9
le_female	100	67.51	9.834987	42.2	79.5
le_w	100	65.688	9.171269	39.8	77.3
le_wmale	100	63.143	8.503954	37.1	74.6
le_wfemale	100	68.434	9.797167	43.2	80
le_b	100	56.033	12.48937	30.8	71.4
le_bmale	100	53.589	11.4569	29.1	67.8
le_bfemale	100	58.567	13.5409	32.5	74.8

Analyzing the relationship between life expectancy (`le`) and time (`year`) via listing raw data values is not a realistic possibility. One should also consider that these data contain a natural ordering of the variable (`year`), which makes them naturally suited for a line graph.

The command to create line plots is `line`. Its basic operation involves calling `line` and specifying y and x variables. Stata will then proceed to graph y versus x in a line plot form, as demonstrated in figure 14.1.

Creating figure 14.1

```
. * A basic line plot
. line le year
```

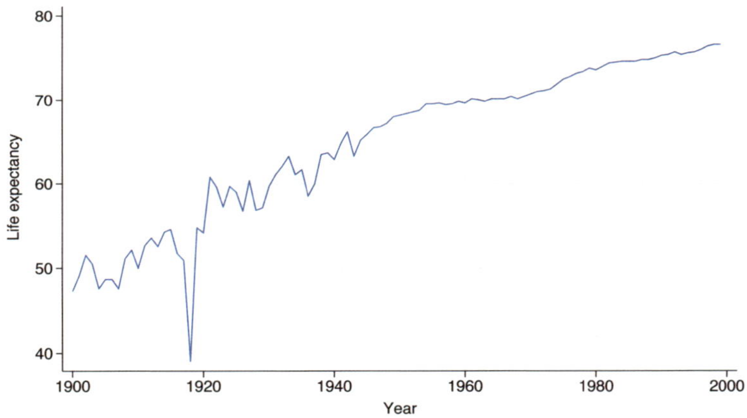

Figure 14.1. A basic line plot

Figure 14.1 exemplifies a basic line plot in a suitable data environment. There is a clear ordering of the x variable (time in years), and the y variable (life expectancy) behaves smoothly without too much year-on-year variation. Life expectancy increases gradually over time, although improvements are subject to more variation pre-World War II than post-World War II. Extreme events like the 1919 flu epidemic are clearly shown to be outliers. Note that `line` provides no data marker symbols by default. Users cannot easily identify individual data values. In addition, only 100 data points underpin this graph; therefore, a large proportion of the total line is interpolated between the data points via direct line connections. In other words, aspects of the data shown on this graph are subject to speculation, and general caution should be applied to line graphs with few data points in the ordered variable.

14.2.1 Important line options

Like scatterplots, line plots are versatile and have access to many options that allow users to customize their lines' shapes, sizes, and colors. Users should always examine the relevant help file in detail, but the most important `line` options are `lpattern()`, `lwidth()`, and `lcolor()`. These options allow users to determine the line patterns (solid, dash, dot, etc.), widths, and colors. The choice of thickness for lines is relatively simple; one can use custom thicknesses provided by the [*#] syntax, where values larger than 1 lead to thicker lines and values between 0 and 1 lead to thinner lines. This avoids having to memorize the *linewidthstyle* (see `help linewidthstyle`) choices. Colors are standard choices accessed via *colorstyle* (see `help colorstyle`). Lines are solid by default, but other patterns exist and can be chosen from *linepatternstyle* (see `help linepatternstyle`). Basic styles include `dash`, `dot`, `dashdot`, `shortdash`, and `longdash`. In addition, a unique custom line pattern style can be created using a "*formula*" as shown in `help linepatternstyle`.

Two examples are shown below in figures 14.2 and 14.3. Figure 14.2 creates a dashed forest-green line that is twice the thickness of a standard line, while figure 14.3 creates a custom line pattern using the "*formula*" syntax consisting of three short dots, three long dots, three short dots, and a blank space.

Creating figure 14.2

```
. * Line plot with markers and options
. line le year, lpattern(dash) lwidth(*2) lcolor(forest_green)
```

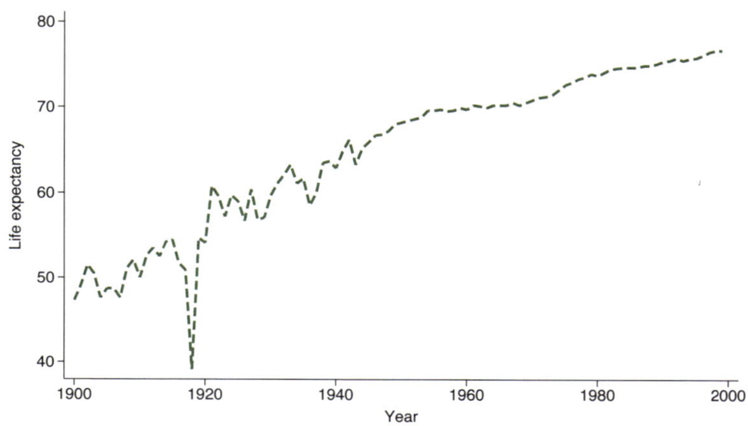

Figure 14.2. Line plot with dashed line pattern

Creating figure 14.3

```
. * Line plot with custom pattern
. line le year, lpattern("...---...#")
```

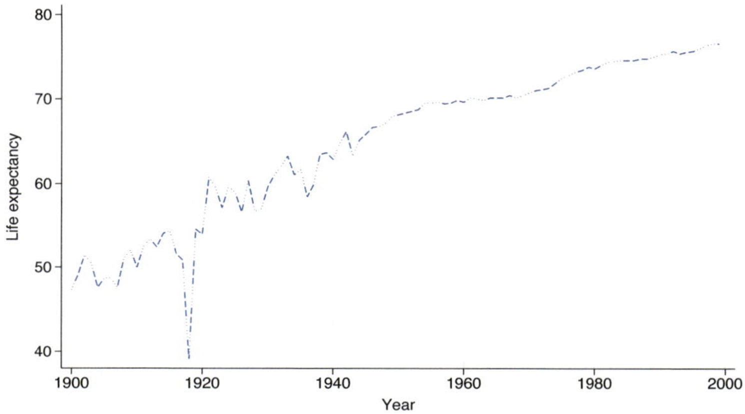

Figure 14.3. Line plot with custom line pattern

14.2.2 Multiple line plots

Like the `scatter` command, `line` can receive a longer variable list than just one y and one x variable. `line` can plot multiple y variables against one x variable all within the same command syntax by specifying three or more variables. All variables except the last variable will be considered as y variables. This makes it quick and easy to plot multiple lines on a line plot. `line` is also part of the two-way family of plots, giving it access to the `twoway` command. As with `scatter`, the `twoway` command is useful for overlaying other types of graphs, although it can also overlay additional line plots.

More conveniently, when using the multiple-variable syntax of `line`, users can specify multiple `line`-level options within one option, with each option choice corresponding to the relevant x–y combination. This makes it faster to specify different `line` options by specifying these choices within one option. In figures 14.4 and 14.5 below, two different methods to create multiple lines are shown: 1) the multivariable method via `line` and many y variables and 2) the `twoway` overlay method. In the first method, the line color options are specified within one `lcolor()` option. In the second method, two `lcolor()` options are required, one for each subgraph. Both commands produce the same graph.

Creating figure 14.4

```
. * Multiple line plots via line
. line le_male le_female year, lcolor(gs12 gs4)
```

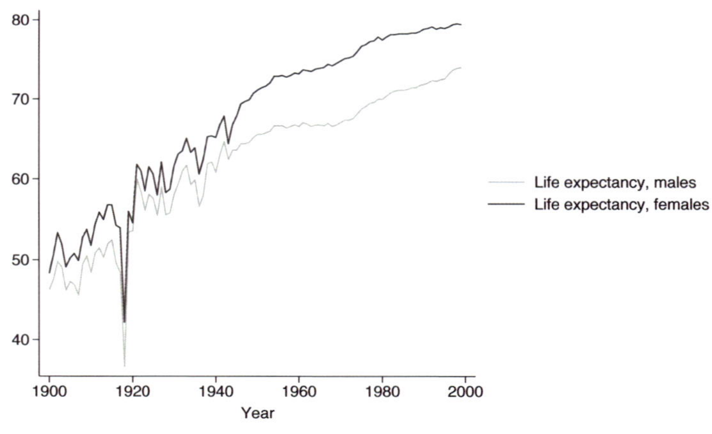

Figure 14.4. Multiple line plots via many *yvars*

Creating figure 14.5

```
. * Multiple line plots via twoway
. twoway (line le_male year, lcolor(gs12)) ///
> (line le_female year, lcolor(gs4))
```

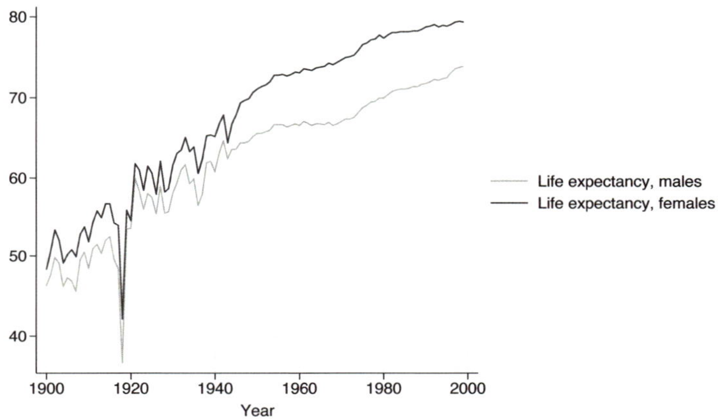

Figure 14.5. Multiple line plots via `twoway`

14.2.3 Line plots with markers

The primary advantage of using the `twoway` code structure is to superimpose other plottypes. A typical plot to overlay is the scatterplot via the `scatter` command. Line plots do not make identifying individual data points on the graph easy. Often, changes in the direction of the line identify an individual data value. Still, this method is not 100% reliable, because data markers can sit within a straight line when the underlying trend has not changed in that local space. To highlight data markers and provide viewers with a better understanding of the raw underlying data, scatterplots can be overlaid on line graphs within the `twoway` wrapper command. Figure 14.6 demonstrates an example and includes various `scatter` options that change the shade, size, and color of the data markers. Note that if users want only a basic line graph with data points, then they can also use the `twoway connected` command, which draws connected scatterplots, or the `twoway scatter` command with the `connect(l)` option. All three methods are suitable for displaying data markers in line plots. Here only the lengthier approach with multiple overlays is demonstrated because it builds the graph more gradually and provides a clearer understanding of how each layer contributes to the final visualization.

Creating figure 14.6

```
. * Line plots with markers
. twoway (line le year) (scatter le year, msymbol(D) msize(*0.5) ///
> mcolor(%50))
```

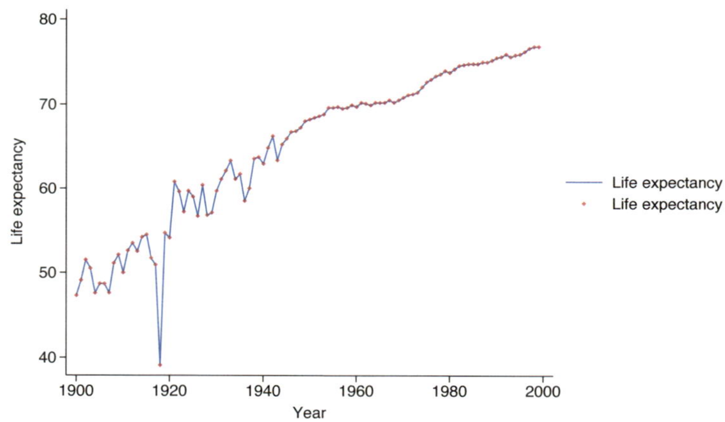

Figure 14.6. Line plot with markers

14.2.4 Different connecting styles

Figure 14.6 displays the actual underlying data of the line plot. In doing so, it also conveniently demonstrates the interpolation aspects of line plots. Data markers are connected via a series of straight lines. This is the default method, but other methods can be accessed via the `connect()` option and the *connectstyle* choices. The most common alternatives are the **`stairstep`** and **`stepstair`** connection styles, which connect markers either flat then vertical or vertical then flat, respectively. These kinds of connection styles are useful if it is known that the y-variable transition is discrete with respect to the x variable and occurs instantaneously before or after a data point. Figure 14.7 demonstrates three common connection types.

Creating figure 14.7

```
. * Different connection styles
. twoway (line le year, connect(stairstep))  ///
> (line le year, connect(stepstair))         ///
> (line le year, connect(direct))
```

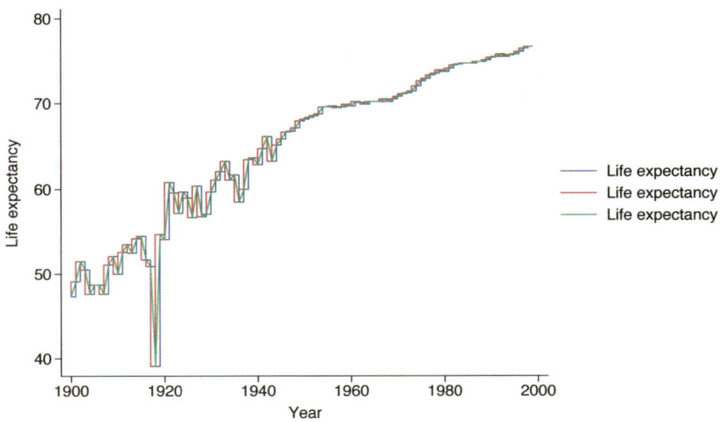

Figure 14.7. Line plot with different connection styles

The choice of connection style remains with the user; in most cases, the direct method is preferred. Moreover, the higher the number of data points, the less these methods deviate. Nondirect connection types should be considered when it is known that data changes occur in particular discrete intervals and when users wish to ensure that no interpolation interpretation errors occur (because staircase connection methods make transitions between data points very obvious).

14.2.5 Multiple y axes

Like scatterplots, care must be taken when using the built-in ability of `line` to plot many y variables. `line` does not automatically place multiple y variables on multiple y axes, resulting in a shared axis for all y variables. Scaling differences across variables can make this a visual problem. In the below code, the difference between female and male life expectancies is computed via the variable `diff`. The variable `diff` has a lower level and less variation than the `le_female` and `le_male` life expectancy variables.

```
. * Demonstrating axis options
. generate diff = le_female - le_male
. label variable diff "Difference"
. summarize diff le_female le_male
```

Variable	Obs	Mean	Std. dev.	Min	Max
diff	100	5.208	1.852145	1	7.800003
le_female	100	67.51	9.834987	42.2	79.5
le_male	100	62.302	8.436369	36.6	73.9

Graphing all three variables with one `line` command results in a line graph that contains one line that may be more difficult to examine because it sits farther away from the other two lines. This is exemplified in figure 14.8.

Creating figure 14.8

```
. * Without axis options
. twoway (line le_female le_male year) (line diff year)
```

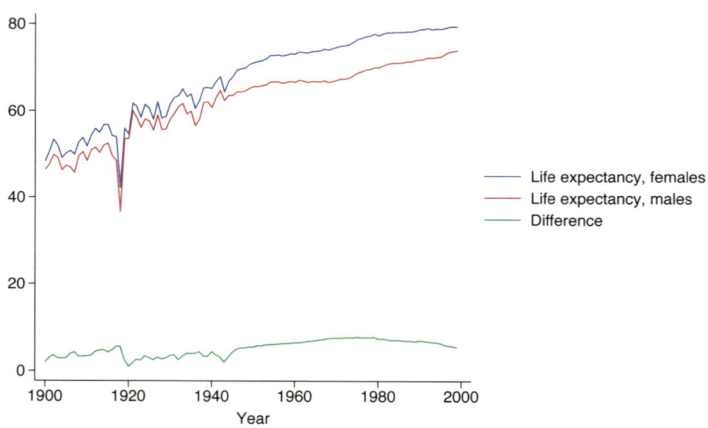

Figure 14.8. Line plots with scale differences

If users wish to bring the third line more into the center of the graph, perhaps to better examine it's exact values, then they can specify multiple y axes via the axis-choice options combined with the `twoway` command. This allows users to specify different y axes for different y variables. In figure 14.9 below, the `yaxis()` option is used to create two different y axes. The variables `le_female` and `le_male` are placed on y-axis 1, while the variable `diff` is placed on y-axis 2. Additional line options are used to better distinguish between these lines. The `diff` line has now become easier to identify and interpret.

Creating figure 14.9

```
. * With axis options
. twoway (line le_female le_male year, yaxis(1))       ///
> (line diff year, yaxis(2) lwidth(*2) lcolor(%40))
```

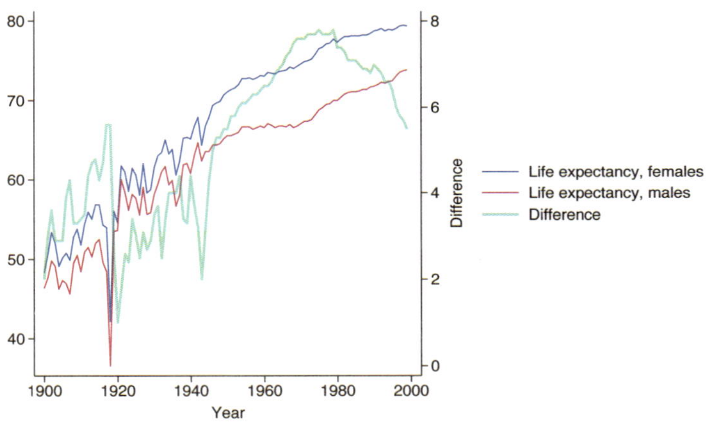

Figure 14.9. Line plots with scale differences and `yaxis()` option

14.2.6 Unsorted, sorted, and missing data

Line graphs require sorting on either the x or the y variable. Most frequently, the sorting is on the x variable because this results in a horizontal scan of the data (often from left to right). In most applications, this is preferable because the human eye is naturally trained to scan horizontally. Sorting on the y variable results in a vertical scan, which is less natural. Nonetheless, there are applied uses of vertical sorting, especially when graph real estate is limited, such as in mobile applications, where screen sizes are much larger in the vertical. Whichever way the sorting takes place, if the data are not sorted, then the result is often a "messy" plot with little analytical value.

Given this importance, the `line` command has a built-in `sort()` option, which means that users are not required to use the `sort` command prior to executing the `line` command. Another useful option is `cmissing()`, which defines how missing values are treated. By default, missing values are ignored, and a connection line is drawn from one marker to the next available marker. If missing values are not ignored, then gaps in the line graph will be generated to indicate missing data more clearly. `cmissing()` can be an important option when users want to clarify that data series are incomplete.

Figure 14.10 demonstrates all four concepts in a combined graph (`graph combine`). The code below randomly sorts the data (on a new variable, `random`, that was created with the `generate` command and the random-number function `runiform()`, causing it to be unsorted) and then replaces 10% of the data with missing values. The `set seed` command ensures perfect replication.

```
. * Unsorted and missing data
. * Shuffling data
. set seed 1234
. generate random = runiform()
. sort random
. replace le = . if random < 0.10
(13 real changes made, 13 to missing)
```

Four separate `line` commands then generate 1) an unsorted line graph, 2) a y-variable-sorted line graph, 3) an x-variable-sorted line graph, and 4) an x-variable-sorted line graph with missing values identified.

Creating figure 14.10

```
. * Sorting in plots
. * Messy plot
. line le year, title(Unsorted) name(g1, replace)
. * Sort on le
. line le year, sort(le) title(Sorted on life expectancy) name(g2, replace)
. * Sort on year
. line le year, sort(year) title(Sorted on year) name(g3, replace)
. * Sort on year with missing values identified
. line le year, sort(year) cmissing(n)     ///
> title(Sorted on year not ignoring missing) name(g4, replace)
. * Combine graphs
. graph combine g1 g2 g3 g4
```

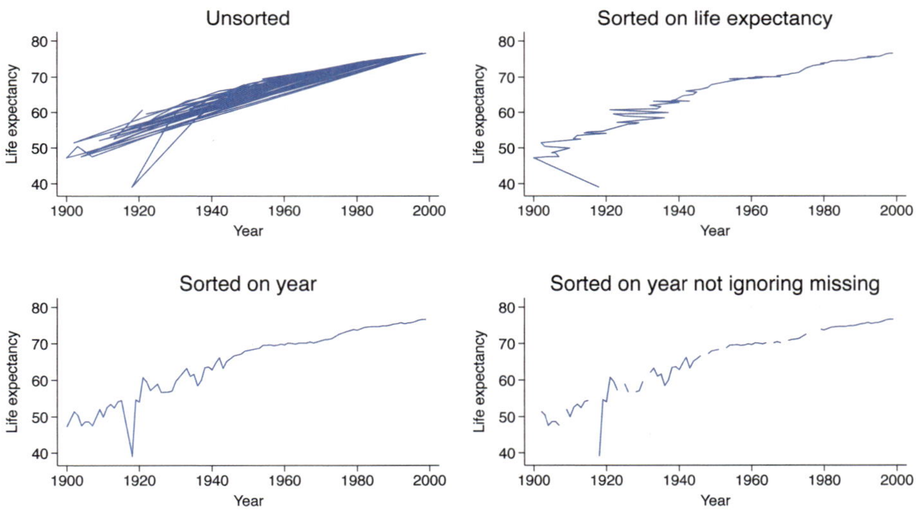

Figure 14.10. Unsorted, y-sorted, x-sorted, and missing-value line plots

Figure 14.10 demonstrates the importance of sorting and dealing with missing values in line graphs. The first panel with unsorted data clearly has little analytical value, and the second panel emphasizes life expectancy as the running variable and requires readers to scan the chart vertically. In such graphs, one can use the `aspectratio()` option to plot a graph where the height is taller than the width. This better helps the eye vertically track the line. The third panel demonstrates a standard line graph sorted on the x variable, and the fourth shows what happens when missing data are actively displayed and not interpolated as in the previous panels.

14.2.7 Line graphs with arrows

Users may wish to add arrows to line graphs to better emphasize the causal direction of the sorted variable. Arrows also have the advantage of better highlighting sudden shifts in trends. Unfortunately, there is no simple arrow option in the `line` command. Therefore, it is not possible to add arrows to basic line plots. The solution is provided by `twoway pcarrow`, which is a specialized graph command that creates paired-coordinate plots with single-direction arrows. Bidirectional arrows can be created with the sister command `twoway pcbarrow`. Both commands require four variables to be specified in the following order: `twoway pcarrow` *y1var x1var y2var x2var*. Each of these variables provides `twoway pcbarrow` with two x sets of Cartesian $[x, y]$ coordinates, which draw the start point and the endpoint of the arrow or arrows.

The `twoway pcarrow` command is generally not used for line graphs, but it can be adapted thanks to the sorted and hence quasilongitudinal nature of the data. The key to using it is to declare the data to be time-series data via the `tsset` command. Specify `tsset` followed by the sorting variable, which is `year` in the below example. Then call the `twoway pcarrow` command, and the starting coordinates of each arrow are the lagged values of the x and y variables. This uses the `l.` prefix before the two y and x variables. The second pair of variables specifies the end coordinates of each arrow, the nonlagged y and x variables. The final result of this code is a line graph that consists of a series of arrows connecting each data point, as demonstrated in figure 14.11.

```
. * Set time-series data
. sysuse uslifeexp, clear
(US life expectancy, 1900-1999)
. tsset year
Time variable: year, 1900 to 1999
        Delta: 1 unit
```

Creating figure 14.11

```
. * Arrow line plots
. twoway (pcarrow l1.le l.year le year)
```

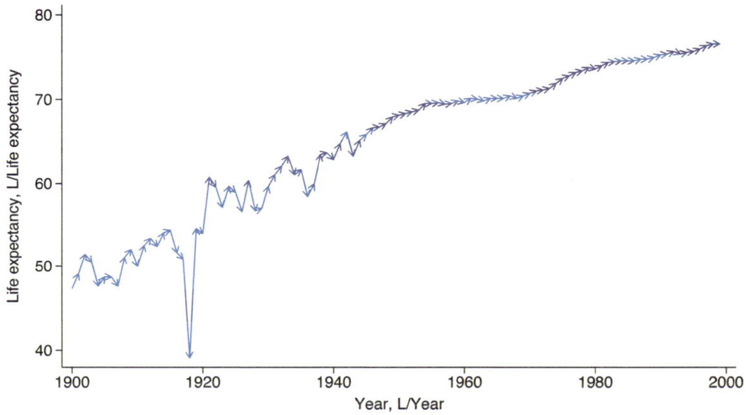

Figure 14.11. Line graph with arrows

Figure 14.11 presents a line graph constructed using the **twoway pcarrow** command. The arrows indicate the ordering of the x variable. Because **pcarrow** is part of the **twoway** family, multiple graphs can be overlaid, and various options are available to adjust the size of the arrow and fill color. A second example, which demonstrates multiple arrow line graphs with small arrowheads, is shown below in figure 14.12. In addition, the **position()** and **row()** suboptions of **legend()** are used to place the line legends on two separate rows at the bottom of the graph.

Creating figure 14.12

```
. * Multiple arrow line plots
. twoway (pcarrow l1.le_male l.year le_male year, msize(*0.5)) ///
> (pcarrow l1.le_female l.year le_female year, msize(*0.5)    ///
> legend(position(6) row(2)))
```

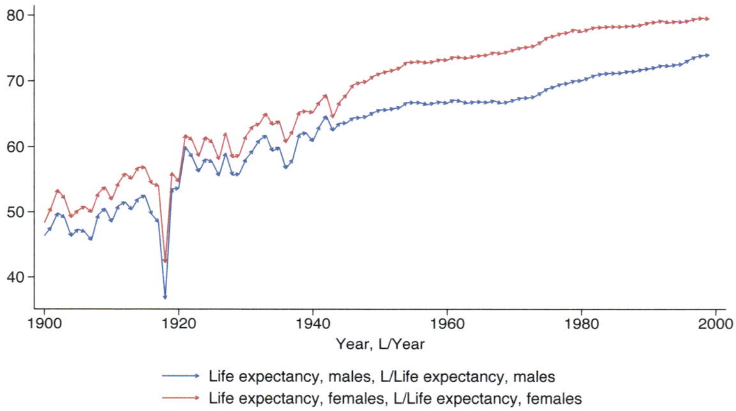

Figure 14.12. Multiple line graphs with arrows

14.3 Sparkline plots

A common user objective is to plot and examine multiple line graphs. When the focus of such comparison is many line graphs and only general trends, including maximum and minimum values, a sparkline plot is an attractive alternative to overlaying many line graphs on a single plot. A sparkline plot is used to graph multiple y variables or multiple categories in one y variable against one x variable. Note that plots are stacked vertically for comparison, and labeling is minimized to allow better focus on the lines themselves. Sparkline plots are particularly useful when there is a lot of overlap between lines, making individual line identification difficult. Sparkline plots are minimalist plottypes that reduce statistical information and can often take on an almost qualitative interpretation.

The advantages of a sparkline plot are exemplified with **grunfeld.dta**, which is available from the Stata website by using the **webuse** command. A summary of the dataset is shown below.

```
. * Load data
. webuse grunfeld
. summarize
    Variable |        Obs        Mean    Std. dev.       Min        Max
-------------+---------------------------------------------------------
     company |        200         5.5    2.879489          1         10
        year |        200      1944.5    5.780751       1935       1954
      invest |        200    145.9583    216.8753        .93     1486.7
      mvalue |        200    1081.681     1314.47      58.12     6241.7
      kstock |        200    276.0172    301.1039         .8     2226.3
-------------+---------------------------------------------------------
        time |        200        10.5    5.780751          1         20
```

grunfeld.dta contains financial information (investment, market value, and capital stock) of 10 companies over about 20 years. A user wishing to examine how the financial investment of these 10 companies evolves by each company over time would probably first resort to a standard line plot. In this case, an overlaid line plot would be created using the **twoway** command to graph 10 **line** plots, each with an **if** qualifier that conditions the plot to 1 company. Figure 14.13 shows an example of this.

Creating figure 14.13

```
. * Many line plots
. twoway (line invest year if company == 1)                ///
> (line invest year if company == 2)                       ///
> (line invest year if company == 3)                       ///
> (line invest year if company == 4)                       ///
> (line invest year if company == 5)                       ///
> (line invest year if company == 6)                       ///
> (line invest year if company == 7)                       ///
> (line invest year if company == 8)                       ///
> (line invest year if company == 9)                       ///
> (line invest year if company == 10, legend(off))
```

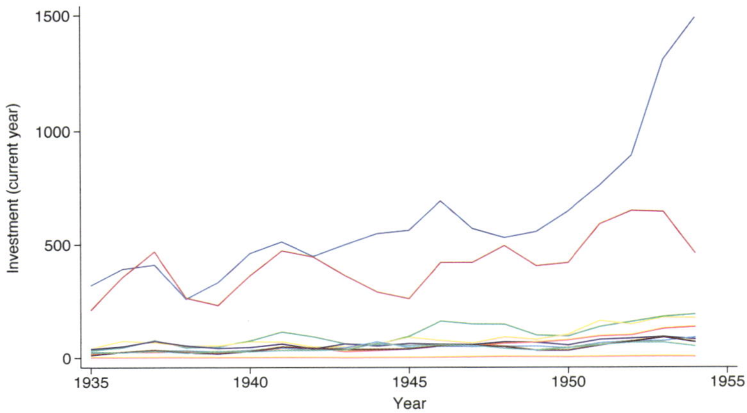

Figure 14.13. Overplotted line plot

Figure 14.13 is not a great plot. It requires lengthy and repetitive code,[1] and eight of the line plots are very hard to distinguish from each other, even with color coding. A more useful alternative is a sparkline plot, which can be created with the community-contributed `sparkline` command (Cox 2013). Similarly to `line`, the `sparkline` command requires one or more y variables to be specified and one x variable. Importantly, an `over()` option also must be specified to split the graph into subgraphs of the specified `over()` variable, in this case `company`. A basic example is demonstrated in figure 14.14.

Creating figure 14.14

```
. * Sparkline plot
. sparkline invest year, over(company)
```

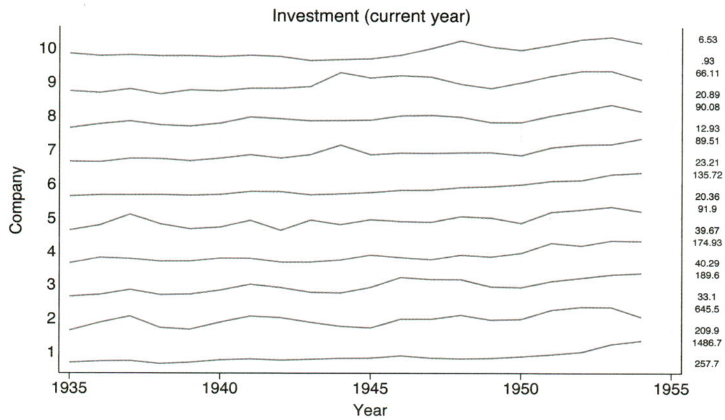

Figure 14.14. Sparkline plot

Figure 14.14 demonstrates a basic sparkline plot. By default, `sparkline` produces the most common kind of sparkline plot, showing several wider-than-tall line graphs stacked vertically. This lets users quickly identify general trends and where maximums and minimums might be located. However, comparison across lines is made difficult by the wide aspect ratio. The y-axis labeling can also be initially confusing. y-axis 1 enumerates the individual groups, while y-axis 2 labels the individual statistics. Information in y-axis 2 is crucial to identifying large or small trend changes, yet only the maximum and minimum values of each series are identified. Additionally, the actual values are scaled to (value − minimum)/(maximum − minimum) so that they lie in a [0, 1] range, and each subset is assigned the same vertical space. Scaling differences are thus hidden.

1. Note that as of the April 30, 2024 release of StataNow, users can also specify the `colorvar()` option to avoid long code with many `if` conditions. The `colorvar()` option colors the lines to reflect different groups of a discrete variable specified in `colorvar()`. This would reduce the complexity of the code in figure 14.13 to

```
line invest year, colorvar(company) colordiscrete coloruseplegend
```

where option `colordiscrete` treats the variable specified in `colorvar()` as discrete instead of continuous and the option `coloruseplegend` ensures that the legend is also discrete.

Many options are available in **sparkline** to customize graphs. Users should examine the help file (see **help sparkline**), but some important options are **extremes**, **extremeslabel**, and **ylabel()**, all of which ultimately modify the information in y-axis 2. In figure 14.15 below, all three options are demonstrated. The option **extremes** adds marker symbols to each line, highlighting maximum and minimum values. The option **extremeslabel** adds the labels of y-axis 2 into the graph, and **ylabel(none, axis(2))** completely removes the second y axis for a cleaner look. The result is a more compact graph that presents all key information within each line graph and not on the second y axis.

Creating figure 14.15

```
. * Sparkline plot with y-axis labels suppressed and minimum and
. * maximum values
. sparkline invest year, over(company) extremes extremeslabel ///
> ylabel(none, axis(2))
```

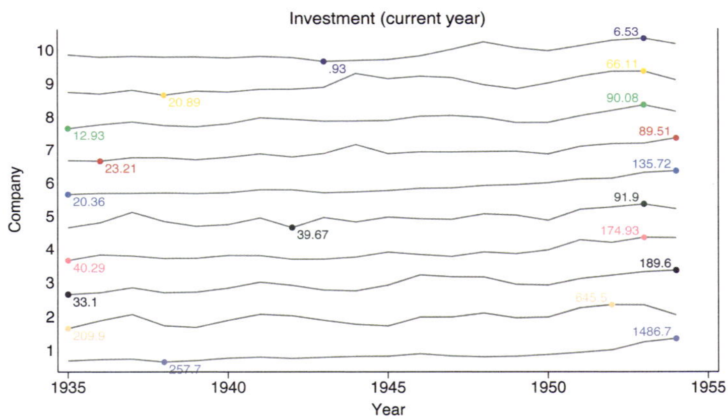

Figure 14.15. Sparkline plot with minimum and maximum values

14.3.1 Multiple sparkline plots

Finally, for those wishing to plot multiple different y variables over many groups, the by() option exists. This allows sparkline plots to be placed on separate graphs that are merged together. In addition, such graphs can be spread over multiple columns with the suboption cols(), as opposed to the single column when using the over() option. Moreover, multiple y variables (**invest**, **mvalue**, and **kstock**) can be added to the command **sparkline**, with the final variable (**year**) being considered the x variable. This results in figure 14.16, which demonstrates a sparkline plot with three sparklines per group spread over two columns. Overall, 30 different line plots are plotted in this graph, and it is easy to identify maximum and minimum peaks in each series. This demonstrates the versatility of sparkline plots.

Creating figure 14.16

```
. * Sparkline plot with multiple y variables
. sparkline invest mvalue kstock year, by(company, cols(2)) extremes ///
> ylabel(none, axis(2))
```

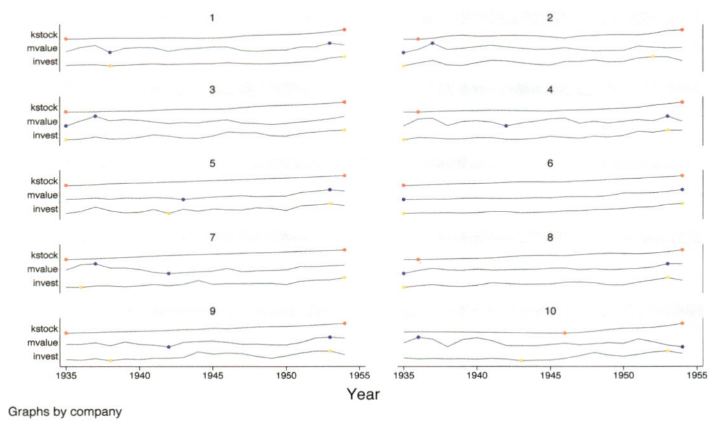

Figure 14.16. Multiple sparkline plots

15 Area and shaded range plots

Advantages

1. Are similar to line plots but visually more appealing for single plots
2. Have shaded areas that quickly convey how far data points are to a reference line
3. Are great at comparing how two line plots evolve across the x axis

Disadvantages

1. Become less useful when many line plots on one graph are needed
2. May result in overlapping colors that can cause confusion
3. Are less effective if there are missing values in the data series
4. Require considerations of accessibility and color choices

Required community-contributed package

To install the community-contributed command that is required to follow along with this chapter completely, type

```
. ssc install pairplot
```

15.1 Introduction

Area and shaded range plots are alternatives to the line plots discussed in the previous chapter. Line plots are convenient data visualization methods when the relationship between a y and an x variable is of interest and there is ordering in one of the variables. This ordering often has a time component, but any logical ordering is acceptable. A potential disadvantage of basic line plots is that they can leave a lot of empty data space in the visualization. Depending on the background color of the plot, this can mean that the majority of the visualization is often white. This can often be appropriate, but there may be instances where additional information can be visually added to line plots to convey the relationship between y and x better.

Area plots take the line plot operation one step further by coloring in the area underneath the graph and the reference line. This leads to a graph with much more color intensity and provides a more intuitive understanding of how far data points are from the reference line (often 0, but it can be any value). Ultimately, the area shading provides a stronger and more visual narrative of how far the bivariate relationship is from a reference point. The general x–y relationship is still shown, although it is slightly deemphasized because of the often strong coloring underneath the line.

Area plots are generally appropriate when users wish to emphasize how far data points are from an important reference point and when there is a lot of variation around a reference line.

A close cousin of the area plot is the shaded range plot, which modifies the area plot so that the shaded area is not one permanent reference point but instead a second bivariate relationship that varies constantly. Shaded range plots excel at highlighting differences between two line plots and how such differences evolve as a function of the x variable. For example, they are often used in finance, where they might denote a stock price's open and closing values. Larger shaded areas denote larger differences between two line plots, while smaller shaded areas denote smaller differences.

15.2 A basic area plot

The basic problem addressed by area and shaded range plots is the same as the basic problem for line plots: users wish to answer questions about the nature of one or more bivariate relationships of underlying data. The key difference from line plots is that 1) a particular reference number is of interest or 2) the data contain multiple y variables that must be compared.

The below data summary is from `uslifeexp.dta`, which contains 100 observations of various groups' life expectancy values for the United States from 1900 to 1999. Analyzing the relationship between life expectancy and time simply by listing raw data values is not analytically sensible. Moreover, there is a natural ordering of the `year` variable, and multiple variables measuring life expectancy are of interest. For example, `le_male` measures male life expectancy and `le_female` measures female life expectancy. Lastly, some reference values, such as 0, 40, or 65 years of age, may be important to this analysis.

```
. * Load data
. sysuse uslifeexp
(US life expectancy, 1900-1999)
. summarize
```

Variable	Obs	Mean	Std. dev.	Min	Max
year	100	1949.5	29.01149	1900	1999
le	100	64.829	9.158628	39.1	76.7
le_male	100	62.302	8.436369	36.6	73.9
le_female	100	67.51	9.834987	42.2	79.5
le_w	100	65.688	9.171269	39.8	77.3
le_wmale	100	63.143	8.503954	37.1	74.6
le_wfemale	100	68.434	9.797167	43.2	80
le_b	100	56.033	12.48937	30.8	71.4
le_bmale	100	53.589	11.4569	29.1	67.8
le_bfemale	100	58.567	13.5409	32.5	74.8

The command to create area plots is the **twoway area** command. Unlike **line**, **area** requires **twoway** to be specified. Thus, it cannot be executed on its own for a convenient shortcut. Its basic operation involves calling the command **twoway area** and specifying a y and an x variable. Stata will then plot y versus x as a line graph, with the area underneath the line shaded. Figure 15.1 demonstrates.

Creating figure 15.1

```
. * A basic area plot
. twoway (area le year)
```

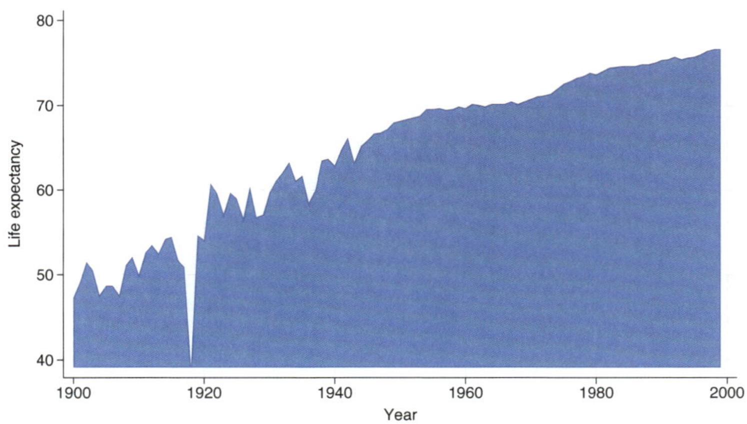

Figure 15.1. A basic area plot

Figure 15.1 displays a basic area plot. A relationship between life expectancy and time identifies increasing life expectancy during the 20th century, albeit with some dips and one traumatic event from the post-World War I flu epidemic. Importantly, the actual x–y relationship is somewhat deemphasized compared with a basic line graph, and more visual emphasis is placed on the distance between a reference point and the x–y relationship at various points of x. By default, area plots set the reference point to the lowest value of the y variable, which is 39.1 years in this case. All shading is thus relative to this point.

Note that interpolation issues are still present. Only 100 data points underpin this graph, one for each year. Therefore, a large portion of this graph is interpolated via direct lines between years. Reading of values of life expectancy that relate to midyear estimates will be subject to some error.

15.2.1 Different types of area plots

Stata offers different types of commands to create area-type graphs. In most cases, the `twoway area` command is used. It shades the entire area underneath the line and provides a strong visual narrative. The `twoway bar` command produces a bar graph that provides a similar visual appearance but removes the spikiness that results from the direct line connections that area plots use. The horizontal bar connections will create a stepladder-type visual appearance, which can be a better option if there is a lot of variation from point to point because it reduces the problem of sharp angular turns in the graph. The `twoway spike` command produces spike plots that are similar to bar charts but reduce the size of the bar width. This leaves more white space between data points, thereby reducing the visual impact of the area shading, which allows for better comparison between adjacent data points. In addition, an area focus is still provided by the narrow vertical bars. Spike plots are helpful alternatives if there are many data points and the emphasis is on allowing users to examine data points closely while providing a visual clue to a reference point. Finally, the `twoway dropline` command creates drop line plots that are similar to spike plots but with marker points shown. This reemphasizes the global x–y relationship but, similarly to the spike plot, gives users better point-to-point comparisons than a standard area plot.

Figure 15.2 exemplifies all four graph types below. The `title()` and `name()` options are used to title and store each graph before being merged into one graph with the `graph combine` command.

Creating figure 15.2

```
. * Four basic area-type plots
. * Basic area plot
. twoway (area le year, title(Area) name(g1, replace))
. * Basic bar plot
. twoway (bar le year, title(Bar) name(g2, replace))
. * Basic spike plot
. twoway (spike le year, title(Spike) name(g3, replace))
. * Basic dropline plot
. twoway (dropline le year, title(Dropline) name(g4, replace))
. * Combine graphs
. graph combine g1 g2 g3 g4
```

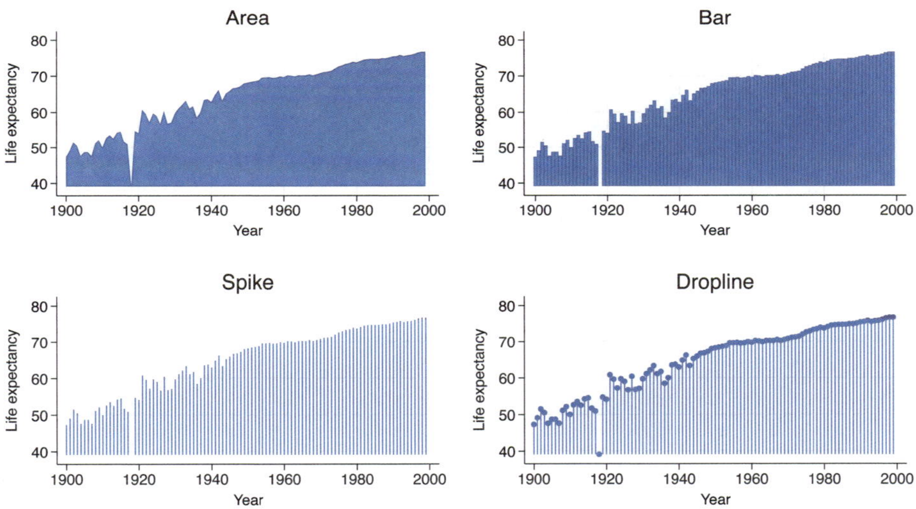

Figure 15.2. Different area-type plots

Figure 15.2 demonstrates four different area-type plots. The top panels better emphasize the area underneath the line plot and are therefore better suited when the distance between a reference point and the x–y relationship matters. The area plot in the first panel provides a more natural progression of the line plot, while the bar graph in the second panel allows for better point-to-point identification. The two bottom panels put slightly less emphasis on the area component underneath the line plot and emphasize the point-to-point transitions. This gives users an improved ability to identify minute numerical changes in local areas of the graph.

15.2.2 Important area plot options

Options matter, especially for area-type plots. A basic understanding of how to modify reference lines and graph colors (especially opacity) is essential for getting the most out of area plots. Users should examine the help file for each graph individually because not all options exist across each `twoway` graph. Below, some of the most important options for each area-type graph are demonstrated.

Many of the below graphs can use the `horizontal` option, which produces a horizontal plot. In most cases, this will not be useful because `horizontal` rotates the x–y relationship into the vertical and can make identification more difficult; however, for mobile or other nonstandard applications (such as posters), the `horizontal` option can provide the means to create a waterfall-type chart where data are examined from top to bottom. The `aspectratio()` and `yscale()` options can also be used to better customize the graph to read data from top to bottom in a vertical format.

The `base()` option sets the reference line for area shading. The life-expectancy data have a minimum value of 39.1 that is used as the base, but other values can be specified. An example with the value 0 is shown below in the second graph.

Colors matter greatly for area plots because so much of the graph is dominated by one shaded area. Users should read about area options (`help area_options`) to examine the available choices. Below is an example with the `fcolor()` option, which affects only the shaded area and not the line. A forest-green color with an opacity of 40% is demonstrated.

The `twoway bar` command has many shared options, but the `barwidth()` option is unique and modifies the width of bars. Values larger than 1 lead to overlapping bars, while values less than 1 lead to bars with gaps. The below example demonstrates this with the `lcolor()` and `fcolor()` options, which adjust bar line and bar fill color, respectively. The result is a series of overlapping opaque bars.

Spike plots are essentially very thin bar plots and have no unique options available to them. The long spiked lines to the baseline can be modified using the `lpattern()` option, which allows users to implement dashed, dotted, or customized line patterns.

The `dropline` command also uses markers, which gives it access to *marker_options* (`help marker_options`). These can be used to change marker visuals, as shown below with the `mfcolor()` and `mlcolor()` options, which change the marker fill and marker outline color. Also exemplified is the `base()` option with a nonzero value and the `yline()` option, which adds a horizontal line at specified y values for better reference.

Creating figure 15.3

```
. * Important options
. * Area plot with options
. twoway (area le year, title(1) horizontal aspectratio(2)                    ///
> yscale(reverse) name(g1, replace))
. twoway (area le year, title(2) base(0) ylabel(0(20)100)                      ///
> name(g2, replace))
. twoway (area le year, title(3) fcolor(forest_green%40)                       ///
> lcolor(black) name(g3, replace))
. * Bar plot with options
. twoway (bar le year, title(4) horizontal barwidth(1.5) lcolor(black)        ///
> fcolor(%30) name(g4, replace))
. * Spike plot with options
. twoway (spike le year, title(5) lpattern(longdash) name(g5, replace))
. * Drop line plot with options
. twoway (dropline le year, title(6) base(65) yline(65)                        ///
> mfcolor(ltblue%80) mlcolor(black%80) name(g6, replace))
. * Combine graphs
. graph combine g1 g2 g3 g4 g5 g6, row(2)
```

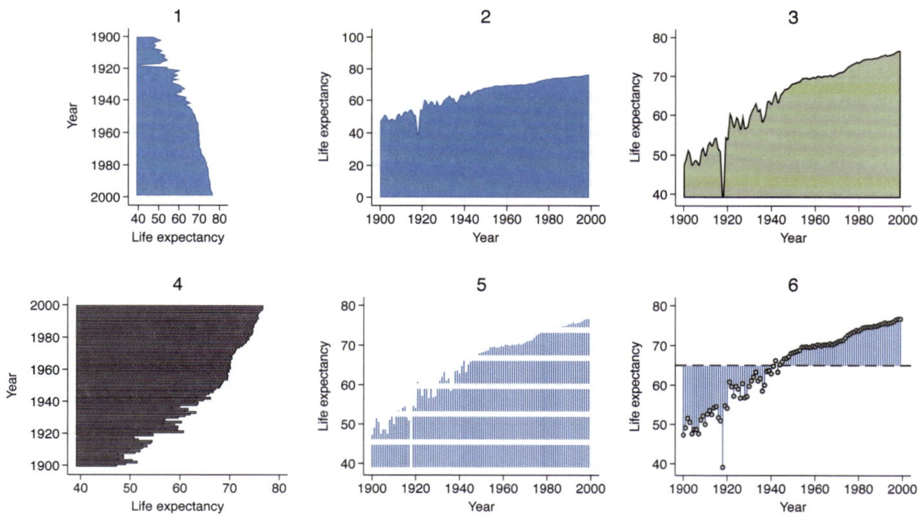

Figure 15.3. Important area-type plot options

Figure 15.3 presents the combined plot of previously specified area-type plots and their associated options. There are many other options, but the above graphs are designed to demonstrate some of the most important options that, in many cases, can be mixed and matched across the different area-type plots.

15.3 Data with variation

The different styles of area plots become particularly useful when there is a lot of variation in the data. This is often the case with finance data; an example is demonstrated below. The S&P 500 dataset (`sp500.dta`) is loaded, which contains stock market price information. A summary is provided below.

```
. * Load data
. sysuse sp500
(S&P 500)

. summarize

    Variable |        Obs        Mean    Std. dev.       Min        Max
-------------+---------------------------------------------------------
        date |        248    15156.85    105.3335      14977      15340
        open |        248    1194.884    87.12808      965.8    1373.73
        high |        248    1204.044    86.53465     984.54    1383.37
         low |        248    1183.334    87.47576     944.75    1364.66
       close |        248    1194.179    86.79468      965.8    1373.73
-------------+---------------------------------------------------------
      volume |        248    12320.68    2585.929       4103    23308.3
      change |        247   -.5473282    15.84464  -53.77002   64.29004
```

The variable `change` measures the change in closing price from the previous day. These data are highly variable and consistently flip around the value 0 over time. The code below demonstrates graphing these data using 1) a basic line plot and 2) three types of area plots. The `twoway dropline` command is excluded because it is similar to a spike plot but with added marker symbols. The `graph combine` command combines all four saved graphs.

Creating figure 15.4

```
. * Data with variation around an important value
. * Basic line plot
. twoway (line change date in 1/60, title(Line) yline(0) name(g1, replace))
. * Area plot
. twoway (area change date in 1/60, title(Area) yline(0) name(g2, replace))
. * Bar plot
. twoway (bar change date in 1/60, title(Bar) yline(0) name(g3, replace))
. * Spike plot
. twoway (spike change date in 1/60, title(Spike) yline(0)  ///
> name(g4, replace))
. * Combine graphs
. graph combine g1 g2 g3 g4, row(2)
```

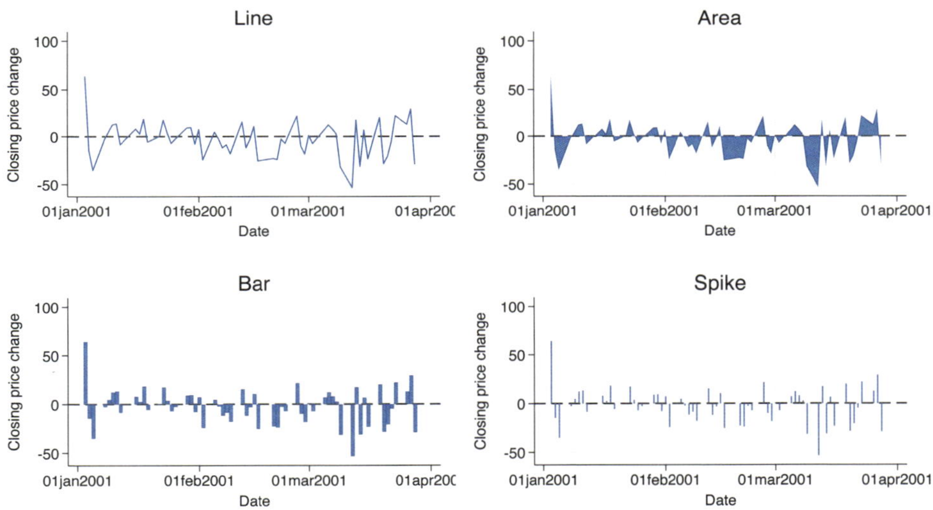

Figure 15.4. Four graphs showing deviation from zero

Figure 15.4 exemplifies the use of area shading in line graphs. All plots provide visualizations of stock market changes over time. Changes can be positive and negative, and the basic line plot in the first panel reveals much day-to-day variation in how the stock market evolved. The line plot does not reveal a stable pattern, and identifying deviation from a baseline is not easy because there are so many small deviations. The other three area-type plots do much better jobs of showing how the stock market evolved by emphasizing the distance between the change and a reference line. Positive and negative events are extremely easy to identify, and consecutive positive and negative days are more tractable. Overall, the area-type plots better illustrate this volatile time series.

15.4 Shaded range plots

Shaded range plots are identical to area plots, with the only difference being that the area in question is not a fixed point on the *y* axis but another line plot that varies continuously. Similarly to area plots, shaded range plots excel at highlighting differences between two different lines and how such differences evolve across the range of the *x* axis. For example, they are often used in finance, where they denote opening and closing values of stock market indices. Larger shaded areas denote bigger differences between two data series, while smaller areas denote smaller differences. The key advantage of shaded range plots is that the area shading between two lines makes it much easier for a layperson to identify differences in the evolution of both lines.

`uslifeexp.dta` is reloaded to demonstrate shaded range plots because it contains multiple time series. Of specific interest is the comparison of life expectancy between different groups, such as men (`le_male`) and women (`le_female`).

```
. * Load data
. sysuse uslifeexp
(US life expectancy, 1900-1999)
. summarize
    Variable |        Obs        Mean    Std. dev.       Min        Max
-------------+---------------------------------------------------------
        year |        100      1949.5    29.01149       1900       1999
          le |        100      64.829    9.158628       39.1       76.7
     le_male |        100      62.302    8.436369       36.6       73.9
   le_female |        100       67.51    9.834987       42.2       79.5
        le_w |        100      65.688    9.171269       39.8       77.3
-------------+---------------------------------------------------------
    le_wmale |        100      63.143    8.503954       37.1       74.6
  le_wfemale |        100      68.434    9.797167       43.2         80
        le_b |        100      56.033    12.48937       30.8       71.4
    le_bmale |        100      53.589     11.4569       29.1       67.8
  le_bfemale |        100      58.567     13.5409       32.5       74.8
```

A basic shaded range plot can be created with the `twoway rarea` command. Similarly to `twoway area`, it requires that `twoway` be specified. All shaded range plots require two *y* variables and one *x* variable to be specified in the form of `twoway rarea` *y1var y2var xvar*. Figure 15.5 below demonstrates a basic shaded range plot using the female and male life expectancy variables from `uslifeexp.dta`.

Creating figure 15.5

```
. * A basic shaded range plot
. twoway (rarea le_female le_male year)
```

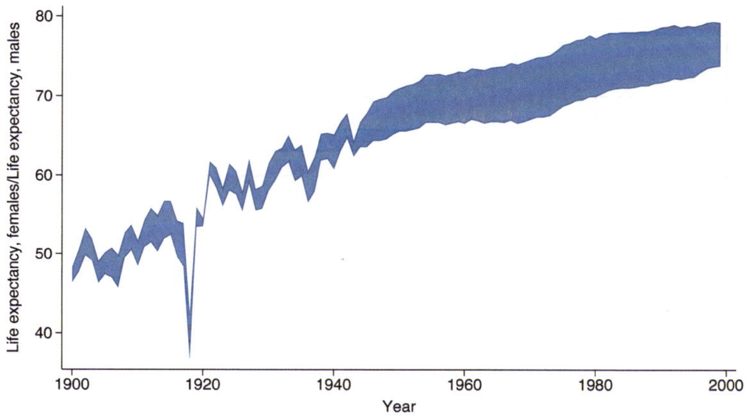

Figure 15.5. A basic shaded range plot

Figure 15.5 presents a shaded range plot of two individual line plots. This allows users to quickly identify where the lines and therefore both x–y relationships diverge or converge. For example, it is easy to see that life expectancy between males and females was relatively close in the 1920s, while larger divergence occurred after 1945, with females having higher life expectancies than men. A narrowing of this gap appeared to occur only very late in the 20th century.

15.4.1 Different types of shaded range plots

By default, Stata offers eight different range-type plots. These are demonstrated in the code below, although realistically, only a subset of these are commonly used. Note that all the following commands require that `twoway` be specified. `rline` plots two lines and forgoes any area shading between the lines. Instead of using `twoway rline`, users may prefer to overlay two individual `line` plots using `twoway`. The `rarea` plottype provides area shading between both lines, while the `rbar` plottype plots ranges with bars. `rspike` offers the same visual narrative as `rbar` but with thinner spikes. A combination approach is offered by `rcap`, which creates ranged spike plots but with capped marker symbols that resemble the top of bars. `rcapsym` takes the previous command one step further by offering users the ability to attach custom markers to the top and bottom of spikes. By default, these are circular markers, but they can be modified. Similarly to `rline`, `rscatter` and `rconnected` remove the area shading between the lines while emphasizing marker values. `rscatter` removes the line connections, while `rconnected` leaves the relevant line connections. Like `rline`, users may prefer to use `twoway` to overlay multiple `scatter` or `connected` plots instead of using the `rscatter` or `rconnected` plottype.

Creating figure 15.6

```
. * Different range plots
. * Basic range line plot
. twoway (rline le_female le_male year, title(1)                    ///
> ytitle("") name(g1, replace))
. * Basic shaded range plot
. twoway (rarea le_female le_male year, title(2)                    ///
> ytitle("") name(g2, replace))
. * Basic range bar plot
. twoway (rbar le_female le_male year, title(3)                     ///
> ytitle("") name(g3, replace))
. * Basic range spike plot
. twoway (rspike le_female le_male year, title(4)                   ///
> ytitle("") name(g4, replace))
. * Basic range capped spikes plot
. twoway (rcap le_female le_male year, title(5)                     ///
> ytitle("") name(g5, replace))
. * Basic capped spikes with symbols plot
. twoway (rcapsym le_female le_male year, title(6)                  ///
> ytitle("") name(g6, replace))
. * Basic range scatterplot
. twoway (rscatter le_female le_male year, title(7)                 ///
> ytitle("") name(g7, replace))
. * Basic range connected scatterplot
. twoway (rconnected le_female le_male year, title(8)               ///
> ytitle("") name(g8, replace))
. * Combine graphs
. graph combine g1 g2 g3 g4 g5 g6 g7 g8, row(3)
```

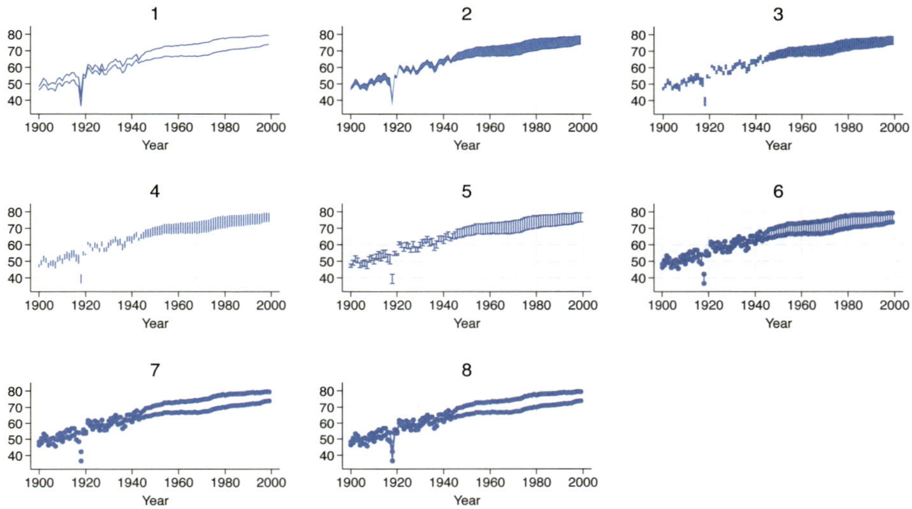

Figure 15.6. Different types of range plots

Figure 15.6 displays the various range plots that Stata offers. Each individual plot was created using the `ytitle("")` option to remove the clutter of repeated y-axis titles. Plots 1, 7, and 8 (`rline`, `rscatter`, and `rconnected`) contain no area shading and are generally not used when the objective is to demonstrate how two data series relate to each other. Plots 2, 3, 4, 5, and 6 all offer some area shading between the two data series. Which plottype is best suited will depend on the user's objective. Plots 5 and 6 (`rcap` and `rcapsym`) place heavier emphasis on the individual data markers and less on the area-shaded component. This is suitable for when detailed data comparisons are needed. Plots 3 and 4 (`rbar` and `rspike`) remove the line connections from the data and emphasize the data series as discrete bars. This makes marker-to-marker comparison easier, although individual data points are not as heavily emphasized as in plots 5 and 6. Plot 2 (`rarea`) is often the standard range plot; it uses both heavy area shading and line connections while removing emphasis on markers. This kind of plot is appropriate when users wish to concentrate on both the data series and its global evolution.

15.5 Pair plots

The community-contributed command `pairplot` (Cox 2003b) also creates a ranged area visualization between two data series while offering some useful options. The `pairplot` command graphs paired data for two y variables by using parallel lines to illustrate the differences or ratios between values. Its basic operation requires two y variables and an x variable to be specified in the command, similarly to other range commands. This is demonstrated in the code to create figure 15.7.

Creating figure 15.7

```
. * A basic pair plot
. pairplot le_male le_female year
```

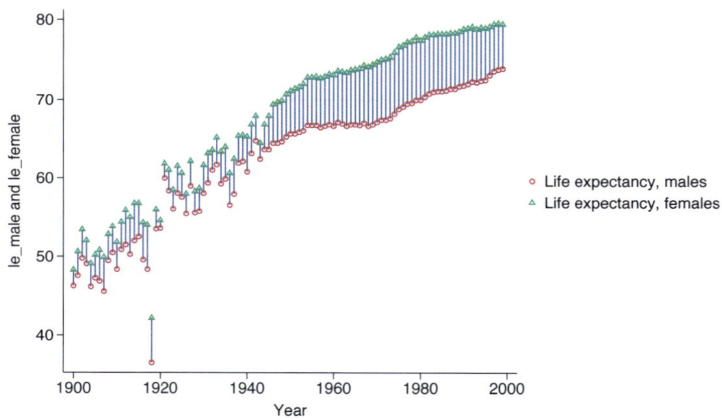

Figure 15.7. Pair plot

Figure 15.7 produces a visualization resembling that of the `rcapsym` command in style; however, it produces an arguably more distinct and visually attractive graph by default. Different marker symbols identify the two different variables; color and symbol variations make it obvious which markers belong to which variable. Area shading is provided thanks to connecting bars, and overall, the visualization produces a neat and consistent visualization of the difference between male and female life expectancies.

Two convenient options of `pairplot` are `diff` and `ratio`, which specify that the y axis should show the difference of *yvar1* − *yvar2* or the ratio of *yvar1*/*yvar2*. This means that the difference between the two series can be graphed without further computation. This turns `pairplot` from a shaded range plot into an area plot; similarly to other area plots, when `pairplot` is used in this configuration, the `base()` option becomes available, which allows users to modify the reference line. An example with the `ratio` option is shown in figure 15.8.

Creating figure 15.8

```
. * Pair plot ratios
. pairplot le_male le_female year, ratio base(0.95)
```

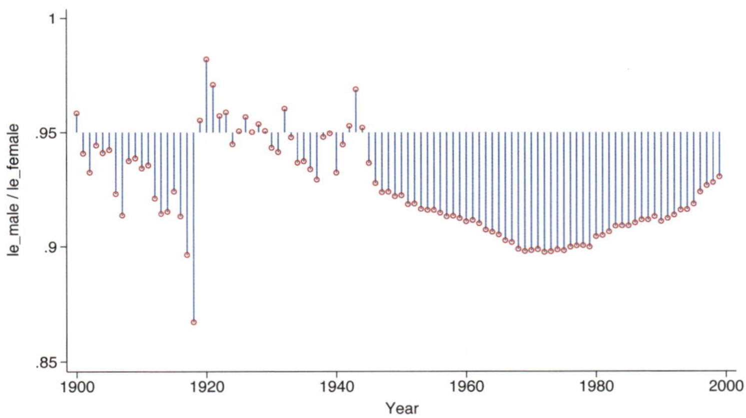

Figure 15.8. Pair plot ratios

15.6 Multiple area and shaded range plots

Area plots create a lot of visual impact because of the often large graph areas that they shade. This makes plotting multiple area or shaded range plots on a single graph problematic because area shading may overlap, thereby making identification of data difficult. This can be mitigated to some extent by using color opacity (the inverse of transparency). A color's opacity percentage determines how much of that color hides the background. A color that is 100% opaque completely hides the background, and a color with 0% opacity is entirely transparent. Opacity in area and range graphs can be adjusted with the `%` modifier when using either the `color()` or the `fcolor()` option (the first changing line and area coloring and the second changing only area coloring). Figure 15.9 shows an example with two overlaid area plots and transparent colors. The option `color(%50)` is used in both overlaid area plots.

Creating figure 15.9

```
. * Two area plots overlaid
. twoway (area le_male year, color(%50))    ///
> (area le_female year, color(%50) legend(row(2)))
```

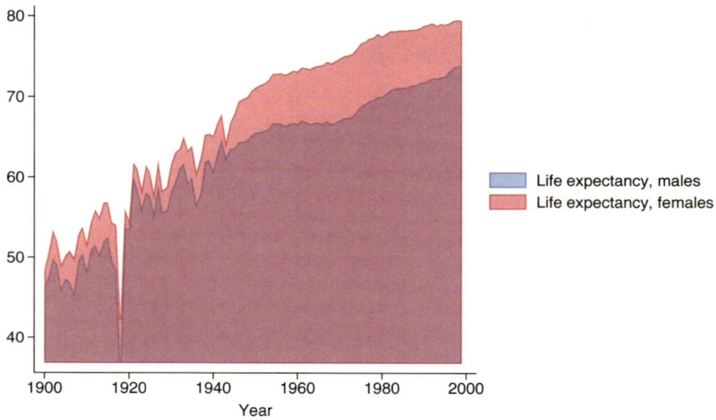

Figure 15.9. Two area plots

Figure 15.9 demonstrates why it is generally a poor idea to overlay multiple area plots. While both area plots can be identified in this particular case, it requires some effort by the user. Two simple line plots would have done a much better job. The area components underneath the two line plots are marred by color mixing. This can cause confusion in more complex graphs, so users should generally not overlay multiple area plots.

The same narrative applies to shaded range plots, although there is more scope for multiple range plots on a single graph here. This is because shaded range plots do not area shade to a reference line but to other line plots. Depending on the trending of both shaded range plots, there may be less area overlap. A positive example is shown below in figure 15.10. Note the use of the `legend()` option with the suboptions `position()` and `row()` to place the lengthy legend text at the bottom of the graph.

Creating figure 15.10

```
. * Two shaded range plots overlaid
. twoway (rarea le_wmale le_wfemale year, color(%50))  ///
> (rarea le_bmale le_bfemale year, color(%50) legend(position(6) row(2)))
```

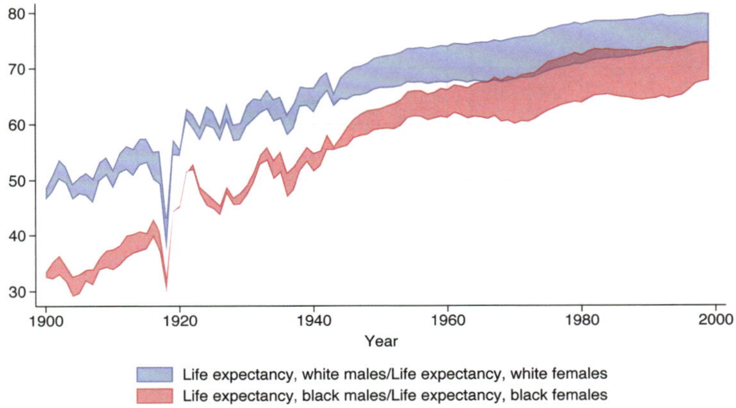

Figure 15.10. Two shaded range plots

Figure 15.10 is an example where multiple shaded range plots on a single graph work and make narrative sense. The ranges do not overlap much, and it is easy to identify both area shadings. This allows the visualization to present four separate lines and signify the relationship between these variables. However, caution must be advised. Overloading shaded range plots with too many plots will lead to plots that offer little value. A negative example is shown in figure 15.11.

Creating figure 15.11

```
. * Limit reached with four shaded range plots
. twoway (rarea le_male le_female  year, color(%50))    ///
> (rarea le_w le_b year, color(%50))                    ///
> (rarea le_wmale le_wfemale year, color(%50))          ///
> (rarea le_bmale le_bfemale year, color(%50) legend(position(6) row(4)))
```

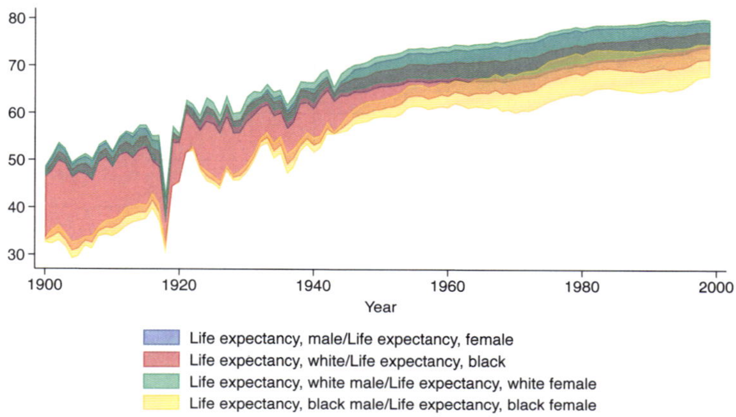

Figure 15.11. A negative example: Four shaded range plots

Figure 15.11 is a clear example of a graph with too many overlaid shaded range plots. It becomes difficult to identify different data series and different area shadings. Users should generally not put multiple area or shaded range plots on a single graph.

16 Lines of best fit

Advantages

1. Quickly overview the bivariate relationship
2. Are great for large samples or overloaded scatterplots
3. Can accommodate group comparison by plotting multiple lines
4. Allow for statistical eyeball testing via inclusion of confidence intervals
5. Can help identify linear and nonlinear trends more easily

Disadvantages

1. Often need a scatterplot for better data contextualization
2. Can compete to be the "correct" line
3. If parametric, require prior assumptions on the general relationship
4. If nonparametric, require computational choices and cannot be easily transferred
5. May oversimplify complex relationships

Required community-contributed packages

To install the community-contributed commands that are required to follow along with this chapter completely, type

```
. ssc install combomarginsplot
. ssc install supsmooth
. net install https://www.stata-journal.com/software/sj15-4/gr0021_2
. ssc install ellip
```

16.1 Introduction

Unlike connected plots, lines of best fit do not connect individual data points. They attempt to express the relationship between all data points on a graph via a line that best meets certain conditions. The outcome is that such lines show a trend, or correlation, between two variables. This concept is closely related to regression analysis, which performs such computations in a multidimensional space and outputs slope parameters of independent variables, highlighting their relationships to the dependent variable or variables. In a practical environment, regression methods are often most appropriate when many independent variables need to be jointly analyzed with one or more dependent variables. However, in a graphical setting, line-of-best-fit plots are generally employed in bivariate cases where there is only one y variable and one x variable.

A key advantage of lines of best fit is that they clearly indicate the general relationship between y and x. Lines of best fit excel at quickly providing an overview of how two continuous variables are related (correlated). This is especially useful if a standard scatterplot reveals an ambiguous relationship or if one has so much data that overplotting occurs.

Many techniques for fitting lines have been developed, and they are generally split into parametric and nonparametric methods. Parametric methods impose one or multiple parameters onto the data relationship. In simple terms, the underlying line of best fit has a global equation that allows it to be reproduced outside the data space. This allows parametric lines of best fit to be transferable outside the dataset and to extrapolate relationships outside data boundaries. Nonparametric lines of best fit let the data speak for themselves and apply less restrictive parameters that are not global. Usually, a bandwidth parameter determines how much data are looked at. In all cases, nonparametric lines look at subportions of the data and then stitch the relevant relationships together to make a global relationship. The advantage of nonparametric fits is that users need to make fewer assumptions about initial relationships in the data. A disadvantage is that such lines of best fit are not transferable.

A good rule of thumb is that nonparametric lines of best fit should be used for descriptive analysis or basic data exploration. No priors are needed about potential relationships. This makes them very suitable for graphical analysis. Parametric methods are best employed in analyses of multiple variables (such as regression) or in cases where there is a prior theoretical foundation about the global relationship between y and x. Parametric methods should also be used where extrapolation beyond known data points is needed or when relationships need to be taken outside a given dataset and applied or tested in another dataset.

16.2 A basic best fit plot

The basic problem that lines of best fit try to answer is similar to that of scatterplots and other relationship plots: they attempt to provide a visual summary of how two (continuous) variables are related. In scatterplots, users must imagine the overall relationship between the variables from the numerous points drawn on the graph. This can be problematic when 1) the scatterplot is overloaded with data points or 2) the relationship between y and x is ambiguous (or complex).

An example is shown in figure 16.1 below. auto.dta is loaded, and the two continuous variables price and mpg are plotted against each other on a basic scatterplot.

```
. * Load data
. sysuse auto
(1978 automobile data)
. summarize
```

Variable	Obs	Mean	Std. dev.	Min	Max
make	0				
price	74	6165.257	2949.496	3291	15906
mpg	74	21.2973	5.785503	12	41
rep78	69	3.405797	.9899323	1	5
headroom	74	2.993243	.8459948	1.5	5
trunk	74	13.75676	4.277404	5	23
weight	74	3019.459	777.1936	1760	4840
length	74	187.9324	22.26634	142	233
turn	74	39.64865	4.399354	31	51
displacement	74	197.2973	91.83722	79	425
gear_ratio	74	3.014865	.4562871	2.19	3.89
foreign	74	.2972973	.4601885	0	1

Creating figure 16.1

```
. * The relationship between car price and miles per gallon
. scatter price mpg
```

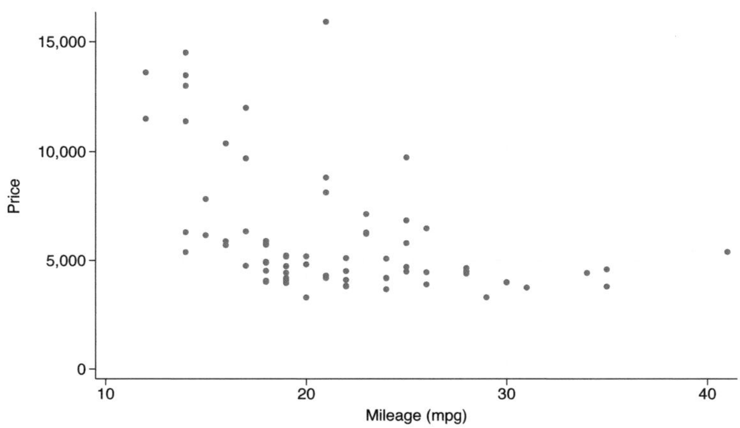

Figure 16.1. A basic scatterplot

The result is a relationship that appears negative in aggregate, but some local parts of this relationship appear very negative and others even appear positive. It is not entirely clear how the relationship between `price` and `mpg` should be interpreted.

Lines of best fit attempt to overcome the ambiguity problem of figure 16.1 by providing a clear visual depiction of how these data are related via a best fit modeling approach. The final relationship depicted will be subject to whatever best fit equation is used to model the data. Still, in all cases, a clear visual line depicting the relationship between y and x is presented, making relationship interpretation faster and easier than examining individual data points.

The basic line-of-best-fit plot is a linear fit that shows whether an x–y relationship is positively or negatively sloped using a single slope parameter and a single intercept parameter. Such graphs can be created with the `twoway lfit` command. This command requires a *yvar* and an *xvar* to be specified. Unlike `scatter`, the `lfit` plottype cannot be used as a standalone command; the full `twoway lfit` command name is required. A simple example is displayed in figure 16.2.

Creating figure 16.2

```
. * A basic linear fit
. twoway (lfit price mpg)
```

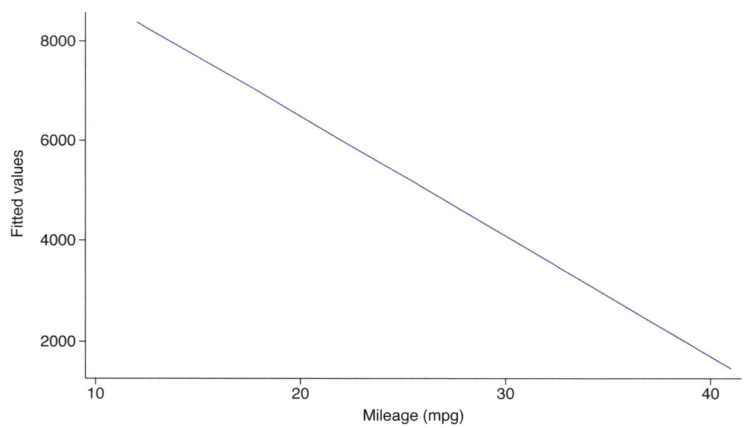

Figure 16.2. A basic linear fit plot

Figure 16.2 demonstrates the relationship between the price of cars and their mileage. It is a somewhat naïve example because the scaling of the y axis is such that the relationship looks strongly negative; because no data points are overlaid on the plot, the y-axis range is relatively limited and does not encompass the range of data points for the y axis. The y axis is the variable `price`, whose data values lie in the range of

[3,291, 15,906], which is not reflected in this plot. This, in turn, gives a false impression of a strongly negative relationship. This is why fit plots are nearly always used with other `twoway` plottypes, primarily scatterplots.

When a fit plot and a scatterplot are overlaid, `scatter` should be specified at the first plot, and fit plots should be specified subsequently. This means that marker symbols will, by default, not block the visualization of the lines of best fit. An example is demonstrated below in figure 16.3.

Creating figure 16.3

```
. * Scatterplot with linear fit
. twoway (scatter price mpg) (lfit price mpg)
```

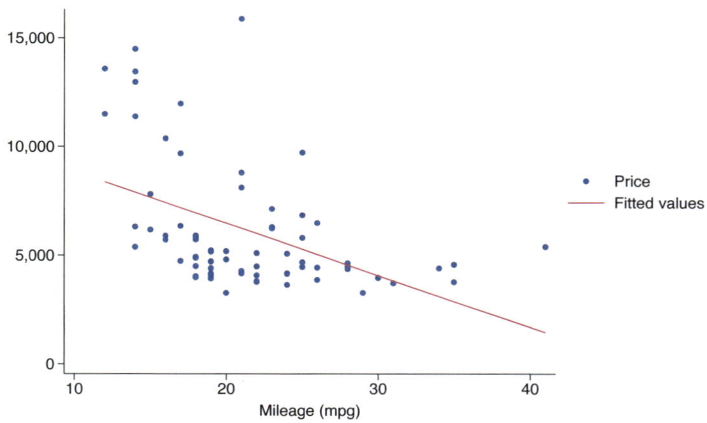

Figure 16.3. Scatterplot with linear fit

Figure 16.3 is a basic linear best fit plot. The linear best fit summarizes the data relationship so that the total distance between all the individual data points and the line is minimized. Any other line through these data would result in more distance between the data points and the line of best fit. A key advantage of this linear fit plot is that it clearly indicates to a user which way the data are sloping. In this case, the relationship between car prices and car mileage is negative.

16.2.1 Predicting out of range

An important capability of parametric fit plots is that they can predict relationships outside the relevant data range. This is because parametric fits are subject to parameterized equations that can easily be extended toward negative and positive infinity. While caution is advised whenever lines of best fit are extended beyond the relevant data range (because the extrapolated data relationship may not hold in this space), one can easily do so with the `range()` option. Two values must be specified within this option that indicate the minimum and maximum prediction values. Figure 16.4 demonstrates an example where the range of the variable `mpg` is extended to 0 and 50.

Creating figure 16.4

```
. * Linear fit extrapolation
. twoway (scatter price mpg) (lfit price mpg, range(0 50))
```

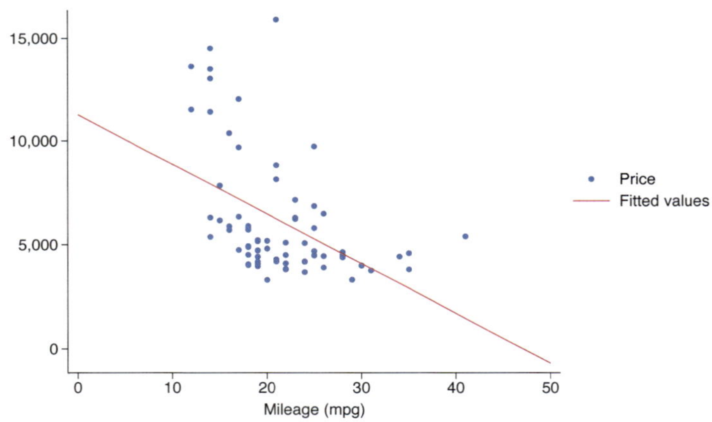

Figure 16.4. Linear fit extrapolation

Figure 16.4 demonstrates a line of best fit between `price` and `mpg` with an out-of-bounds extension of the line of best fit beyond the minimum and maximum `mpg` values. Examining the average price of cars with mileage values between 0 and 50 is now possible. However, caution is always urged. In this case, the line of best fit predicts that car prices will be negative at 50 miles per gallon, suggesting that car sellers will pay people to take their cars. Clearly, such predictions make little sense.

The `range()` option is available to all default parametric fit plots, including `qfit` and `fpfit`.

16.3 Quadratic fits

Stata also offers quadratic fit plots via the `twoway qfit` command. Quadratic fit plots fit lines of best fit with two slope parameters and one intercept parameter. This, in turn, allows them to depict relationships that have one inflection. Depending on the estimated slope parameters, lines of best fit will be either convex or concave. The `twoway qfit` command works identically to the previous `twoway lfit` command, and an example is demonstrated in figure 16.5 below.

Creating figure 16.5

```
. * Quadratic fit
. twoway (scatter price mpg) (qfit price mpg)
```

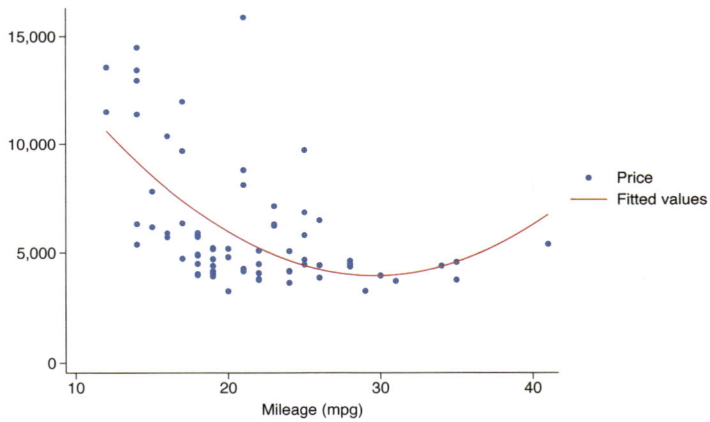

Figure 16.5. Quadratic fit

16.4 Fractional polynomial fits

A more complex parametric fit can be achieved via `twoway fpfit`, which fits fractional-polynomial lines of best fit. Fractional polynomials differ from regular polynomials by allowing logarithmic transformation, noninteger powers, and repeated powers in their polynomial expansions. The underlying estimation is relatively complex, and more information can be found in the help file of the `fp` prefix command (see `help fp`), which is the estimation counterpart of `twoway fpfit`. While graphs can be drawn after estimation with the `fp` prefix, the `twoway fpfit` command omits the information output. It proceeds straight to the graphical element of fractional polynomial fitting. Fitted graphs of this nature generally produce highly fitted curves that can have multiple inflection points but continue to be characterized by equations with relatively few parameters. An example is shown below in figure 16.5.

Note that while `twoway fpfit` is relatively easy to implement, it contains much more complex options than its `twoway lfit` and `twoway qfit` counterparts. The default degrees (two polynomial terms) and default power choices (-2, -1, -0.5, $\log(x)$, 0.5, 1, 2, and 3) can both be modified using the `estopts()` option. This can lead to substantially different shapes when customized, but users should become familiar with the underlying `fp` prefix command before attempting to modify these default options.

Creating figure 16.6

```
. * Fractional polynomial fit
. twoway (scatter price mpg) (fpfit price mpg)
```

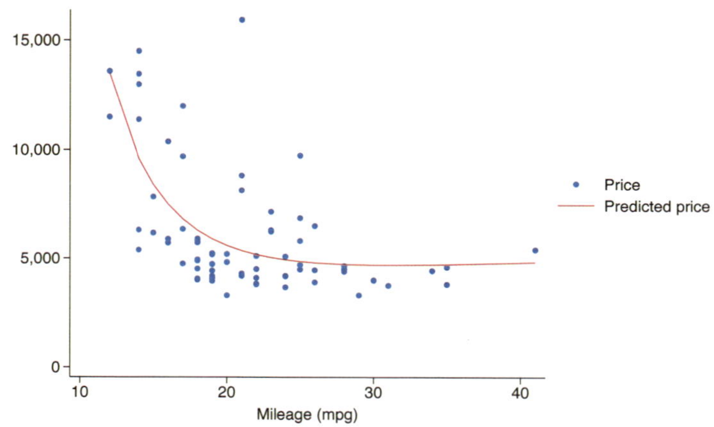

Figure 16.6. Fractional polynomial fit

Figure 16.6 is an example of a fractional polynomial fit. An advantage of fractional polynomial fitting is that low-order polynomials provide a restrictive set of possible shapes, and high-order polynomials exhibit poor fit at extreme value ranges, causing edge or wave effects. Fractional polynomials provide an extended family of curves using only a small set of predefined powers, making them powerful but flexible choices for advanced parametric fits to data relationships. However, the underlying equations are not simple to understand, and it can be difficult to relate empirical fits from fractional polynomial powers to theoretical relationships. To some extent, they offer a halfway point between parametric and nonparametric fit methods in Stata.

16.5 Multiple lines of best fit

There are two primary ways to implement multiple lines of best fit. The first is via the `twoway` command, which allows users to overlay additional plottypes. Users should specify a base scatterplot first and then overlay additional fit plots. Figure 16.7 shows this example, which overlays three different fit plots over one scatterplot. A disadvantage of this approach is that it often requires postgraph editing because the labels will not appropriately signify what each plottype is. To avoid confusion with added legend options, the legend below has been switched off with the `legend(off)` option.

Creating figure 16.7

```
. * Multiple lines of best fit
. twoway (scatter price mpg) (lfit price mpg)      ///
> (qfit price mpg) (fpfit price mpg, legend(off))
```

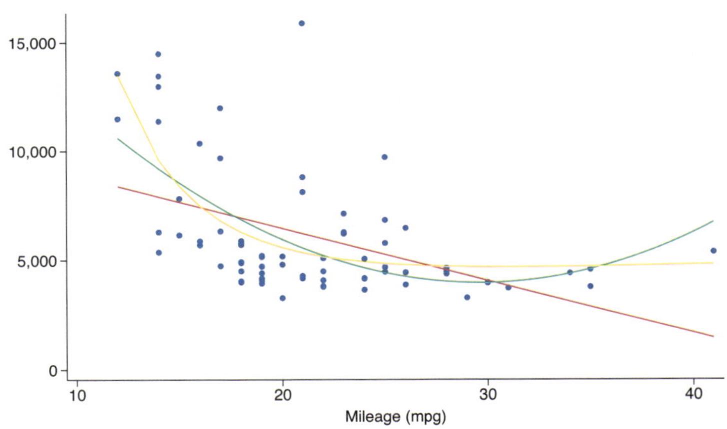

Figure 16.7. Multiple lines of best fit with `twoway`

The second method to generate multiple lines of best fit is by using the `by()` option, which is available to all Stata fit commands. The `by()` option draws separate graphs by a variable list (often one categorical variable, but there can be multiple) and merges these separate graphs into a combined graph. An example using the `twoway lfit` command is shown in figure 16.8. The `by(rep78, compact)` option repeats the linear fit plot by each category of the variable `rep78`, and the suboption `compact` compacts the graph style. The `range()` option is also used to extrapolate the results into an out-of-bounds data space.

Creating figure 16.8

```
. * Lines of best fit with the by() option
. twoway (scatter price mpg)    ///
> (lfit price mpg, by(rep78, compact) range(12 41))
```

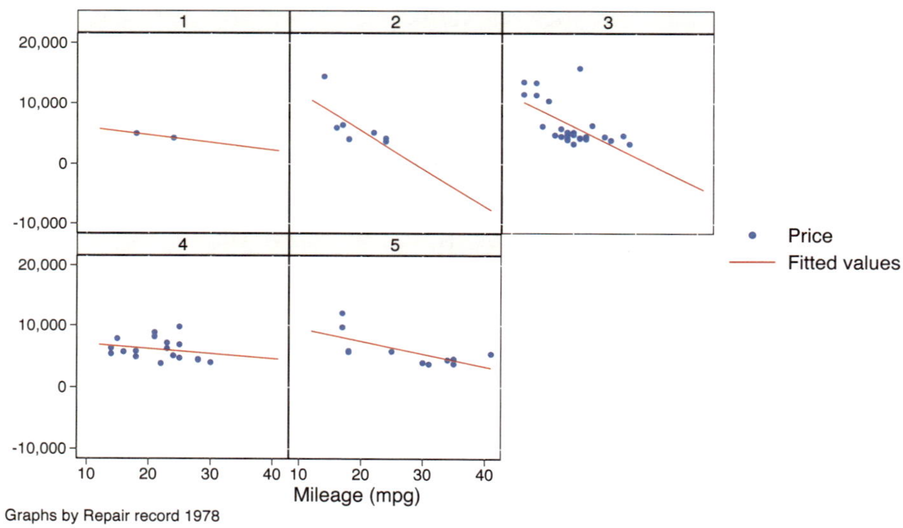

Figure 16.8. Multiple lines of best fit with `by()`

16.6 Custom polynomial fits

Astute readers will have noticed that no cubic or other higher-order polynomial fit commands have been presented. This is because no official graph command plots high-order polynomial fits, likely for good reason; such fits are often specialized and are hard to reconcile with theory. However, it is relatively simple to produce custom polynomial fits of any order by first estimating the relevant relationship in a regression using the `regress` command. This estimation command can easily fit high-order interaction terms via Stata's factor-variable expansion capabilities (see `help fvvarlist`). Variables can be factorially interacted with each other via the `##` operator. The `c.` variable prefix identifies variables as continuous (categorical variables use the `i.` prefix). For example, the code `c.mpg##c.mpg##c.mpg` will result in three polynomial terms being created: `mpg`, `mpg2`, and `mpg3`. Inserting this line of code into the `regress` command and setting the dependent variable as `price` will result in a regression model of car prices that has a linear `mpg` variable, a quadratic `mpg` variable, and a cubic `mpg` variable. For each variable, a coefficient is estimated, and together, these coefficients model a cubic relationship between `price` and `mpg`. The results can then be used to compute predictions with the `margins` command. In the example below, `margins` takes the most recent estimates (those of `regress`) and computes the average value of `price` at values of `mpg` that range between 0 and 50 in increments of 2.

```
. regress price c.mpg##c.mpg##c.mpg

      Source |       SS           df       MS      Number of obs   =        74
-------------+----------------------------------   F(3, 70)        =     14.34
       Model |   241701668         3  80567222.7   Prob > F        =    0.0000
    Residual |   393363728        70  5619481.83   R-squared       =    0.3806
-------------+----------------------------------   Adj R-squared   =    0.3540
       Total |   635065396        73  8699525.97   Root MSE        =    2370.5

------------------------------------------------------------------------------------
             price | Coefficient  Std. err.      t    P>|t|     [95% conf. interval]
-------------------+----------------------------------------------------------------
               mpg |  -3990.861   1301.467    -3.07   0.003    -6586.556   -1395.166
                   |
       c.mpg#c.mpg |   135.6783   53.59796     2.53   0.014      28.7805     242.576
                   |
 c.mpg#c.mpg#c.mpg |  -1.503489   .7007825    -2.15   0.035    -2.901155   -.1058222
                   |
             _cons |   43105.62   10054.97     4.29   0.000     23051.62    63159.63
------------------------------------------------------------------------------------
```

```
. margins, at(mpg = (0(2)50))

Adjusted predictions                                    Number of obs = 74
Model VCE: OLS

Expression: Linear prediction, predict()
1._at:  mpg =  0
2._at:  mpg =  2
3._at:  mpg =  4
4._at:  mpg =  6
5._at:  mpg =  8
6._at:  mpg = 10
7._at:  mpg = 12
8._at:  mpg = 14
9._at:  mpg = 16
10._at: mpg = 18
11._at: mpg = 20
12._at: mpg = 22
13._at: mpg = 24
14._at: mpg = 26
15._at: mpg = 28
16._at: mpg = 30
17._at: mpg = 32
18._at: mpg = 34
19._at: mpg = 36
20._at: mpg = 38
21._at: mpg = 40
22._at: mpg = 42
23._at: mpg = 44
24._at: mpg = 46
25._at: mpg = 48
26._at: mpg = 50
```

	Margin	Delta-method std. err.	t	P>|t|	[95% conf.	interval]
_at						
1	43105.62	10054.97	4.29	0.000	23051.62	63159.63
2	35654.59	7679.647	4.64	0.000	20338.01	50971.16
3	29216.81	5698.44	5.13	0.000	17851.63	40581.99
4	23720.12	4078.497	5.82	0.000	15585.82	31854.43
5	19092.36	2787.671	6.85	0.000	13532.53	24652.19
6	15261.35	1795.462	8.50	0.000	11680.42	18842.29
7	12154.93	1075.646	11.30	0.000	10009.62	14300.24
8	9700.936	613.1166	15.82	0.000	8478.114	10923.76
9	7827.193	403.1501	19.42	0.000	7023.135	8631.251
10	6461.536	369.7162	17.48	0.000	5724.16	7198.912
11	5531.798	372.5949	14.85	0.000	4788.681	6274.915
12	4965.811	368.1946	13.49	0.000	4231.471	5700.152
13	4691.409	384.626	12.20	0.000	3924.296	5458.521
14	4636.423	449.6667	10.31	0.000	3739.591	5533.255
15	4728.686	550.7482	8.59	0.000	3630.253	5827.118
16	4896.03	656.5638	7.46	0.000	3586.555	6205.505
17	5066.289	747.4985	6.78	0.000	3575.45	6557.128
18	5167.294	833.3894	6.20	0.000	3505.151	6829.437
19	5126.879	972.3569	5.27	0.000	3187.574	7066.183
20	4872.875	1266.535	3.85	0.000	2346.851	7398.899
21	4333.115	1804.911	2.40	0.019	733.3345	7932.896
22	3435.433	2630.614	1.31	0.196	-1811.161	8682.026
23	2107.659	3766.48	0.56	0.578	-5404.348	9619.667
24	277.6279	5235.424	0.05	0.958	-10164.1	10719.35
25	-2126.829	7064.082	-0.30	0.764	-16215.7	11962.04
26	-5177.879	9281.851	-0.56	0.579	-23689.95	13334.19

The resulting predictions from the estimation analysis are then used by the command **marginsplot**, which plots the predicted values on a graph. The **noci** option suppresses confidence intervals, and the **addplot()** option adds a scatterplot of **price** versus **mpg** as an overlay. Because **scatter** is overlaid on **marginsplot**, the marker colors of the scatterplot are made transparent using the (%50) modifier in the **mcolor()** option. In addition, legends are suppressed to avoid clutter in the graph.

Creating figure 16.9

```
. * Cubic polynomial
. marginsplot, noci addplot(scatter price mpg, mcolor(%50) legend(off))
```

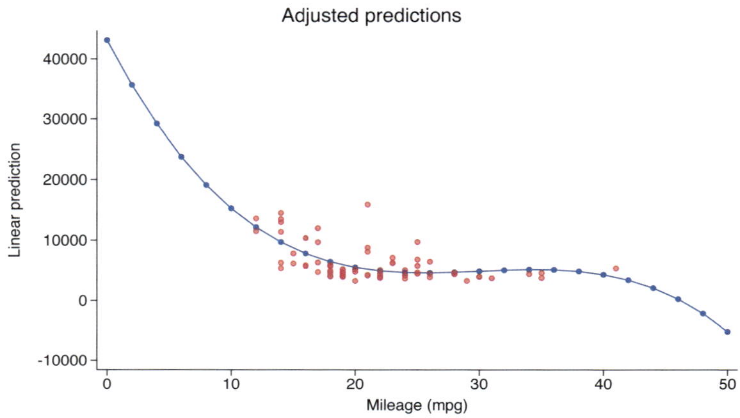

Figure 16.9. Custom polynomial fit

Figure 16.9 demonstrates how to create a cubic polynomial fit using the `regress` estimation command and the plotting of the subsequent predictions. Note that figure 16.9 predicts out of bounds, which was achieved via the `margins` command. Modifying the above regression to higher nth-degree polynomials is a relatively simple process of adding more `##c.mpg` terms to the regression code. However, this is unlikely to be a good idea in most cases because few data processes follow very high polynomial patterns.

16.7 Multiple custom polynomial fits

If multiple custom polynomial plots are desired, one must instead replace the official command `marginsplot` with the community-contributed command `combomarginsplot` (Winter 2014). `marginsplot` is not able to (easily) plot more than one `margins` prediction on a single plot, but `combomarginsplot` helps users achieve this without complex code. The `combomarginsplot` command requires users to save their predictions from multiple `margins` commands and then specify these saved predictions within the `combomarginsplot` code.

An example is shown below in figure 16.10. Two polynomial regressions are estimated (a cubic and a quartic), and predictions from these are then stored via the `margins` command with the `saving()` option. The `quietly` prefix command is used to suppress terminal output.

```
. * Regressions with interaction terms
. regress price c.mpg##c.mpg##c.mpg

      Source |       SS           df       MS      Number of obs   =        74
-------------+----------------------------------   F(3, 70)        =     14.34
       Model |  241701668         3  80567222.7   Prob > F        =    0.0000
    Residual |  393363728        70  5619481.83   R-squared       =    0.3806
-------------+----------------------------------   Adj R-squared   =    0.3540
       Total |  635065396        73  8699525.97   Root MSE        =    2370.5

------------------------------------------------------------------------------
       price | Coefficient  Std. err.      t    P>|t|     [95% conf. interval]
-------------+----------------------------------------------------------------
         mpg |  -3990.861   1301.467    -3.07   0.003    -6586.556   -1395.166
             |
 c.mpg#c.mpg |   135.6783   53.59796     2.53   0.014      28.7805     242.576
             |
 c.mpg#c.mpg#|
       c.mpg |  -1.503489   .7007825    -2.15   0.035    -2.901155   -.1058222
             |
       _cons |   43105.62   10054.97     4.29   0.000     23051.62    63159.63
------------------------------------------------------------------------------

. quietly margins, at(mpg = (12(1)41)) saving(file1, replace)
. regress price c.mpg##c.mpg##c.mpg##c.mpg

      Source |       SS           df       MS      Number of obs   =        74
-------------+----------------------------------   F(4, 69)        =     12.27
       Model |  263997367         4  65999341.7   Prob > F        =    0.0000
    Residual |  371068029        69  5377797.53   R-squared       =    0.4157
-------------+----------------------------------   Adj R-squared   =    0.3818
       Total |  635065396        73  8699525.97   Root MSE        =      2319

------------------------------------------------------------------------------
       price | Coefficient  Std. err.      t    P>|t|     [95% conf. interval]
-------------+----------------------------------------------------------------
         mpg |  -14640.63    5383.09    -2.72   0.008     -25379.6   -3901.657
             |
 c.mpg#c.mpg |   818.6972   339.5205     2.41   0.019     141.3724    1496.022
             |
 c.mpg#c.mpg#|
       c.mpg |  -20.03977   9.129399    -2.20   0.032    -38.25242   -1.827116
             |
 c.mpg#c.mpg#|
 c.mpg#c.mpg |   .1799426   .0883742     2.04   0.046     .0036408    .3562444
             |
       _cons |   102336.3   30707.67     3.33   0.001     41076.16    163596.4
------------------------------------------------------------------------------

. quietly margins, at(mpg = (12(1)41)) saving(file2, replace)
```

The user must then specify both saved prediction files (`file1` and `file2`) in the `combomarginsplot` command. The remaining options specified are identical to those accessible by `marginsplot`. The result is an overlaid graph with 1) a cubic fit, 2) a quartic fit, and 3) a basic scatterplot.

Creating figure 16.10

```
. * Multiple polynomials with combomarginsplot
. combomarginsplot file1 file2, noci                          ///
> addplot(scatter price mpg, mcolor(%50) legend(off))  ///
> title(Adjusted predictions)
```

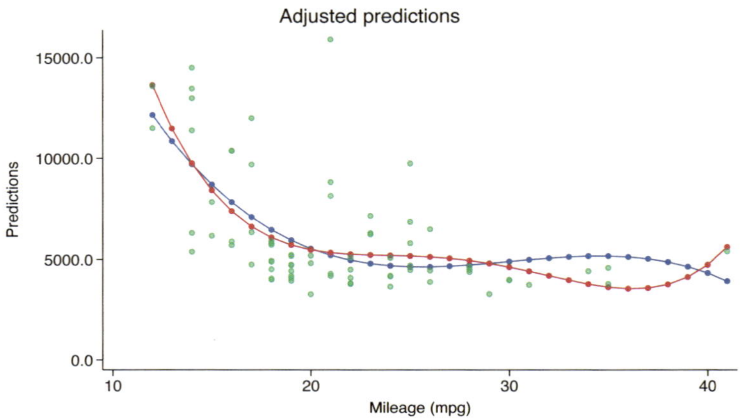

Figure 16.10. Multiple custom polynomial fits

16.8 Custom polynomial fits with functions

Finally, for users wishing to access the underlying parametric equation of their custom polynomial or maybe for users who are not content with displaying polynomial fit via linear connected intervals, the `twoway function` command can specify custom equations on plots. The concept here is that coefficient estimates from the previous estimation command (`regress`) are converted directly onto the graph via the `twoway function` command, leaving out the intermediate step of predictions using the `margins` command.

The `twoway function` command requires a specification of the nature $[y =]f(x)$, where $f(x)$ is a function of x. The trick is to realize that estimates from the regression can be directly accessed from the `e(b)` matrix through the use of system variables such as `_b[mpg]`. See `help _variables` for more information. All users need to do is specify a function in the form of the previous regression (in this example, a cubic polynomial) and insert the relevant `_b[varname]` as the coefficient for each polynomial. The resulting code, which produces figure 16.11, is shown below. At first glance, it can seem a somewhat unwieldy code setup, but this is mainly a product of the long variable names caused by the "on-the-fly" interaction expansion. The logic behind the code construction is relatively simple. A function in the form of $y = a + bx + cx^2 + dx^3$

is specified and drawn over a numerical range of 12 to 41 (representing the range of the mpg values), where the values for a, b, c, and d are automatically taken from the previous regression estimates.

```
. * Regressions with interaction terms
. regress price c.mpg##c.mpg##c.mpg

      Source |       SS           df       MS      Number of obs   =        74
-------------+----------------------------------   F(3, 70)        =     14.34
       Model |   241701668         3  80567222.7   Prob > F        =    0.0000
    Residual |   393363728        70  5619481.83   R-squared       =    0.3806
-------------+----------------------------------   Adj R-squared   =    0.3540
       Total |   635065396        73  8699525.97   Root MSE        =    2370.5

------------------------------------------------------------------------------
       price | Coefficient  Std. err.      t    P>|t|     [95% conf. interval]
-------------+----------------------------------------------------------------
         mpg |  -3990.861   1301.467    -3.07   0.003    -6586.556   -1395.166
             |
 c.mpg#c.mpg |   135.6783   53.59796     2.53   0.014      28.7805     242.576
             |
 c.mpg#c.mpg#|
       c.mpg |  -1.503489   .7007825    -2.15   0.035    -2.901155   -.1058222
             |
       _cons |   43105.62   10054.97     4.29   0.000     23051.62    63159.63
------------------------------------------------------------------------------
```

Creating figure 16.11

```
. * Drawing a custom function
. twoway (scatter price mpg) (function y = _b[_cons] + _b[mpg]*x +     ///
> _b[c.mpg#c.mpg]*x^2 + _b[c.mpg#c.mpg#c.mpg]*x^3, range(12 41))
```

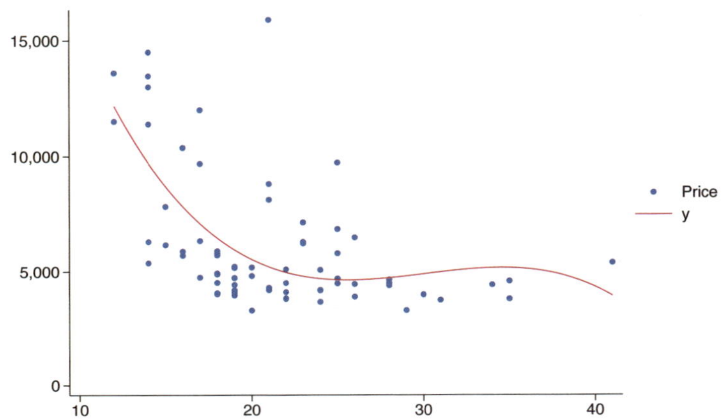

Figure 16.11. Custom polynomial fit with twoway function

Figure 16.11 repeats figure 16.9 but through a different method. The **marginsplot** method has more steps but generally requires less complex code, while the **twoway function** method requires fewer steps but more complex code. Note that the most common mistake when using the **twoway function** command to plot polynomial fits is forgetting to add the constant (`_b[_cons]`) into the function specification.

16.9 Constrained lines of best fit

It is also possible to fit constrained lines of best fit, which constrain one or more parameters in a best fit plot. Similarly to custom polynomial plots, no graph command immediately plots constrained lines of best fit. One will need to use the **cnsreg** estimation command to fit constrained linear regression models and then plot the relevant predictions. Prior to running **cnsreg**, users will need to define one or more constraints using the **constraint** command, which then enters the **cnsreg** command via the **constraints()** option. Users should examine the **constraint** help file (see **help constraint**) to familiarize themselves with the various **constraint** options.

In the below example, a simple linear fit is estimated with the constraint that the value of **price** is 10,000 when **mpg** is 0. This is done by defining a constraint (*#1*) that sets the coefficient on the variable **_cons** to 10,000. **cnsreg** then estimates a best linear fit between **price** and **mpg** with that constraint. Subsequently, the **margins** and **marginsplot** commands are used to 1) predict the estimates over a range of x and then 2) plot the predicted estimates. The previous example with the **twoway function** command could also have been used. The final result is figure 16.12.

```
. constraint 1 _cons = 10000
. cnsreg price mpg, constraints(1)
Constrained linear regression                   Number of obs =        74
                                                F(1, 73)      =    176.85
                                                Prob > F      =    0.0000
                                                Root MSE      = 2626.2650
 ( 1)  _cons = 10000
------------------------------------------------------------------------------
       price | Coefficient  Std. err.      t    P>|t|     [95% conf. interval]
-------------+----------------------------------------------------------------
         mpg |  -184.0503   13.84009   -13.30   0.000    -211.6336   -156.4671
       _cons |      10000  (constrained)
------------------------------------------------------------------------------
. quietly margins, at(mpg = (0(2)50))
```

Creating figure 16.12

```
. * Constrained line of best fit
. marginsplot, noci addplot(scatter price mpg, mcolor(%50))
```

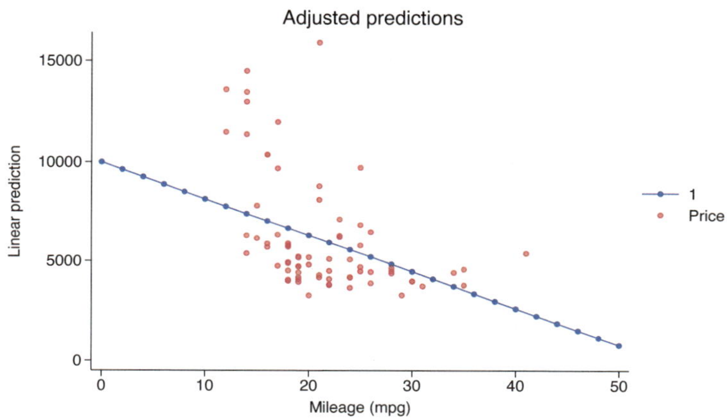

Figure 16.12. Constrained line of fit

Figure 16.12 is an example of a simple constrained fit. The slope coefficient remains free to be estimated, while the intercept is set to 10,000. Clearly, a wide range of possible constraints are feasible that allow users to set custom slopes and custom intercepts while best fitting the remaining parameters.

16.10 Nonparametric fits (smoothers)

There are various default nonparametric fit plots (sometimes called bivariate smoothers) available in Stata. In addition, several community-contributed commands have been developed over time. The exact mechanics of each bivariate smoother will vary, and users should examine individual help files. In almost all cases, a bandwidth choice underpins each nonparametric fit plot. A bandwidth determines the local space in which a nonparametric estimator estimates a data relationship. Multiple bandwidth estimates are then merged to form a global picture of the data relationship. Most commands will have methods to determine a default starting bandwidth (for example, Silverman's rule), but varying the bandwidth will likely lead to substantially different shapes and relationships. Users must understand the concept of bandwidth and explore how their graphs vary by adjusting bandwidths up or down to match their needs. Small bandwidths should be used when fine and granular data relationships matter, and large bandwidths should be used when coarse and global relationships matter more.

Different commands may also have other options that determine the shape of nonparametric fit plots. Kernel functions or polynomial functions are additional options that change the underlying mechanics of how data points are treated within a given bandwidth range. These can also alter the graphed relationship, although adjusting these parameters is less important than bandwidth choices in practice.

16.10.1 Local polynomial smoother

One of the most versatile nonparametric smoothers is the `twoway lpoly` command, which performs kernel-weighted local polynomial regression. This technique fits a polynomial curve to a subset of the data points within a specified bandwidth around each target point, using a weighting scheme (typically a kernel function) to give more importance to data points closer to the target point. The fitted values from this regression are then plotted as a smoothed curve. Local polynomial regression smoothing is an extension of local mean smoothing techniques, in which the amount of smoothing is controlled only by a bandwidth parameter, and choosing bandwidth values often involves a bias-versus-variance tradeoff. This problem is minimized in local polynomial regression-based smoothing, where higher-order polynomials have better bias properties than zero-degree local polynomial (that is, mean-based) smoothers. They often require no bias adjustment at the boundary values.

An example of a simple local polynomial smoother is shown below. In this case, the local polynomial smoother is superimposed on a scatterplot, and no options are specified. This results in figure 16.13.

Creating figure 16.13

```
. * Local polynomial lines of best fit
. twoway (scatter price mpg) (lpoly price mpg)
```

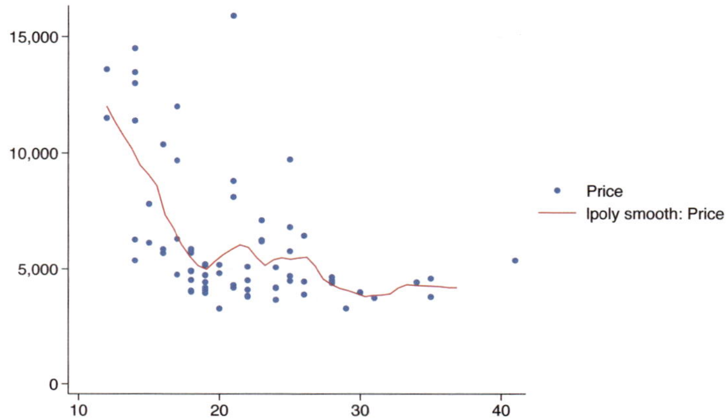

Figure 16.13. Local polynomial smoother with degree 0

Figure 16.13 demonstrates the usefulness of nonparametric smoothers. Rather than imposing a linear or quadratic fit via `lfit` or `qfit`, the `lpoly` command suggests that the relationship between price and miles per gallon is at first highly negative but then changes into a "flat" relationship once the value of miles per gallon increases past 20. The great advantage here is the avoidance of imposing a prior parametric relationship on the data and letting them speak for themselves. A disadvantage of such plots is that nonparametric lines of best fit can look substantially different depending on what kind of bandwidth is chosen.

Users of nonparametric fit plots should experiment with different bandwidth settings because doing so can result in significantly different relationship visualizations. Undersmoothing or oversmoothing can be easy mistakes to make. To adjust the bandwidth, users can specify the `bwidth()` option. Higher numbers will lead to smoother lines of best fit, while lower numbers will lead to more detail but with local variations (spikes). `twoway lpoly` has two further important options that adjust the estimation of the bivariate smoother. The first is the `kernel()` option, which specifies the kernel function (the default is `kernel(epanechnikov)`). The second is the `degree()` option, which specifies the degree of the polynomial smoother (the default is `degree(0)`). Note that the default degree of 0 means that the `twoway lpoly` command without any options is a local mean smoother and not a local regression smoother.

Figure 16.14 shows an example of all three options. A scatterplot with three overlaid polynomial smoothers is drawn. The first smoother uses a rectangular kernel (`rectangle`) and a bandwidth of 1. The second uses a normally distributed kernel (`gaussian`) and a bandwidth of 3. The third uses the default kernel (`epanechnikov`) and default bandwidth but with a polynomial of degree 2, implying local quadratic regression fits.

Creating figure 16.14

```
. * Local polynomial fits with various kernel and bandwidth choices
. twoway (scatter price mpg)                                                ///
> (lpoly price mpg, kernel(rectangle) bwidth(1))                            ///
> (lpoly price mpg, kernel(gaussian) bwidth(3))                             ///
> (lpoly price mpg, degree(2) legend(label(2 "Rectangular kernel, bw1")    ///
> label(3 "Gaussian kernel, bw3") label(4 "Default kernel, degree2")))
```

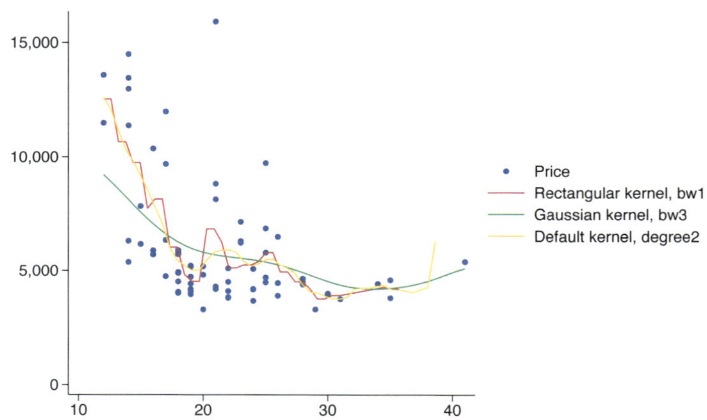

Figure 16.14. Local polynomial smoother options

The aim of figure 16.14 is to show how `twoway lpoly` option choices can result in different plot relationships. Arguably, the overall relationship of all three `twoway lpoly` fits tells the same global story; the relationship between price and miles per gallon is negative, with a steeper negative relationship in the early range of miles per gallon. Which of these three smoothers should ultimately be chosen remains a question that only users can answer.

16.10.2 Other bivariate smoothers

Several other bivariate smoothers exist, both within official Stata and from the wider user base via downloadable commands. Within official Stata, the `twoway mband` command computes cross medians, illustrating them as a line plot. This generally results in a piecewise-linear fit plot that contains many straight lines. The convenience of `twoway mband` is its methodological simplicity. The `twoway mspline` command extends this concept further by first calculating cross medians and then using them as knots to fit a cubic spline. The advantage of a cubic spline fit is that it interpolates a set of data points (the knots) and guarantees smoothness at the data points. This results in a smoothed fit that is similar to `twoway mband` but without the sharp angular breaks. `twoway lowess` is the `twoway` version of the standalone `lowess` command, which implements weighted linear least-squares regression smoothing. It is similar in character to `twoway lpoly` but has fewer options for methodological customization. Figure 16.15 demonstrates all plots in an overlaid fashion.

Creating figure 16.15

```
. * Other nonparametric fits
. twoway (scatter price mpg) (mband price mpg) ///
> (mspline price mpg) (lowess price mpg)
```

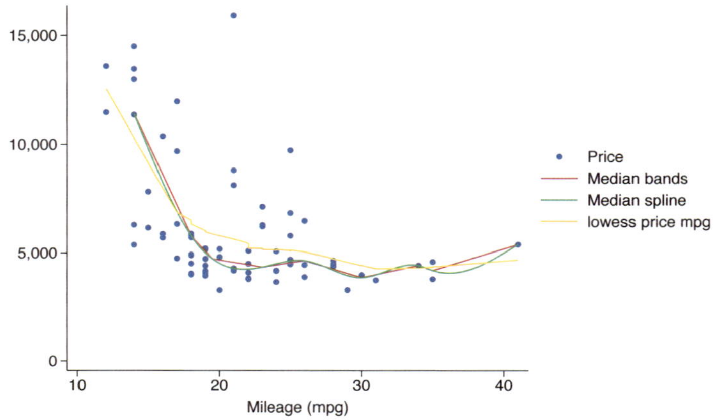

Figure 16.15. Other default smoother fits

Figure 16.15 displays the various other types of nonparametric fit plots that exist within official Stata. Users should experiment with various fit plots and their associated options. In most cases, the `twoway lpoly` command is the most powerful and flexible option with which to create visualizations of nonparametric fits.

One can use the `search` command within Stata to identify other nonparametric fit plots that might be useful. A small example selection is shown below in figure 16.16. The community-contributed command `supsmooth` (Luedicke 2015) implements a smoother that has varying bandwidths and therefore tries to overcome the constraint that users often need to select one bandwidth. `doublesm` (Cox 2005b) provides two distinct nonparametric fits of y versus x and x versus y. This can be useful if there is no assumed causal flow between the y and x variables or when y and x measure similar properties. `polarsm` (Cox 2005b) implements a smoother on polar-transformed coordinates of the bivariate data. This smoother is then mapped back into Cartesian coordinates, resulting in a closed circular smoother that defines the central (often half) part of the data. Often, ellipse-type curves are the result, providing a different visual feel of the data relationship compared with traditional line smoothers. Comparing the length versus the width of such an elliptical smoother can give a better idea of the correlation between the two variables. A parametric version of this idea is provided by the `ellip` command (Alexandersson 2004). All four custom smoothers are demonstrated below.

Creating figure 16.16

```
. * Super smoother
. supsmooth price mpg, alpha(1) title(Super smoother) name(g1, replace)
. * Double smoother
. doublesm price mpg, title(Double smoother) name(g2, replace)          ///
> legend(position(6))
. * Polar smoother
. polarsm price mpg, title(Polar smoother) smooth(lcolor(black))        ///
> name(g3, replace)
. * Confidence ellipse
. ellip price mpg, title(Confidence ellipse) plot(scatter price mpg)   ///
> name(g4, replace) legend(position(6))
. * Combine graphs
. graph combine g1 g3 g2 g4, ycommon xcommon
```

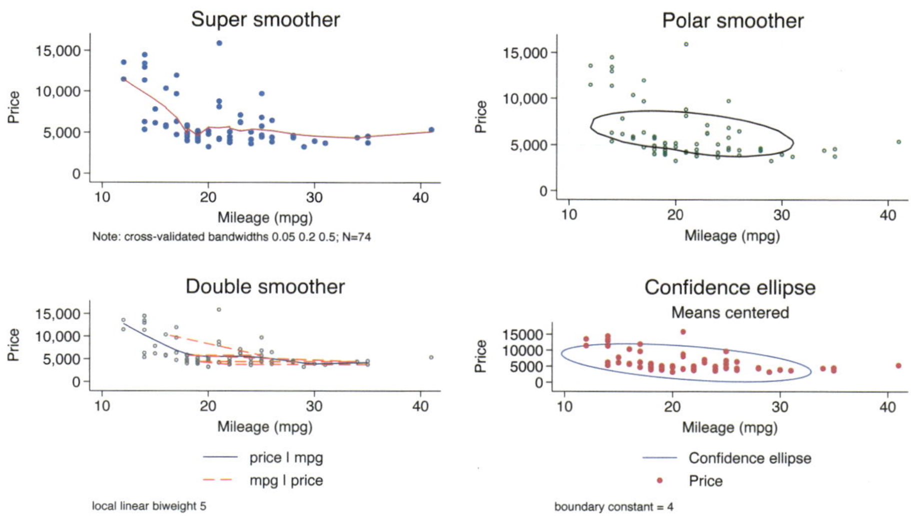

Figure 16.16. Other community-contributed smoother fits

Figure 16.16 demonstrates the versatility of Stata's user base. Each command will have various options that allow further permutation of these example graphs.

16.11 Lines of best fit with confidence intervals

The default line-of-best-fit commands (both parametric and nonparametric) all have confidence interval counterparts that allow users to plot lines of best fit with confidence intervals. This can be useful if statistical testing is required as part of the graphical analysis because confidence intervals give users an indication of how precisely a statistic is estimated. The default confidence intervals for these commands are for the mean prediction. Thus, users can test whether the mean prediction of car price at `mpg` = 20 is £5000 or not. The relevant commands all end with the suffix `ci`. Table 16.1 shows the twin command for each default fit command.

Table 16.1. Line-of-best-fit confidence interval commands

Basic line-of-best-fit command	Confidence interval counterpart
`lfit`	`lfitci`
`qfit`	`qfitci`
`fpfit`	`fpfitci`
`lpoly`	`lpolyci`

Like other fit plots, confidence interval fit plots are best used with scatterplots. An example is shown below in figure 16.17, which overlays a linear fit with confidence intervals onto a scatterplot. The option `level()` is used to control the confidence level, in this case 90%. The default is `level(95)`.

Creating figure 16.17

```
. * Linear fit with 90% confidence intervals
. twoway (scatter price mpg) (lfitci price mpg, level(90))
```

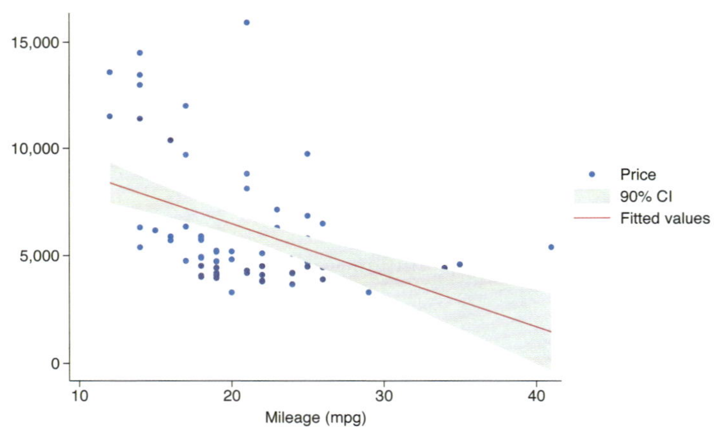

Figure 16.17. Linear fit plot with confidence intervals

Figure 16.17 shows a linear line of best fit with 90% confidence intervals. The confidence intervals can be used to gauge the overall statistical relationship of the linear fit. In this case, because the upper confidence interval on the right side of the graph is below the lower confidence interval on the left side of the graph, it is feasible to argue that the global linear relationship is statistically significant at the 90% level.

16.12 Multiple lines of best fit with confidence intervals

Note that confidence intervals introduce area shading into the graphs. One should either use transparent coloring or set the confidence interval plot as the first plottype. This avoids too many scatterplot markers being hidden by the area shading of the confidence interval. An example of making the confidence interval shading transparent is demonstrated in figure 16.18. The option `acolor()` is used to change the outline, fill color, and opacity of the confidence interval shading. In addition, two best fit confidence plots are overlaid (linear and quadratic). Because multiple plottypes are overlaid, the legend is suppressed with the `legend(off)` option to avoid clutter.

Creating figure 16.18

```
. * Multiple fits with confidence intervals
. twoway (scatter price mpg) (lfitci price mpg, acolor(eltgreen%30))  ///
> (qfitci price mpg, acolor(erose%30) legend(off))
```

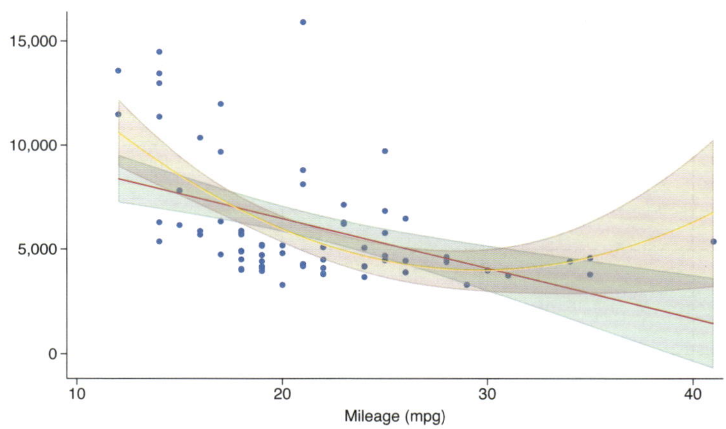

Figure 16.18. Multiple fit plots with confidence intervals

Figure 16.18 demonstrates how transparent colors can be used with multiple lines of best fit that include confidence intervals. Each graph element is recognizable, although users should avoid overlaying too many confidence interval plots onto one graph.

16.13 Lines of best fit with recast confidence intervals

Finally, all confidence interval plots can be recast as another plottype to offer a different visual feel to the graph. This can be done with the `ciplot()` option, which allows users to select a different plottype. The range-type plots (`rarea`, `rbar`, `rspike`, `rcap`, `rcapsym`, `rscatter`, `rline`, and `rconnected`) from the `twoway` plot family will often be most appropriate for confidence intervals. Figure 16.19 below demonstrates a local polynomial smoother with confidence intervals recast into a range plot with capped spikes (`rcap`). The recast confidence interval area is colored black via the `acolor()` option.

Creating figure 16.19

```
. * Local polynomial fit with confidence intervals as spikes
. twoway (scatter price mpg) (lpolyci price mpg, ciplot(rcap) acolor(black))
```

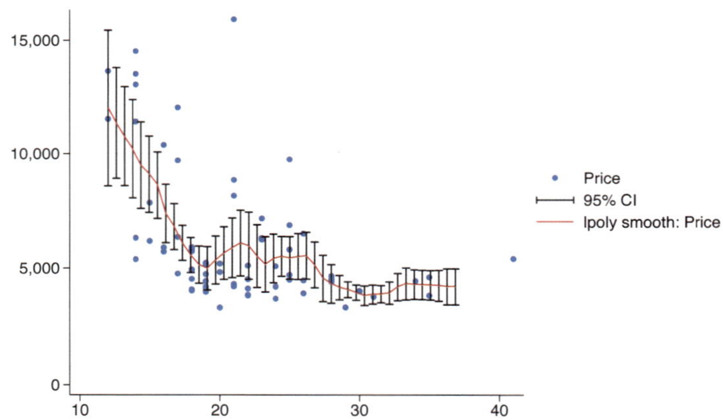

Figure 16.19. Line-of-best-fit plot with recast confidence intervals

17 Jitter plots

Advantages

1. Are useful for seeing how two discrete variables are jointly distributed
2. Are intuitive to understand
3. Help to reduce overplotting

Disadvantages

1. Do not reflect true data values with their data markers
2. Struggle when discrete variables have a very high or very low number of categories
3. Also struggle with very large or very small data samples
4. May require explanation for audiences unfamiliar with jittering techniques

Required community-contributed packages

No additional installations are required to follow along with this chapter completely.

17.1 Introduction

Scatterplots are the mainstays of bivariate data analysis. No graphical visualization is simpler and more intuitive than plotting data values from two variables on a 2D plane where each axis represents one variable. Assuming a small- or medium-size dataset, the resulting scatterplot will often neatly show the form of the relationship between x and y. A precondition of a successful scatterplot is that the values underpinning x and y are continuous (or quasicontinuous with many different discrete categories). When the x and y variables are binary or discrete with few categories, the resulting plot often has little analytical value because most data points "sit" on each other.

There is a way to make scatterplots work with such discrete data. The secret is jittering the relevant markers. Jitter plots, sometimes called jittered scatterplots, randomly displace the marker values away from the actual values. This is generally done spherically (360 degrees around the data point), although it can be reduced to more limited circular displacement (such as 180 or 90 degrees) for edge markers. In addition, the random displacement distance from the origin can be controlled, allowing users to "explode" data markers to varying degrees. The result is that more data markers can be graphed, which in turn can reveal data densities and allow inference of how discrete variables are related.

17.2 A basic jitter plot

A typical application of jitter plots is to reveal discrete joint distributions for small or medium-sized datasets. Figure 17.1 below presents a naïve scatterplot using `auto.dta`, which contains two discrete variables: `rep78`, which has five repair categories, and the binary variable `foreign`, which measures foreign or domestic car status.

```
. * Load data
. sysuse auto
(1978 automobile data)
. summarize rep78 foreign
    Variable |        Obs        Mean    Std. dev.       Min        Max
-------------+---------------------------------------------------------
       rep78 |         69    3.405797    .9899323          1          5
     foreign |         74    .2972973    .4601885          0          1
```

Creating figure 17.1

```
. * Basic scatterplot
. scatter foreign rep78
```

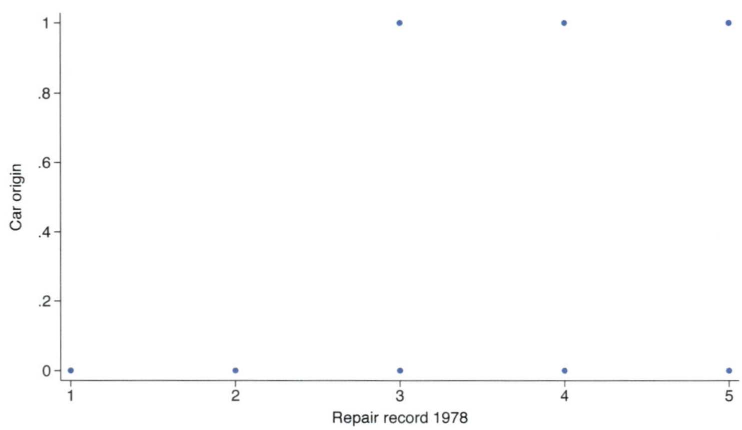

Figure 17.1. A basic scatterplot of two discrete variables

The resulting scatterplot presents only seven individual data markers and is not very informative. The majority of the data markers sit on top of each other, and little can be done with such a plot.

To obtain some usability from this scatterplot, one can randomly jitter the data markers; readers could then make some inferences about the data density in each of these 10 bivariate categories (5×2 variable categories). This can be done with the `jitter()` option, which specifies the size of the random displacement as a percentage of the graph area. Larger values will move data markers a greater random distance away from the origin; smaller values keep the "explosion" closer to the origin. Figure 17.2 shows an example with a jitter value set to 12.

Creating figure 17.2

```
. * Scatterplot with discrete data and jitter()
. twoway (scatter foreign rep78, jitter(12))
```

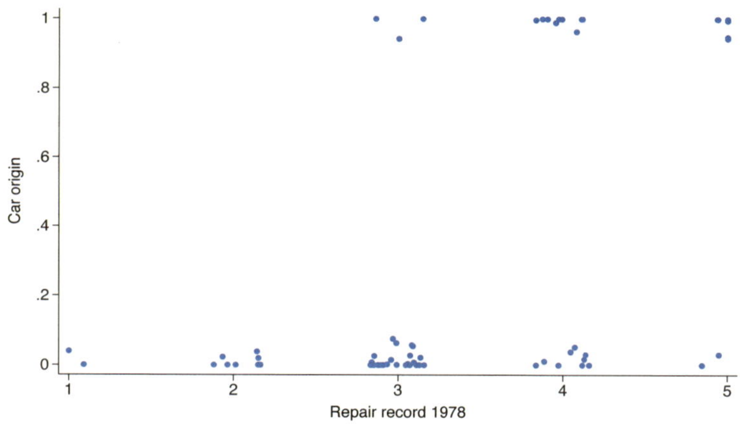

Figure 17.2. A basic jitter plot of two discrete variables

Figure 17.2 demonstrates the usefulness of a jitter plot; the individual markers are moved away from their true values to random values, which, in turn, allows users to gauge how many markers were in each category. The resulting visualization allows users to easily understand how the data are distributed across these two discrete variables.

The random displacement of the markers is a function of the original Stata seed conditions (which reset with every start-up). To replicate a specific jitter plot, one can use the `jitterseed()` option, in which a number between 0 and 2,147,483,647 is specified. Below is an example with the seed set to 1,234; this code will re-create the same graph every time.

Creating figure 17.3

```
. * Different jitter seed
. twoway (scatter foreign rep78, jitter(12) jitterseed(1234))
```

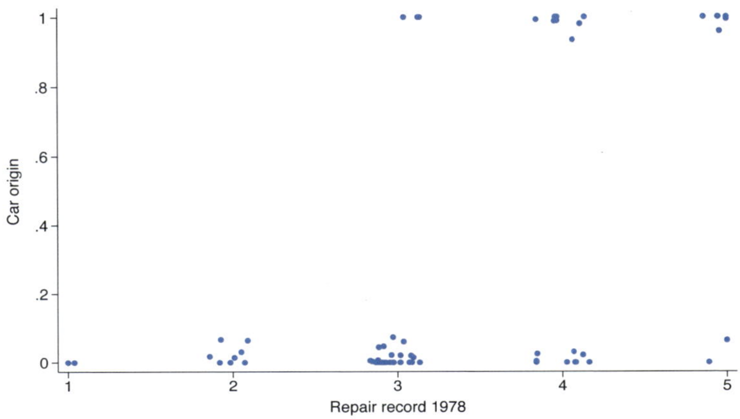

Figure 17.3. Jitter plot with specified seed

17.3 A complex jitter plot

Although the previous example shows the concept of a jitter plot well, it is unlikely to be useful in practice. The sample count of `auto.dta` is low, and both variables used have few categories. A basic two-way table would likely have been a better representational option. A more useful example can be presented with `nlsw88.dta`. Here the variables `occupation` and `industry` measure 13 and 12 different categories, respectively, and their sample sizes are both over 2,000 observations.

```
. * Load data
. sysuse nlsw88
(NLSW, 1988 extract)

. summarize occupation industry

    Variable |        Obs        Mean    Std. dev.       Min        Max
-------------+---------------------------------------------------------
  occupation |      2,237    4.642825    3.408897          1         13
    industry |      2,232    8.189516    3.010875          1         12
```

A basic scatterplot of these two variables looks as follows:

Creating figure 17.4

```
. * Scatterplot of many discrete categories
. twoway (scatter occupation industry,                         ///
> ylabel(1(1)13, valuelabel angle(45))                         ///
> xlabel(1(1)12, valuelabel angle(45)))
```

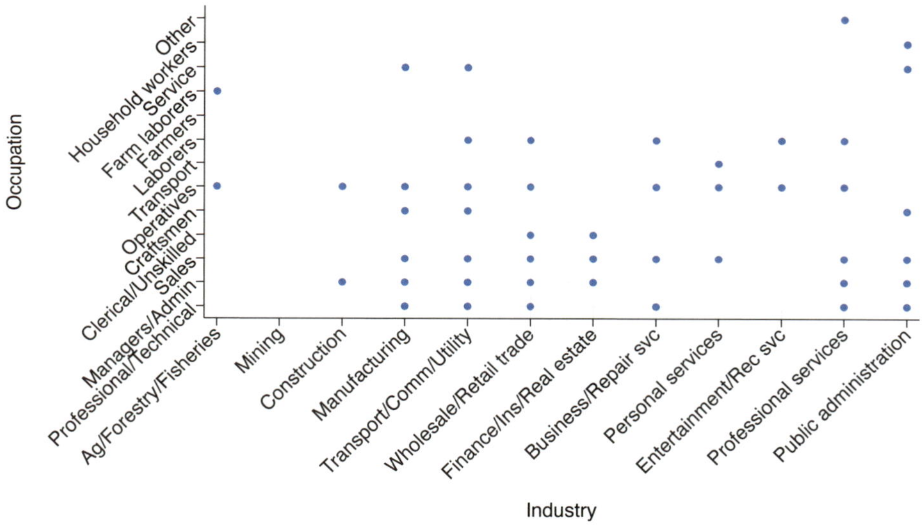

Figure 17.4. Scatterplot of two complex discrete variables

At first glance, it might seem that a jitter plot of this graph with so many observations and discrete values is likely to be unmanageable because markers will continue to overlap and high displacement values will lead to merging of markers across categories. However, a surprising amount of flexibility can be found when faced with a large sample size through the `msize()` and `mcolor()` options. Tiny markers and opaque color choices can diminish the problem of having too many data markers. A resulting jitter plot with these options might look as follows:

Creating figure 17.5

```
. * Scatterplot of many discrete categories with jitter()
. twoway (scatter occupation industry,                                    ///
> ylabel(1(1)13, valuelabel angle(45)) xlabel(1(1)12, valuelabel        ///
> angle(45)) jitter(4) msize(tiny) mcolor(%30))
```

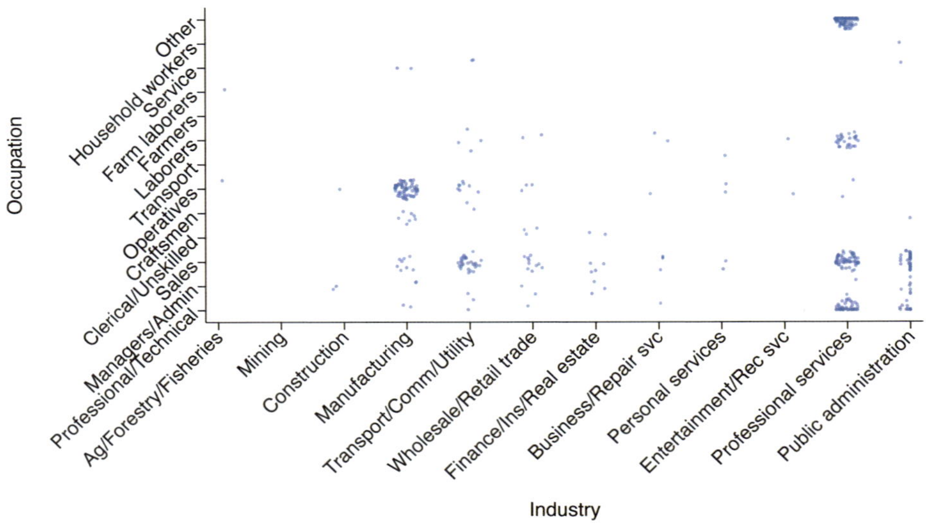

Figure 17.5. Jitter plot of two complex discrete variables

From the resulting jitter plot, it is now possible to see that operative occupations are prevalent in the manufacturing industry, while professional and technical occupations are concentrated in the professional services industry. This complex jitter plot is a good example of how jitter plots can quickly communicate complex bivariate discrete distributions.

17.4 Jitter plots with continuous values

A jitter plot can also be useful when one variable is continuous and the other is discrete, especially when there are many observations and the discrete variable has few values. In figure 17.6 below, the discrete variable `race` (which has only three categories) is scattered against the continuous variable `wage`.

```
. summarize wage race

    Variable |        Obs        Mean    Std. dev.       Min        Max
-------------+---------------------------------------------------------
        wage |      2,246    7.766949    5.755523   1.004952   40.74659
        race |      2,246    1.282725    .4754413          1          3
```

Creating figure 17.6

```
. * One continuous variable with jitter()
. twoway (scatter race wage) (scatter race wage, jitter(15)     ///
> mcolor(%20) legend(label(1 "Original") label(2 "Jittered")))
```

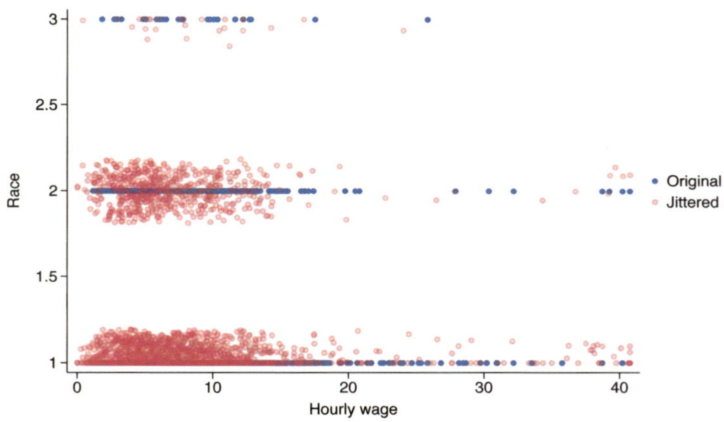

Figure 17.6. Jitter plot with one continuous variable

Figure 17.6 presents both the original plot and the jittered plot. The original plot has little analytical value because only the extreme outliers of the wage distribution are clearly displayed for each group. The jittered plot reveals much more of the data massing, and some inferences can be made about the wage distribution between the three race categories. Unfortunately, boundary effects squash some of the jittered values into the boundary, making the data density of group 1 harder to understand. Thankfully, this can be relatively easily solved with the `ylabel()` option, which can define a longer y axis (in this case, between 0 and 4 in increments of 1). This allows jittered data points more room to go beyond their original bounds. Figure 17.7 exemplifies this, although it remains important to remember that only three categories of actual data exist (values 1, 2, and 3).

Creating figure 17.7

```
. * One continuous variable and extended y axis with jitter()
. twoway (scatter race wage) (scatter race wage, jitter(15)    ///
> mcolor(%20) legend(label(1 "Original") label(2 "Jittered"))  ///
> ylabel(0(1)4))
```

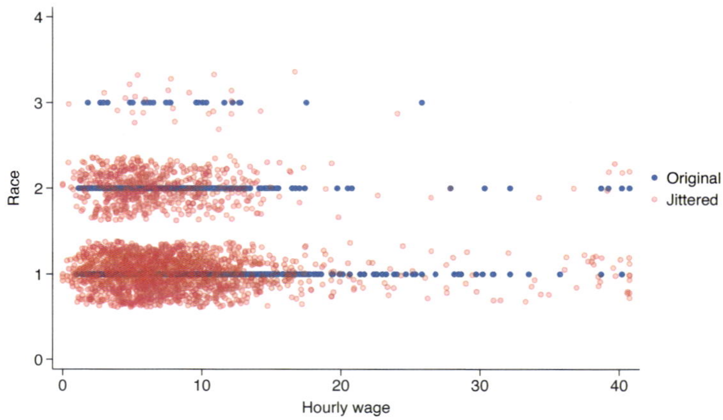

Figure 17.7. Jitter plot with one continuous variable and extended y axis

Jittering can also be applied to two continuous variables. This may seem like an odd choice because a jittered scatterplot of two continuous variables will randomly move all markers away from their data points. However, this can be valuable when data are sensitive and the actual "true" data points must stay hidden. Because of its random spherical displacement of markers, a jittered scatterplot will continue to have the same aggregate properties and visual similarity as the "true" scatterplot, but all individual markers will be anonymized. Figure 17.8 presents an example using the variables `wage` and `ttl_exp`. The code consists of 1) the original scatterplot, 2) the jittered plot, and 3) a linear fit of the original values.

Creating figure 17.8

```
. * Two continuous variables with jitter()
. twoway (scatter wage ttl_exp if race == 2, mcolor(%30))                    ///
> (scatter wage ttl_exp if race == 2, jitter(5) mcolor(%60)                  ///
> mlcolor(black) legend(label(1 "Original") label(2 "Jittered")))            ///
> (lfit wage ttl_exp if race == 2)
```

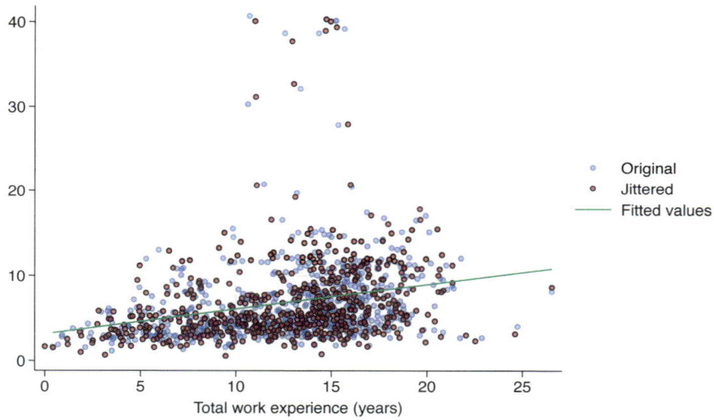

Figure 17.8. Jitter plot with two continuous variables

Figure 17.8 shows the power of using a jitter plot with two continuous variables. The red data markers, in aggregate, look similarly distributed to the original blue data markers. The green line, a linear fit on the original values, matches the red jittered markers well. Therefore, the overall relationship of these data is maintained and visually well presented, but markers do not identify their original true data values, thereby reducing the sensitivity of this plot. Note that the original and jittered markers are overlaid only for comparative purposes; only the jittered markers would be shown in an applied setting.

18 Table plots

Advantages

1. Resemble the traditional table setup most closely
2. Are intuitive to understand
3. Have many visualization options
4. Enable easy comparison across different categories

Disadvantages

1. Are less useful for small tables with few categories
2. Make it difficult to compute precise statistics between cells
3. Require careful design to maintain clarity and readability

Required community-contributed package

To install the community-contributed command that is required to follow along with this chapter completely, type

```
. ssc install tabplot
```

18.1 Introduction

Table plots are relatively simple constructs but offer a powerful and intuitive visualization of a standard two-way table. At their core, they replace the numeric cell values in each table cell with a marker symbol that reflects the size of the numeric value in that particular cell. This makes table plots intuitive to understand because the basic table framework is maintained; rows and columns represent different categories of two categorical variables. Cell marker symbols vary in size, allowing readers to quickly identify which cells have many observations and which cells have few.

Table plots are particularly useful when users wish to visualize medium-to-large tables because smaller tables continue to be best represented by numerical tables. A simple 2×2 table generally does not need a pictorial representation. Table plots aim to reduce complex numeric two-way tables into quick visual forms.

18.2 A basic table plot

The basic problem is illustrated by the table below, generated from `nlsw88.dta`. A user wishes to examine the bivariate distribution of two categorical variables with many categories. In this case, the variables `occupation` and `industry` are used, which contain 13 and 12 different categories, respectively. The resulting two-way table is large, contains many numbers, and is complex to navigate. While it provides a lot of statistical detail, it is not useful when a fast overview of the data is required.

```
. * Load data
. sysuse nlsw88
(NLSW, 1988 extract)

. tabulate occupation industry

                     |                     Industry
          Occupation | Ag/Forest     Mining  Construct  Manufactu |     Total
---------------------+--------------------------------------------+----------
Professional/Technica |         1          1          1         24 |       316
      Managers/Admin |         1          0          5         37 |       263
               Sales |         6          3         16         73 |       725
  Clerical/Unskilled |         1          0          1         13 |       102
           Craftsmen |         0          0          0         29 |        53
          Operatives |         2          0          2        185 |       244
           Transport |         0          0          0          0 |        28
            Laborers |         0          0          1          1 |       285
             Farmers |         1          0          0          0 |         1
       Farm laborers |         5          0          0          0 |         7
             Service |         0          0          3          5 |        16
   Household workers |         0          0          0          0 |         2
               Other |         0          0          0          0 |       186
---------------------+--------------------------------------------+----------
               Total |        17          4         29        367 |     2,228

                     |                     Industry
          Occupation | Transport  Wholesale  Finance/I  Business/ |     Total
---------------------+--------------------------------------------+----------
Professional/Technica |         4          7          8         13 |       316
      Managers/Admin |         9         65         56          9 |       263
               Sales |        49        113         95         39 |       725
  Clerical/Unskilled |         2         50         29          4 |       102
           Craftsmen |         5         10          0          1 |        53
          Operatives |        12         23          0          1 |       244
           Transport |         0          0          0          0 |        28
            Laborers |         6         58          4         16 |       285
             Farmers |         0          0          0          0 |         1
       Farm laborers |         0          2          0          0 |         7
             Service |         2          2          0          2 |        16
   Household workers |         0          0          0          0 |         2
               Other |         0          2          0          1 |       186
---------------------+--------------------------------------------+----------
               Total |        89        332        192         86 |     2,228
```

Occupation	Industry Personal	Entertain	Professio	Public ad	Total
Professional/Technica	2	2	208	45	316
Managers/Admin	5	2	56	18	263
Sales	14	4	213	100	725
Clerical/Unskilled	1	1	0	0	102
Craftsmen	1	0	5	2	53
Operatives	12	1	5	1	244
Transport	28	0	0	0	28
Laborers	33	6	155	5	285
Farmers	0	0	0	0	1
Farm laborers	0	0	0	0	7
Service	0	1	0	1	16
Household workers	0	0	0	2	2
Other	0	0	182	1	186
Total	96	17	824	175	2,228

An alternative visualization option is provided by the community-contributed command `tabplot` (Cox 1999b). `tabplot` can graph one-, two-, and three-way tables by replacing numeric cell values with bar charts. Using `tabplot` for one-way tables is not recommended, because the same outcome can be reached with the official command `tab1` or `graph bar`. However, for two- and three-way tables, `tabplot` offers a convenient and fast way to create visualizations of complex tables. Figure 18.1 shows a basic example of a two-way table plot. The basic use syntax for a two-way table is `tabplot` *rowvar colvar*. Because of the large number of categories in this example, the `ylabel()` and `xlabel()` options are used to create nonoverlapping y and x titles. These options are part of the standard `twoway graph` command and are not specific to `tabplot`. The suboption `labsize()` reduces the label text size, and the suboption `angle()` angles the label text, all for better identification of labels.

Creating figure 18.1

```
. * Basic table plot
. tabplot occupation industry, ylabel(, labsize(small) angle(45))   ///
> xlabel(, labsize(small) angle(45))
```

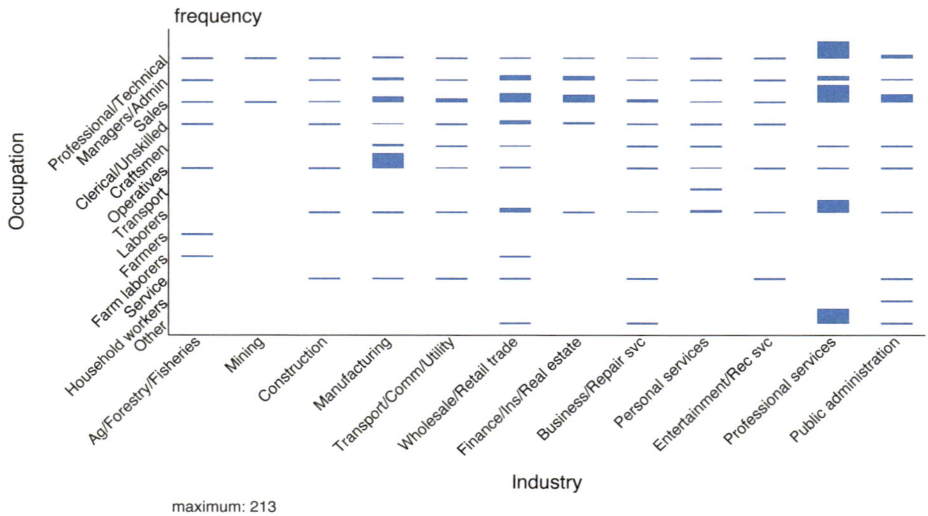

Figure 18.1. A basic table plot

Figure 18.1 demonstrates the strength of a basic table plot. It retains the same cellular setup as the original two-way table, but bar charts now replace the numeric values. Each bar chart is relative in size to a full bar (100%) in the cell with the highest number of observations. The exact frequency that the maximum bar height represents is given in the note of the table, in this case, 213. The primary advantage of this table plot is the rapid identification of large cells and how the data are distributed across the categories.

18.2.1 Horizontal and percentage table plots

For those wishing to create a slightly different visualization or to mimic the visual style of the `graph hbar` command, the `horizontal` option allows users to replace the vertical bars with horizontal bars. The `percent` option converts the frequency measure into a percentage of the total frequency of all values displayed in the graph. This option is often helpful when users want to examine the bivariate data distribution between the categories. Figure 18.2 presents an example.

Creating figure 18.2

```
. * Horizontal table plot
. tabplot occupation industry,                                    ///
> ylabel(, labsize(small) valuelabel angle(45))                   ///
> xlabel(, labsize(small) valuelabel angle(45)) percent missing horizontal
```

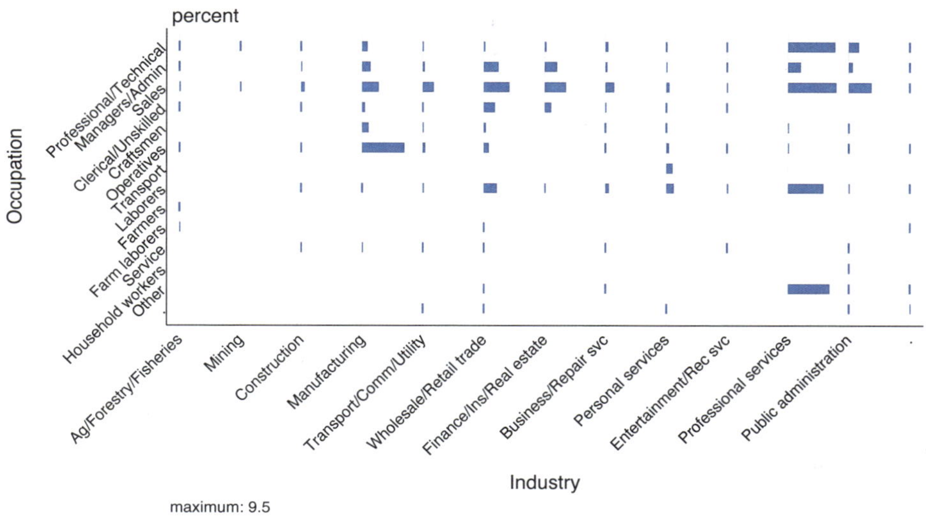

Figure 18.2. Horizontal percentage table plot

Figure 18.2 offers a distinctively different visualization compared with figure 18.1. The bars are now horizontal, and the maximum bar value now represents 9.5% instead of 213 observations. However, the information presented ultimately remains similar.

18.2.2 Complex table plot

A disadvantage shared by figures 18.1 and 18.2 is that the statistical information visualized is greatly diminished compared with the traditional numerical table format. Exact comparison of similarly sized cells is difficult, and individual frequencies or percentages can be difficult to infer. For those wishing for more detail while keeping the bar graph visualization for rapid identification of approximate cell sizes, the `showval()` option offers an important additional element of control. It places the numeric data markers back into each cell, thereby merging both the traditional two-way numeric table and the simpler table plot visualization. Suboptions will need to be used when using `showval()` because the markers will often overlap with the bar graphs and need to be shifted or formatted appropriately. Both cases are shown below in figure 18.3, which uses the `offset()` suboption to shift markers up or down and the `format()` suboption to format the numeric labels. In addition, for this plot and the next plot, the `mlabsize()` option is used to reduce the marker label size. The numbers may appear difficult to read when printed in this size, but when the graph is displayed in a larger size (for example, in a PowerPoint presentation), this combination of bars and numbers is very informative. Finally, the `missing` option is also used, which introduces an extra category to both variables that consists of all missing values, thereby ensuring that the graph reflects observations from the entire dataset.

Creating figure 18.3

```
. * Table plot with values
. tabplot occupation industry, ylabel(, labsize(small) angle(45))  ///
> xlabel(, labsize(small) angle(45)) percent missing               ///
> showval(offset(0.15) format(%2.1f) mlabsize(*0.5))
```

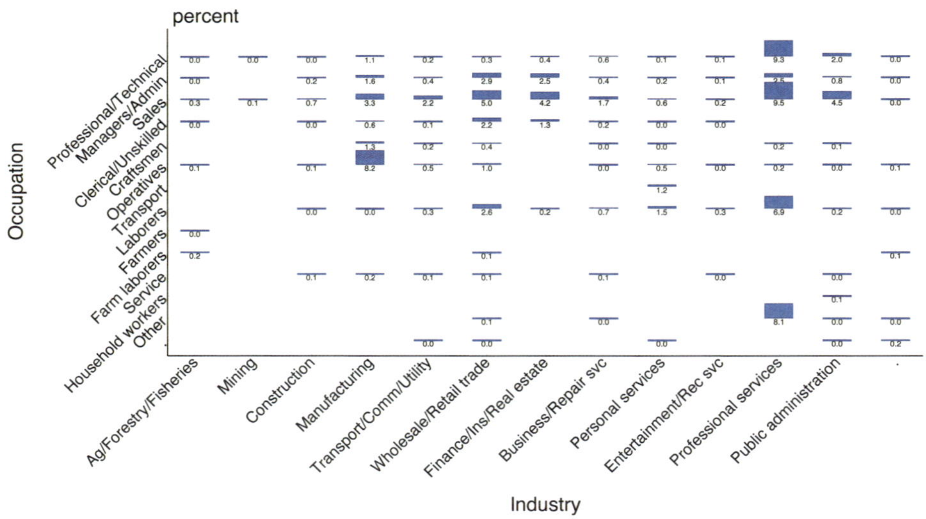

Figure 18.3. Complex table plot

The advantages of figure 18.3 are clear; it allows the rapid identification of the data distribution provided by figures 18.1 and 18.2, while a prolonged examination allows users to identify detailed cell statistics. Note that careful marker placement and sizing are important components of such graphs, and even here some markers are obscured by bars.

18.2.3 Framed table plot

Further customization options are available to improve the quality of complex table plots. In figure 18.4 below, the `frame()` option is used to place a frame around each cell. The numeric value within this option is the height of the frame and should normally be set to the maximum cell value, 9.5% in this case. The framing of each bar greatly improves the ability to identify individual cells, especially for larger tables like the one used in this demonstration. The `frameopts()` option allows for many customization options, although it leaves the frames blank by default. The fill color of the bars can be modified using the standard `graph bar` options. Figure 18.4 shows the addition of the `bfcolor()` option to create opaque color shading that results in better identification of numeric markers.

Creating figure 18.4

```
. * Table plot with frames
. tabplot occupation industry, ylabel(1(1)13, labsize(small) angle(45))  ///
> xlabel(1(1)12, labsize(small) angle(45)) percent missing               ///
> showval(offset(-0.5) format(%2.1f) mlabsize(*0.5)) frame(9.5)          ///
> bfcolor(%30)
```

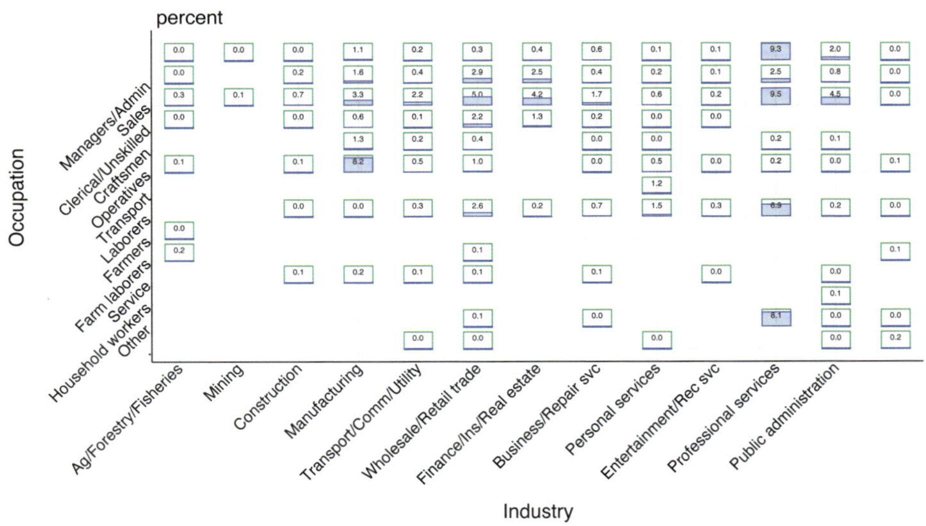

Figure 18.4. Framed table plot

Figure 18.4 shows a powerful table plot that clearly illustrates the matrix of cells and categories, shades areas of cells according to their numerical sizes, and provides detailed statistics via individual numeric markers. `tabplot` also contains many additional visualization options, including ways to change the bar size or recast the bars as another `twoway` plottype, such as spikes.

18.3 Three-way table plot

Three-way table plots attempt to create 3D tables using two dimensions. The lack of a third dimension in many presentational mediums, such as paper or a computer screen, means that these are often pseudo-three-way tables. Usually, this is done by simply repeating a two-way table over the categories of a third variable. This process can be accessed in `tabplot` with the `by()` option, which creates three-way table plots by producing a set of standard two-way table plots over several panels. Each panel corresponds to a category of the variable specified in `by()`. An example is shown in figure 18.5.

Creating figure 18.5

```
. * Three-way table plot
. tabplot occupation industry, by(union) ylabel(1(1)13, labsize(small)  ///
> angle(45)) xlabel(1(1)12, labsize(small) valuelabel angle(45))
```

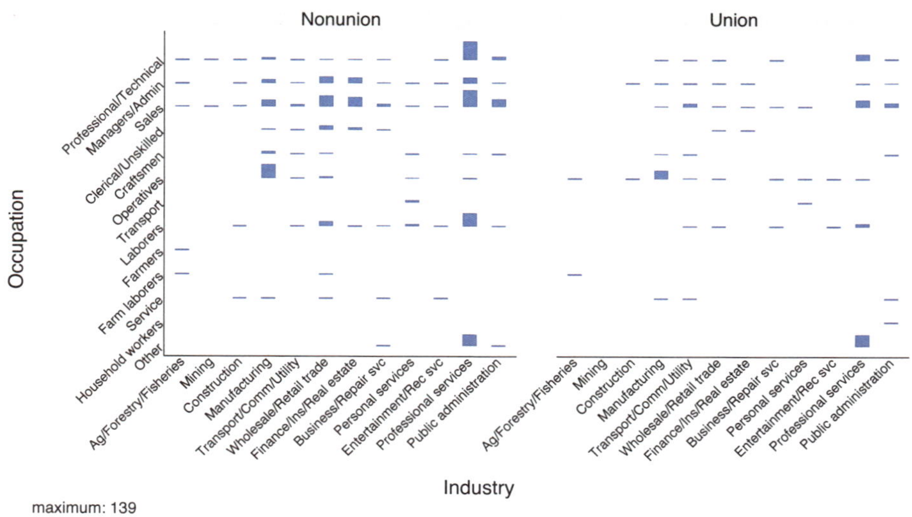

Figure 18.5. Three-way table plot

Figure 18.5 highlights an advantage of three-way table plots compared with a (extremely) large numeric table; many data are shown and reduced into a clear visual image that can be quickly scanned and analyzed.

18.4 Balloon plots

`tabplot` is a convenient and fast community-contributed command to create tabular visualizations. Its speed and ease of use should make it the first port of call for many applications. However, if you wish to deviate away from bar charts as the primary visualization method, then `tabplot` offers limited options. The `recast` option allows some alternative visualizations, but they are mainly confined to a few range-type graphs such as `twoway rspike`. To create custom markers in table plots, users must build their own table plot using official Stata commands.

A popular alternative marker visualization in the table plot is a circle, which results in a graph sometimes known as a balloon plot. To access this graph, users must create their own custom table plot using the `twoway scatter` command and then modify the marker symbols using the `msymbol()` option. To inflate markers to represent the cell quantity's size, one must also use the [*weight*] modifier to inflate marker sizes accordingly. That, in turn, often requires the prior use of the `collapse` command to calculate the relevant cell sums of a table.

In the below code, which creates figure 18.6, all observations are first given a weight of 1. The dataset is then collapsed by the two categorical variables, where the sum of weight for each categorical combination is computed. The new collapsed dataset is then used to underpin a `twoway scatter` command with the [*weight*] modifier.

```
. * Give all observations a weight of 1
. generate weight = 1
. * Collapse observations
. collapse (sum) weight, by(occupation industry)
```

Creating figure 18.6

```
. * Scatterplot with weights
. twoway (scatter occupation industry [weight=weight])
```

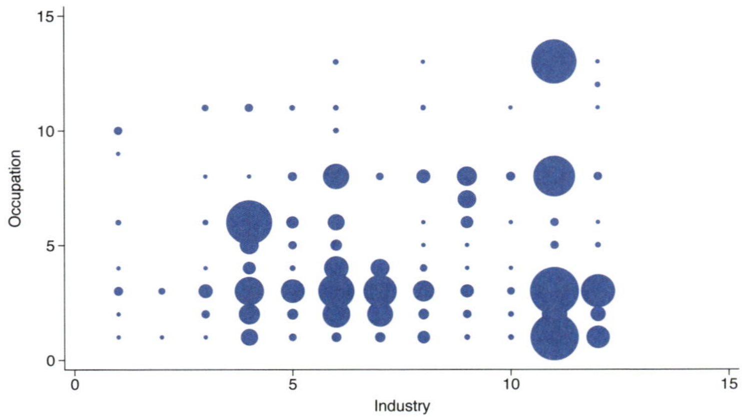

Figure 18.6. Balloon plot

Figure 18.6 presents a first attempt at a balloon plot. The underlying idea is correct, but the graph needs more formatting to be presentable. Many variations are possible, and a suggestion is shown below in figure 18.7. Here the `msymbol()`, `mfcolor()`, `mlcolor()`, and `msize()` options are used to modify the marker symbols, marker fill color, marker line color, and marker size, respectively. The `ylabel()` and `xlabel()` options and their associated suboptions are used as previously and modify the axis labeling.

In addition, a second scatterplot is superimposed. This plot hides its own markers using the `msymbol(i)` option but, more importantly, uses the `mlabel()` option to add the weighted marker labels into the plot. These labels are simply the cell sizes of each cell, so this has the effect of adding the cell sizes into the table plot. This must be done via a second hidden scatterplot because adding this code in the first would cancel the effect of the "ballooning" [*weight*] modifier. Additional marker options are used to create a desired effect; `mlabsize()`, `mlabgap()`, and `mlabcolor()` are used to change marker label sizes, marker offset locations, and marker label colors, respectively. The resulting graph is shown in figure 18.7.

Creating figure 18.7

```
. * Scatterplot with weights, adjusted marker sizes, and cell sizes
. twoway (scatter occupation industry [weight=weight], msymbol(circle)  ///
> mfcolor(eltblue%50) mlcolor(black) msize(*0.4)                          ///
> ylabel(1(1)13, labsize(small) valuelabel angle(45))                     ///
> xlabel(1(1)12, labsize(small) valuelabel angle(45)))                    ///
> (scatter occupation industry, msymbol(i) mlabel(weight)                 ///
> mlabsize(*0.7) mlabgap(2) mlabcolor(maroon) legend(off))
```

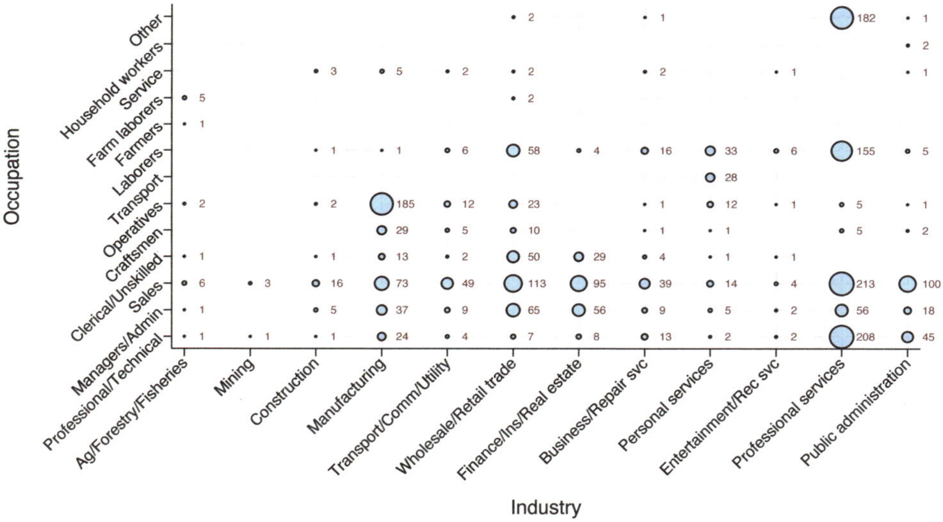

Figure 18.7. Formatted balloon plot

Figure 18.7 presents the resulting graph, which looks similar to the table plots created with the `tabplot` command except that bar symbols have been replaced with circular symbols. In addition, the code requirements are much more extensive.

18.4.1 Custom markers

Once the initial code setup of figure 18.7 is understood and practically applied, users are free to experiment and modify aspects of this code to create custom table plot designs. The `msymbol()` option provides many different marker symbols; changing this option can quickly create a visually different graph. An example with triangles is shown below in figure 18.8.

Creating figure 18.8

```
. * Scatterplot with weights, custom markers, and cell sizes
. twoway (scatter occupation industry [w=weight], msymbol(triangle)     ///
> mfcolor(erose%50) mlcolor(black) msize(*0.4)                         ///
> ylabel(1(1)13, labsize(small) valuelabel angle(45))                  ///
> xlabel(1(1)12, labsize(small) valuelabel angle(45)))                 ///
> (scatter occupation industry, msymbol(i) mlabcolor(black)            ///
> mlabel(weight) mlabsize(*0.7) mlabgap(2) mlabcolor(dknavy) legend(off))
```

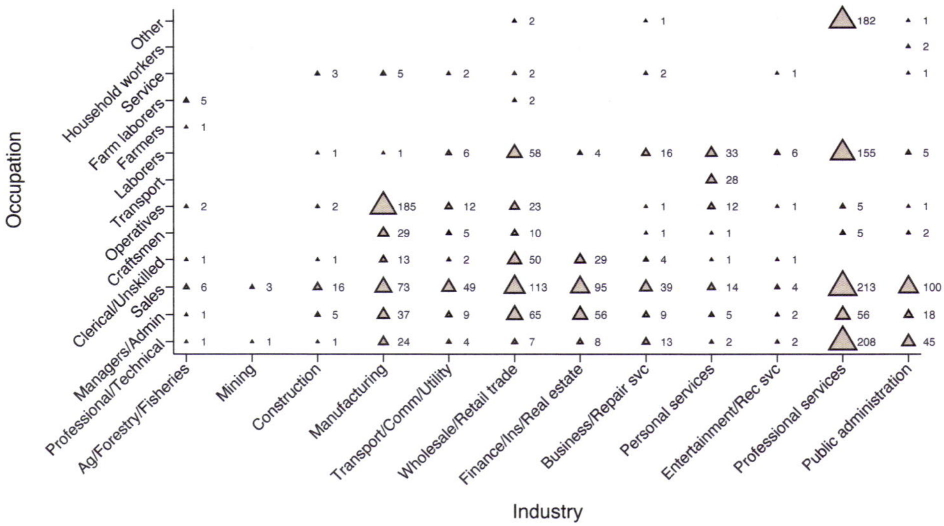

Figure 18.8. Custom marker table plot

Figure 18.8 presents a basic permutation of the previously constructed balloon plot. Users may construct their own table plots using official Stata commands or using the **tabplot** command. In all cases, the core concept of table plots is to visually present large numeric two-way tables.

19 Bivariate bar charts

Advantages

1. Are easy to understand and commonly used
2. Excel for variables with many categories
3. Facilitate easy comparison between categories
4. Can be cast in vertical or horizontal mode

Disadvantages

1. Are less useful for variables with few categories
2. Do not handle extreme number ranges well
3. Require careful axis scaling to avoid false impressions
4. Can be poor at presenting precise statistical values

Required community-contributed packages

No additional installations are required, but a custom dataset is used.

Custom data

A custom dataset taken from Cox (2016) is used in this chapter. The below code creates the data.

```
. * Input data
. input agegroup health freq
. 1 1 243
. 1 2 789
. 1 3 167
. 1 4 18
. 1 5 6
. 2 1 220
. 2 2 809
. 2 3 164
. 2 4 35
. 2 5 6
. 3 1 147
. 3 2 658
. 3 3 181
. 3 4 41
. 3 5 8
. 4 1 90
. 4 2 469
. 4 3 236
. 4 4 50
. 4 5 16
. 5 1 53
. 5 2 414
. 5 3 306
. 5 4 106
. 5 5 30
. 6 1 44
. 6 2 267
. 6 3 284
. 6 4 98
. 6 5 20
. 7 1 20
. 7 2 136
. 7 3 157
. 7 4 66
. 7 5 17
. end
. * Create labels
. label define agegroup 1 "16-24", modify
. label define agegroup 2 "25-34", modify
. label define agegroup 3 "35-44", modify
. label define agegroup 4 "45-54", modify
. label define agegroup 5 "55-64", modify
. label define agegroup 6 "65-74", modify
. label define agegroup 7 "75+", modify
. label define health 1 "Very good", modify
. label define health 2 "Good", modify
. label define health 3 "Regular", modify
. label define health 4 "Bad", modify
. label define health 5 "Very bad", modify
```

```
. * Attach labels
. label values agegroup agegroup
. label values health health
. * Expand data and drop freq variable
. expand freq
. drop freq
. * Save data
. save hage.dta, replace
```

19.1 Introduction

The basic use of bar charts for single or multiple variables has already been described in chapter 9. In this chapter, the discussion will extend to bivariate data, where users are faced with two discrete variables that they wish to examine jointly.

The primary advantages of bar charts are that they remain by far the most common type of data visualization technique and they are tremendously flexible. Most laypeople will have a natural understanding of bar charts. However, bar charts do have some disadvantages, especially when they are used to illustrate discrete bivariate data. When displaying such data, the bar chart may have too many categories. Better methods exist to display bivariate discrete data with many categories, including stacked bar charts or table plots. Nonetheless, it is important to understand how to use this mainstay of data visualization to examine two discrete variables jointly.

19.2 A basic two-way frequency bar chart

The basic problem is that a user wishes to visualize a numerical two-way table like the one below, created from the data that are generated at the beginning of this chapter. This dataset contains two categorical variables that measure health status and age as discrete groups. Both variables have a medium number of categories, and the dataset includes over 6,000 observations. The resulting two-way table is medium-sized with numerous cells.

```
. * Load data
. use hage, clear
. summarize
```

Variable	Obs	Mean	Std. dev.	Min	Max
agegroup	6,371	3.427248	1.867386	1	7
health	6,371	2.284885	.8290198	1	5

```
. table health agegroup
```

	agegroup							
	16-24	25-34	35-44	45-54	55-64	65-74	75+	Total
health								
Very good	243	220	147	90	53	44	20	817
Good	789	809	658	469	414	267	136	3,542
Regular	167	164	181	236	306	284	157	1,495
Bad	18	35	41	50	106	98	66	414
Very bad	6	6	8	16	30	20	17	103
Total	1,223	1,234	1,035	861	909	713	396	6,371

A visualization of this table using bar charts can be created with the **graph bar** command. The basic setup to re-create the above two-way table is to specify **(count)** after the main command and not specify any variables. The variables **health** and **agegroup** are specified using two **over()** options, one for each variable. The **over()** option specifies a categorical variable over which the counts are to be calculated; this variable may be string or numeric. Importantly, the **over()** option can be specified up to three times. This results in a bar chart containing multiple repeating bars of one categorical variable within the other. An example with two discrete variables is shown in figure 19.1.

Creating figure 19.1

```
. * Bar chart over two discrete variables
. graph bar (count), over(health) over(agegroup)
```

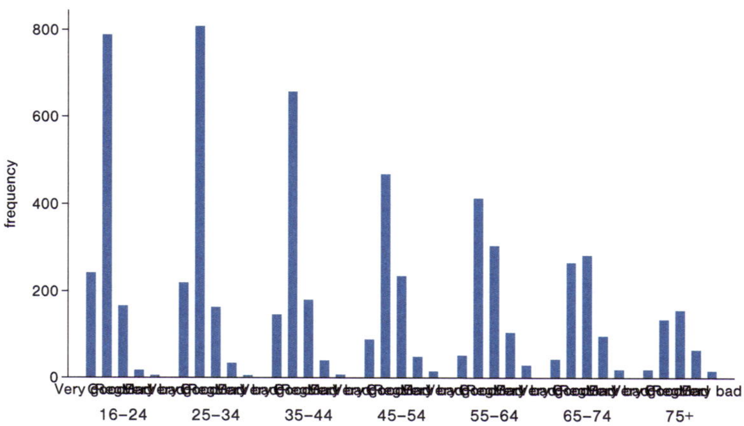

Figure 19.1. A basic two-way bar chart

Figure 19.1 shows a basic bar chart with frequencies for two discrete variables. Note that the discrete categories of the `health` variable are repeated within each `agegroup` category, thereby re-creating the two-way distribution of the previous `table health agegroup` command. Each bar in figure 19.1 represents a cell frequency from the previous two-way table. It is easy to see that most observations are contained in the younger age groups.

Figure 19.1 is hampered by the fact that too many labels are squeezed into too small a space. Two options exist to alleviate this problem. The labels could be changed; for example, they could be made smaller or vertically arranged. Or users could resort to the `graph hbar` command, which generates horizontal bar charts. The second option is shown in figure 19.2 below.

Creating figure 19.2

```
. * Horizontal bar chart over two discrete variables
. graph hbar (count), over(health) over(agegroup)
```

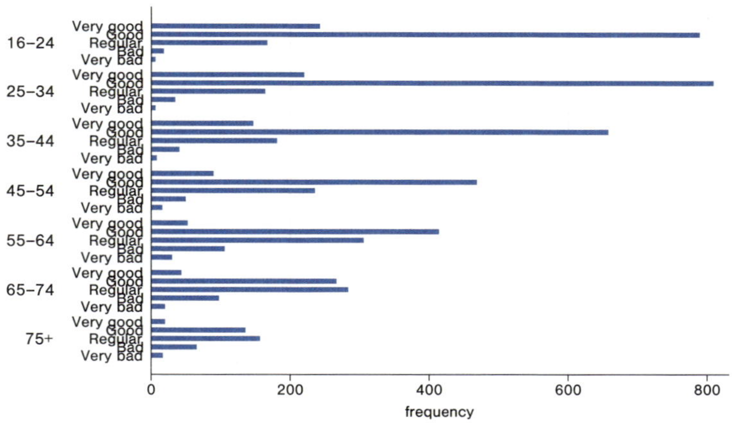

Figure 19.2. Horizontal two-way bar chart

While the labels could still use some further modification, such as adding the suboption `label(labsize(vsmall))` in the option `over(health)`, figure 19.2 is much more user friendly than figure 19.1. The frequencies of the two-way table are maintained, and it is now possible to identify which bars belong to which health categories.

19.2.1 Horizontal two-way frequency bar chart

An alternative visualization effect can be achieved with the `asyvars` option. When the `asyvars` option is specified, the visual results are the same as if the first `over()` group was removed and introduced as a y variable after the command **graph hbar**. The net effect of this option is that the health group is now split into multiple colored bars labeled in the legend instead of unicolored with categorical labeling on the y or x axis. Essentially, whereas the previous y axis identified two categories, it now identifies only one category, and the other is provided by color differentiation. An example is shown in figure 19.3.

Creating figure 19.3

```
. * Horizontal bar chart with legend differentiation
. graph hbar (count), over(health) over(agegroup) asyvars
```

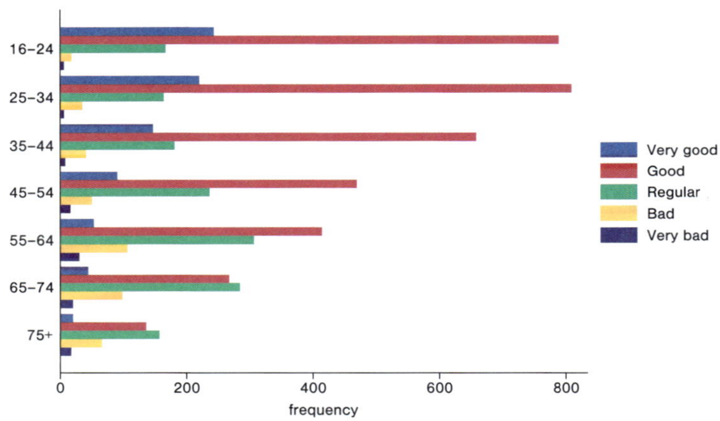

Figure 19.3. Horizontal two-way bar chart with `asyvars()`

Figure 19.3 is identical to figure 19.2 but emphasizes colored bars. The advantage of colored bars is that single health groups are easier to pick out and compare across age categories. A disadvantage is that health groups are more difficult to compare within a particular age category because the labeling is missing from the y axis. It will be up to users and their narratives to decide whether figure 19.2 or figure 19.3 is more appropriate.

19.2.2 Column, row, and cell percentage two-way bar charts

Frequency two-way tables are not the only products of bivariate analysis. Often, percentage statistics are more useful and revealing than frequency statistics. Three sets of percentages are generally of interest to a user examining a two-way table: column, row, and cell percentages. Each highlight to what extent the data in a cell are proportionate to a column total, row total, or "total-total". The result is often a better understanding of how data are distributed and thus potentially correlated across the various discrete categories. Column, row, and cell percentages can be computed for any two-way table using the `tabulate` command with the `column`, `row`, and `cell` options. The `nofreq` option below is also used to suppress frequency statistics.

```
. * Table with multiple cell statistics
. tabulate health agegroup, column row cell nofreq

+-------------------+
| Key               |
|-------------------|
|  row percentage   |
| column percentage |
|  cell percentage  |
+-------------------+

           |                         agegroup
    health |     16-24      25-34      35-44      45-54      55-64 |     Total
-----------+-------------------------------------------------------+----------
 Very good |     29.74      26.93      17.99      11.02       6.49 |    100.00 
           |     19.87      17.83      14.20      10.45       5.83 |     12.82 
           |      3.81       3.45       2.31       1.41       0.83 |     12.82 
-----------+-------------------------------------------------------+----------
      Good |     22.28      22.84      18.58      13.24      11.69 |    100.00 
           |     64.51      65.56      63.57      54.47      45.54 |     55.60 
           |     12.38      12.70      10.33       7.36       6.50 |     55.60 
-----------+-------------------------------------------------------+----------
   Regular |     11.17      10.97      12.11      15.79      20.47 |    100.00 
           |     13.65      13.29      17.49      27.41      33.66 |     23.47 
           |      2.62       2.57       2.84       3.70       4.80 |     23.47 
-----------+-------------------------------------------------------+----------
       Bad |      4.35       8.45       9.90      12.08      25.60 |    100.00 
           |      1.47       2.84       3.96       5.81      11.66 |      6.50 
           |      0.28       0.55       0.64       0.78       1.66 |      6.50 
-----------+-------------------------------------------------------+----------
  Very bad |      5.83       5.83       7.77      15.53      29.13 |    100.00 
           |      0.49       0.49       0.77       1.86       3.30 |      1.62 
           |      0.09       0.09       0.13       0.25       0.47 |      1.62 
-----------+-------------------------------------------------------+----------
     Total |     19.20      19.37      16.25      13.51      14.27 |    100.00 
           |    100.00     100.00     100.00     100.00     100.00 |    100.00 
           |     19.20      19.37      16.25      13.51      14.27 |    100.00 
```

health	agegroup 65-74	75+	Total
Very good	5.39	2.45	100.00
	6.17	5.05	12.82
	0.69	0.31	12.82
Good	7.54	3.84	100.00
	37.45	34.34	55.60
	4.19	2.13	55.60
Regular	19.00	10.50	100.00
	39.83	39.65	23.47
	4.46	2.46	23.47
Bad	23.67	15.94	100.00
	13.74	16.67	6.50
	1.54	1.04	6.50
Very bad	19.42	16.50	100.00
	2.81	4.29	1.62
	0.31	0.27	1.62
Total	11.19	6.22	100.00
	100.00	100.00	100.00
	11.19	6.22	100.00

To visualize the row and column percentage elements in this table via bar charts, users must 1) adjust the order of the `over()` groups and 2) use the `percentages` option, which computes percentages within specified variables. To obtain cell percentages, users must replace the `percentages` option with the `(percent)` statistic.

To create row percentages, repeat the code from figure 19.3 and add the option `percentages`.

Creating figure 19.4

```
. * Row percentages
. graph hbar (count), over(agegroup) over(health) asyvars percentages
```

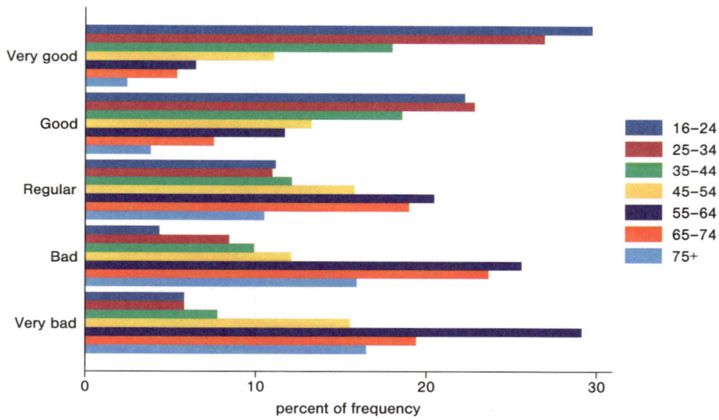

Figure 19.4. Horizontal two-way bar charts with row percentages

To create column percentages, keep the `percentages` option but switch the order of the `over()` groups.

Creating figure 19.5

```
. * Column percentages
. graph hbar (count), over(health) over(agegroup) asyvars percentages
```

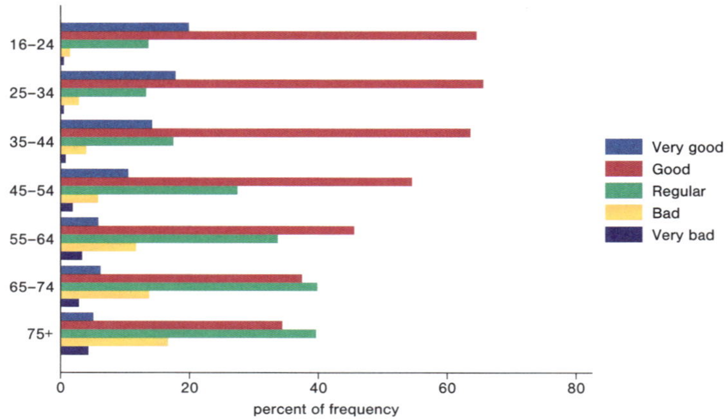

Figure 19.5. Horizontal two-way bar charts with column percentages

Finally, to create cell percentages, remove the `percentages` option, switch back the original order of the `over()` groups, and include the `(percent)` statistic after the main `graph hbar` command.

Creating figure 19.6

```
. * Cell percentages
. graph hbar (percent), over(health) over(agegroup) asyvars
```

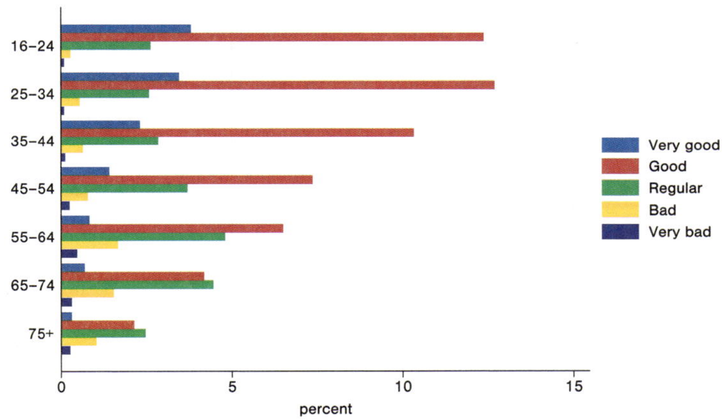

Figure 19.6. Horizontal two-way bar charts with cell percentages

Of course, the same effect can be achieved with `graph bar` for vertical bar representations.

Basic bar charts can be useful for depicting simple two-way tables, though they tend to be visually limited in how well they can convey a complex two-way table. In particular, they miss the $n \times k$ matrix-like structure of tables and instead elongate all $n \times k$ categories onto one axis. This may work well when both discrete variables have only a few categories but may become a major limiting factor when introducing more discrete categories.

20 Stacked bar charts

Advantages

1. Use space efficiently to present complex data
2. Retain bar colors and widths when many bar categories are used
3. Excel at presenting two-way tables with many categories
4. Enable comparison within and across categories
5. Can present an extra frequency dimension via the mosaic variation

Disadvantages

1. Make statistically comparing bars more difficult, especially bars that do not start at the baseline
2. Are not as intuitive or fast to comprehend as basic bar charts
3. Can become cluttered with too many categories

Required community-contributed packages

To install the community-contributed commands that are required to follow along with this chapter completely, type

```
. ssc install slideplot
. ssc install spineplot
```

Custom data

A custom dataset taken from Cox (2016) is used in this chapter. The below code creates the data.

```
. * Input data
. input agegroup health freq
. 1 1 243
. 1 2 789
. 1 3 167
. 1 4 18
. 1 5 6
. 2 1 220
. 2 2 809
. 2 3 164
. 2 4 35
. 2 5 6
. 3 1 147
. 3 2 658
. 3 3 181
. 3 4 41
. 3 5 8
. 4 1 90
. 4 2 469
. 4 3 236
. 4 4 50
. 4 5 16
. 5 1 53
. 5 2 414
. 5 3 306
. 5 4 106
. 5 5 30
. 6 1 44
. 6 2 267
. 6 3 284
. 6 4 98
. 6 5 20
. 7 1 20
. 7 2 136
. 7 3 157
. 7 4 66
. 7 5 17
. end
. * Create labels
. label define agegroup 1 "16-24", modify
. label define agegroup 2 "25-34", modify
. label define agegroup 3 "35-44", modify
. label define agegroup 4 "45-54", modify
. label define agegroup 5 "55-64", modify
. label define agegroup 6 "65-74", modify
. label define agegroup 7 "75+", modify
. label define health 1 "Very good", modify
. label define health 2 "Good", modify
. label define health 3 "Regular", modify
. label define health 4 "Bad", modify
. label define health 5 "Very bad", modify
```

```
. * Attach labels
. label values agegroup agegroup
. label values health health
. * Expand data and drop freq variable
. expand freq
. drop freq
. * Save data
. save hage.dta, replace
```

20.1 Introduction

Stacked bar charts, sometimes known as compound bar charts, stack two or more bars on top of each other. This is in contrast to the traditional bar chart, which presents bars next to each other. The advantage of stacked bar charts is that they make more efficient use of the available space, which is useful when many bars must be presented. Often, traditional bar charts with many bars become overloaded, so bar widths become very thin, making reading the bars more challenging and increasing the complexity of the graph. Stacked bar charts can retain wider bars for much longer, thereby retaining a key feature of bar charts—the color and width of a bar as easy identifiers for users. However, a disadvantage is that because bars are now stacked on top of each other, statistical comparison across bars can be more difficult because bars will have different baselines. This makes fast bar-to-bar comparison a complex endeavour, especially for bars stacked near the top.

20.2 A basic two-way stacked frequency bar chart

Similarly to the previous chapter, the basic problem is that a user wishes to visualize a numerical two-way table like the one below, generated from the dataset provided at the beginning of this chapter. This dataset contains two categorical variables that measure health status and age as discrete groups. The resulting two-way table is a large table with many numbers.

```
. * Load data
. use hage
. summarize
```

Variable	Obs	Mean	Std. dev.	Min	Max
agegroup	6,371	3.427248	1.867386	1	7
health	6,371	2.284885	.8290198	1	5

```
. table health agegroup
```

	agegroup							
	16-24	25-34	35-44	45-54	55-64	65-74	75+	Total
health								
Very good	243	220	147	90	53	44	20	817
Good	789	809	658	469	414	267	136	3,542
Regular	167	164	181	236	306	284	157	1,495
Bad	18	35	41	50	106	98	66	414
Very bad	6	6	8	16	30	20	17	103
Total	1,223	1,234	1,035	861	909	713	396	6,371

Stacked bar charts are created with the `stack` option in the `graph bar` command. Before proceeding with this chapter, users should first become familiar with the basic `graph bar` command. The key to creating a stacked bar chart with two categorical variables is the combined use of the `over()`, `asyvars`, and `stack` options. The `over()` option is used to specify two categorical variables (`health` and `agegroup`) over which the categorical groups are generated, and the `asyvars` option treats the first `over()` group as a y variable. The `stack` option then stacks the y-variable categories; all that is left is defining the statistics that must be generated. For a normal frequency stacked bar chart, this is done by specifying `(count)`, which is added after the main command, `graph bar`. The resulting chart is shown in figure 20.1.

Creating figure 20.1

```
. * Stacked bar chart
. graph bar (count), over(health) over(agegroup) stack asyvars ///
> ytitle(Frequency)
```

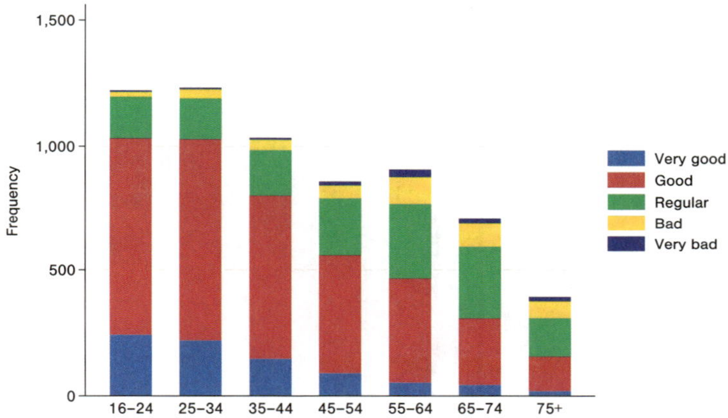

Figure 20.1. Two-way stacked frequency bar chart

Figure 20.1 shows a basic stacked bar chart with frequencies. The plot re-creates the original two-way table by populating each age category with five different health groups and visually depicting the observation count in each health category. It should be relatively easy for any graph user to identify that there are more young people than older people in these data and that young people experience better health outcomes.

20.3 Two-way stacked percentage bar chart

Stacked frequency bar charts are helpful for re-creating basic two-way frequency tables, but the most common application of stacked bar charts is percentage-based graphs. This has the effect of "stretching" each bar to the top of the graph, representing 100%; each stacked subbar then reflects what proportion of that particular x category constitutes the relevant y category. This, in turn, provides an even faster visual overview of how the underlying data are distributed among the two categorical variables. Little modification to the code is needed; only the `percentages` option must be added to the original `graph bar` code.

Creating figure 20.2

```
. * Stacked bar chart with percentages option
. graph bar (count), over(health) over(agegroup) stack asyvars ///
> ytitle(Percent) percentages
```

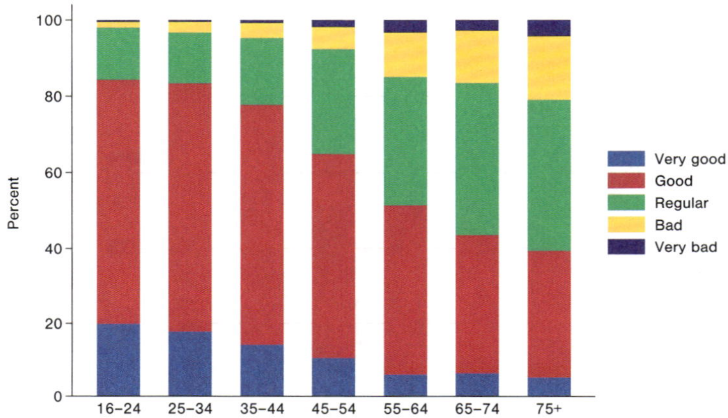

Figure 20.2. Two-way stacked percentage bar chart

Figure 20.2 presents a percentage stacked bar chart of the underlying data, which graphs the percentages of health responses within a given age group. It is now easy to see that the proportion of people with bad or very bad health outcomes is significantly higher in older age groups. Stacked percentage bar charts are great methods for allowing readers to quickly understand how two categorical variables are jointly distributed.

For an alternative visualization option, consider using the `graph hbar` command. This sister command to `graph bar` uses the same syntax but produces horizontal bars instead of vertical bars. This can be especially useful if users have long labels in one of the categorical variables. In the example below, the `blabel()` option is used to add bar labels to the graph. These are colored white (`color(white)`), positioned in the center (`position(center)`), and formatted to have no decimal places (`format(%3.0f)`).

Creating figure 20.3

```
. * Stacked horizontal bar chart with percentages option
. graph hbar (count), over(health) over(agegroup) stack asyvars     ///
> ytitle(Percent) percentages blabel(bar, color(white)              ///
> position(center) format(%3.0f))
```

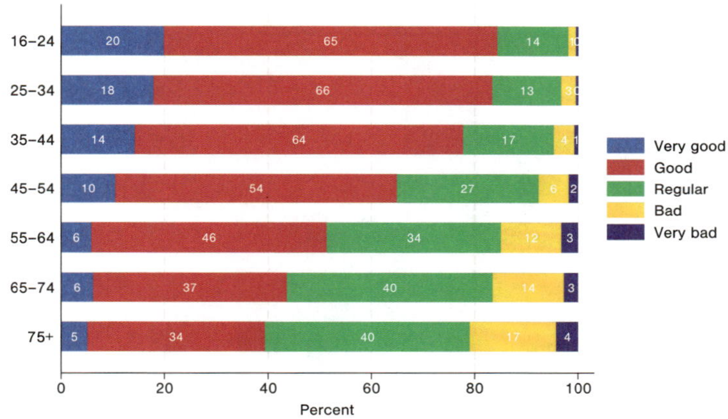

Figure 20.3. Two-way stacked horizontal percentage bar chart

20.4 Slide plots

A derivative of the stacked bar chart is the slide plot. The community-contributed command `slideplot` (Cox 2003c) generates a sliding bar chart that plots the frequencies of categories from one or more variables. Sliding bar charts are useful for naturally ordered categories and when users want to clearly distinguish between such categories.

The idea is that one set of categories represents a positive set of outcomes, while another set of categories represents a set of negative outcomes. The positive and negative categories are anchored on either side of zero, which moves the reference line away from the edge of a graph and toward the middle. This makes it easier to quickly identify "good" or "bad" outcomes across a wide range of categorical groups.

The command `slideplot` comes in two flavors: `slideplot bar` and `slideplot hbar`. The first creates vertical bars, while the second creates horizontal bars. For two categorical variables, the first variable must be treated as a y variable and be specified after the command and the second is included in the `by()` option. Note that the `negative()` and `positive()` options must be specified for the command to work. Include the relevant y-variable category numbers that should be treated as negative and positive. The negative group will be stacked below (to the left of) the origin, and the positive group will be stacked above (to the right of) the origin. Note that in these data, lower numbers are associated with better outcomes in the health variable. The resulting command syntax is shown below, and its outcome is in figure 20.4.

Creating figure 20.4

```
. * Slide plot
. slideplot bar health, negative(5 4) positive(3 2 1) by(agegroup)  ///
> ytitle(Frequency)
```

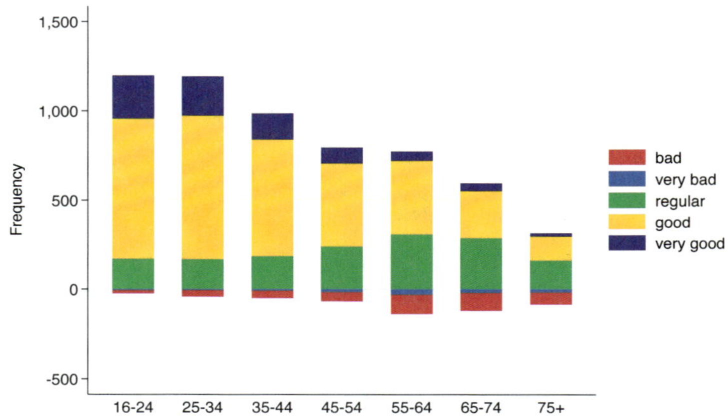

Figure 20.4. Slide plot

Figure 20.4 demonstrates a basic slide plot. The bars continue to be stacked, but the fourth health category (bad) is anchored at zero rather than the first (very good). Users can now better distinguish between "healthy" categories, which sit on top (to the right) of the origin, and "unhealthy" categories, which sit below (to the left of) the origin. It is immediately obvious that good health decreases with age, while poor health increases with age. Note that the colors have reversed across the categories compared with the previous stacked bar charts. A disadvantage is that, by default, the y axis has negative labeling below the origin. Users should consider adding a note to clarify that the y-axis values should be treated as absolute values.

`slideplot` also supports horizontal bar depiction via the sister `slideplot hbar` command. Similarly to other stacked bar charts, a `percent` option exists that converts the frequency statistics to percentage statistics. An example is shown in figure 20.5.

Creating figure 20.5

```
. * Slide plot horizontal
. slideplot hbar health, negative(5 4) positive(3 2 1) by(agegroup)  ///
> ytitle(Percent) percent yline(0)
```

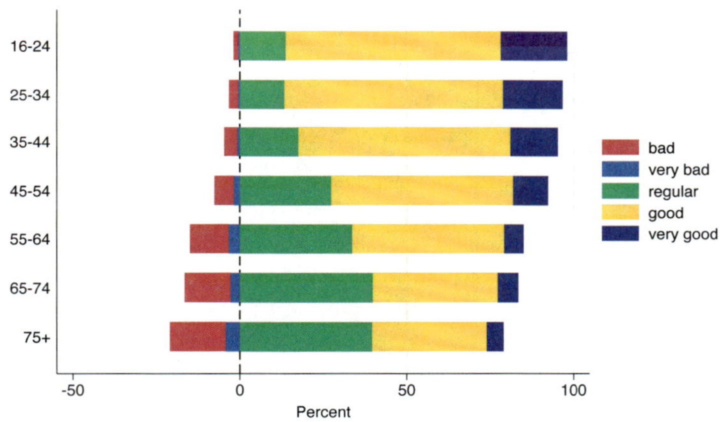

Figure 20.5. Horizontal slide plot

Figure 20.5 converted the previous vertical slide plot into a horizontal slide plot with percentages. The `yline()` option was also used to superimpose a dashed reference line. The graph clearly demonstrates that the incidences of bad and very bad health increase with higher age groups.

20.5 Mosaic plots

A further variation of a stacked bar chart is the mosaic plot, sometimes known as a Marimekko chart or a spine plot. Similarly to a stacked bar chart, a mosaic plot shows multiple bars stacked at each x-axis category, but now the width of each stacked bar is proportional to the total frequency of the entire category. In other words, mosaic plots have two x axes, one indicating the group identifier and another showing the fraction of observations in each category. They are particularly useful when frequency differences between the x-axis groups exist and where a user wishes to highlight these so that readers can form their own opinions on whether particular categories should be deemed more or less important.

In Stata, mosaic plots are created with the community-contributed `spineplot` command (Cox 2007b). The core syntax of `spineplot` is relatively simple; it requires only a categorical y variable and a categorical x variable to be specified after the main command. An example is shown below in figure 20.6.

Creating figure 20.6

```
. * Mosaic plot
. spineplot health agegroup
```

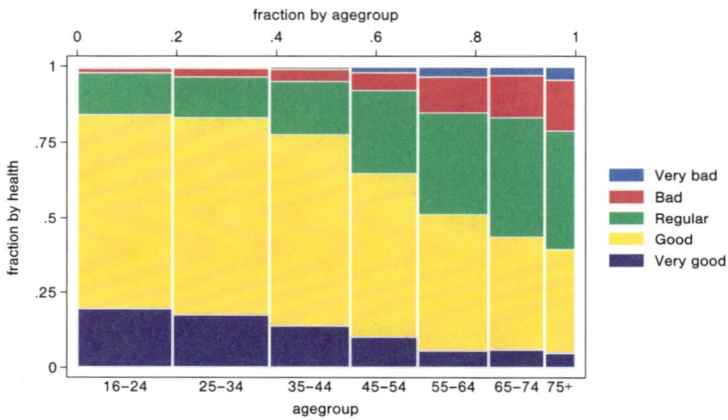

Figure 20.6. Mosaic plot

Figure 20.6 presents a mosaic chart of health versus age. Similarly to the previous stacked bar charts, it is easy to identify that poor health status increases with age group. In addition to this, a second x axis at the top of the graph identifies the proportion of data in each age group. It is now possible to identify that the fraction of observations in the 75+ age group is much less than the fraction of observations in the 25–34 age group.

Mosaic charts carry a lot of information and graphically do an excellent job at presenting the original, numerical two-way table in both its frequency and its percentage forms. Therefore, it can be seen as a "best of both worlds" graph, although the additional complexity may not always be useful for laypeople. In addition, note that the default coloring of the `spineplot` command is the inverse of the `graph bar` command that produced stacked bar charts. Users can fall back on the plethora of options to modify their graph appropriately, and these are not options specific to `spineplot`. An example of a mosaic plot with custom bar colors, bar widths, and axes is shown below in figure 20.7.

Creating figure 20.7

```
. * Mosaic plot with labeling and outline colors
. spineplot health agegroup,                                          ///
> xlabel(0(10)100, axis(1) tposition(outside))                        ///
> percent ylabel(0(10)100, axis(2))                                   ///
> barall(blwidth(*0.2) blcolor(black))                                ///
> bar1(color(maroon)) bar2(color(erose)) bar3(color(eltgreen))        ///
> bar4(color(eltblue)) bar5(color(edkblue))
```

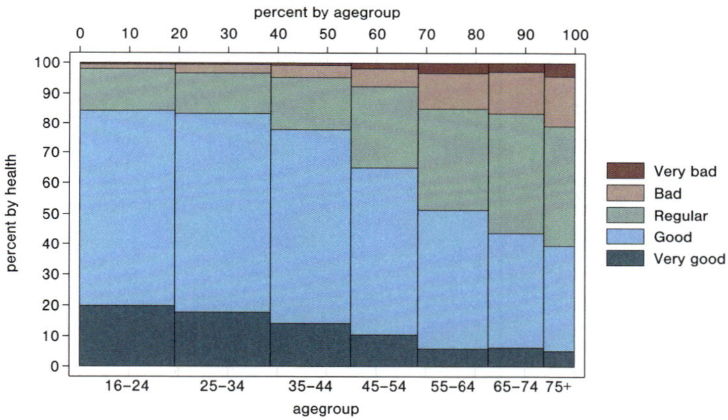

Figure 20.7. Custom mosaic plot

Part III

Graphs for multivariate data

Multivariate analysis is an extension of bivariate analysis where the onus is often on identifying how three or more variables are jointly related. It is mainly done numerically because data visualization has problems extending relationship analysis into three or more dimensions—which is the numerical equivalent of examining multiple variables jointly. However, via innovative methods such as color differentiation or repeated graph patterns, additional (pseudo)dimensions of data can be visualized. This, in turn, gives users powerful tools to examine relationships more carefully and often with less parametric assumptions than the numerical equivalents. Users can thus explore how functional form, data density, coding, or outliers might influence and steer multivariate results.

Multivariate graphs are often specialized graphs that have specific uses. Unlike univariate and bivariate graphs, they are not natural stepping stones for data analysts engaged in data work. The reasons are often in the complexity of communicating such graphs to laypeople, finding the appropriate data environment for such graphs to work, and restricting the analysis to just three, four, or five key variables. These conditions make multivariate graphs somewhat more challenging prospects for standard analysts; however, when used in the right circumstances, they offer a powerful alternative to traditional numerical methods such as multiple regression analysis.

21 Matrix plots

Advantages

1. Can visualize many different bivariate relationships
2. Provide a comprehensive overview of dataset structure
3. Are useful for diagnostic purposes before parametric modeling
4. Can be flexibly used with many different graph types

Disadvantages

1. Can overload users with too many graphs
2. Are limited to many bivariate relationships and incapable of true multivariate analysis
3. Often require a lot of formatting

Required community-contributed packages

To install the community-contributed commands that are required to follow along with this chapter completely, type

```
. ssc install corrtable
. ssc install crossplot
. ssc install trellis
```

21.1 Introduction

Multivariate visualization is difficult. There are not enough dimensions to keep adding more variables to traditional scatterplots. Even graphing three variables jointly requires significant compromise because the third dimension needs to be pseudovisualized in two dimensions. An alternative option to depicting complex multivariate relationships between variables is to use many bivariate plots. Such plots essentially "slice" higher-dimensional plots into 2D subplanes and present many such "slices". This, in turn, allows users to obtain some visual understanding of how data are distributed and how data vary across a wide range of variables.

This approach leads to matrix plots, which present many subgraphs on a single graph, each being related to the overall research question at hand. Their principle is that they condense a large amount of information into many graphs stitched together so that a natural connection between each subgraph exists. This makes it easier to examine such complex graphs because the eye has natural reference points between the various subgraphs.

Matrix plots are often used before parametric modeling such as regression analysis. Many modeling frameworks require users to specify how variables enter a particular statistical model. For example, they might be continuous or categorical. Moreover, continuous variables can take different functional forms, such as linear, quadratic, or polynomial. Alternatively, there might be significant outliers or data bunching in parts of the bivariate distributions. To gauge what kind of transformation variables are needed, users should always visually inspect variables that are about to enter a statistical model. Matrix plots excel at this type of analysis because they can quickly present an entire model's data on a single graph.

There are some caveats. First, multiple bivariate relationships do not imply multivariate relationships. What is true in the bivariate world is not necessarily true in a multivariate world, where other factors may mitigate bivariate relationships. For example, the relationship between a car's price and its mileage may be strongly affected by the weight of the car. As such, a bivariate functional-form analysis of car price and mileage alone may not always conform to what is best for a model. In most cases, though, bivariate relationships translate well to multivariate relationships.

Another caveat is that matrix plots produce many subgraphs by design. This should generally not be the first choice in data visualization, because graphs should be designed to convey the maximum amount of information in one picture. While this is unavoidable when looking at many variables, formatting is key to creating good matrix plots. End users need simple and effective visual markers that aid the eye in scanning the subgraphs. Marker labels and titles should be clear and not repeated. There should be little space between subgraphs. Users must carefully consider what kind of array of subgraphs they need; do they need a 4×5 matrix or a 2×10 matrix?

Finally, there are different types of matrix plots. Many default matrix plots present many scatterplots to visualize many bivariate relationships. Other plottypes also exist, such as trellis plots, which repeat the same graph over many different subgroups. While the first type allows users to quickly analyze how many different variables are related to each other, the second type allows users to examine how two variables are related across many different groups. Both graphs are matrix plots, but the initial analytical emphasis is different.

Matrix plots are very powerful plottypes for understanding how a lot of raw data are distributed between variables. Unfortunately, they are not particularly friendly to the general user, because those outside the immediate data environment will likely have difficulty fully understanding the overall picture. They are best used in operational and analytical environments where users prepare parametric modeling. Their use in presentations or other outside scenarios should generally be limited or well rationalized.

21.2 Correlation matrix

Matrix plots use many subgraphs to visualize a series of variables and their bivariate relationships. `auto.dta` contains 74 observations about car characteristics from 1978. It also contains several variables that relate to car attributes, and a user might be interested in how many of these are related to each other.

```
. * Load data
. sysuse auto
(1978 automobile data)

. summarize

    Variable |        Obs        Mean    Std. dev.       Min        Max
-------------+---------------------------------------------------------
        make |          0
       price |         74    6165.257    2949.496       3291      15906
         mpg |         74     21.2973    5.785503         12         41
       rep78 |         69    3.405797    .9899323          1          5
    headroom |         74    2.993243    .8459948        1.5          5
-------------+---------------------------------------------------------
       trunk |         74    13.75676    4.277404          5         23
      weight |         74    3019.459    777.1936       1760       4840
      length |         74    187.9324    22.26634        142        233
        turn |         74    39.64865    4.399354         31         51
displacement |         74    197.2973    91.83722         79        425
-------------+---------------------------------------------------------
  gear_ratio |         74    3.014865    .4562871       2.19       3.89
     foreign |         74    .2972973    .4601885          0          1
```

The most basic matrix plot is a correlation plot. A tabular version can be easily generated via the `correlate` command, which produces a correlation (or covariance) matrix. A variable list specifies the variables for which the correlation is to be calculated.

```
. * Correlation matrix
. correlate price mpg weight length displacement
(obs=74)

             |    price      mpg   weight   length displa~t
-------------+---------------------------------------------
       price |   1.0000
         mpg |  -0.4686   1.0000
      weight |   0.5386  -0.8072   1.0000
      length |   0.4318  -0.7958   0.9460   1.0000
displacement |   0.4949  -0.7056   0.8949   0.8351   1.0000
```

It is also possible to graph the above tabular results as a graph using the community-contributed command `corrtable` (Cox 2007a), which acts similarly to `correlate`; users must simply insert an appropriate variable list. An example is shown in figure 21.1.

Creating figure 21.1

```
. * Correlation matrix plot
. corrtable price mpg weight length displacement
```

Price	-0.469	0.539	0.432	0.495
-0.469	Mileage (mpg)	-0.807	-0.796	-0.706
0.539	-0.807	Weight (lbs.)	0.946	0.895
0.432	-0.796	0.946	Length (in.)	0.835
0.495	-0.706	0.895	0.835	Displacement (cu. in.)

Figure 21.1. Correlation matrix plot

Figure 21.1 repeats the prior tabular results and provides users with a fast overview of how variables in `auto.dta` are correlated with each other. For example, there is a high correlation between car `weight` and car `length` (which makes sense).

`corrtable` has some useful options that users can invoke to provide more detail in their graphical depiction of the correlation matrix. The option `half` specifies that the graph displays only the lower half of the correlation matrix, thereby removing redundant information. The option `n` displays the sample size of each correlation, and the option `pval` displays the p-values of each correlation. A demonstration of these options is shown in figure 21.2.

Creating figure 21.2

```
. * Half-correlation matrix with options
. corrtable price mpg weight length displacement, half n pval
```

Price				
-0.469 74 0.000	Mileage (mpg)			
0.539 74 0.000	-0.807 74 0.000	Weight (lbs.)		
0.432 74 0.000	-0.796 74 0.000	0.946 74 0.000	Length (in.)	
0.495 74 0.000	-0.706 74 0.000	0.895 74 0.000	0.835 74 0.000	Displacement (cu. in.)

Figure 21.2. Half-correlation matrix with observation count and p-values

Figure 21.2 provides more detail than the previous correlation matrix and allows users to decide which correlations matter. However, to some extent, a basic correlation plot remains unsatisfactory to serious analysts. Correlation coefficients cannot reveal nonlinearities, grouping of data, and outliers within data relationships. Scatterplots excel at that.

21.3 Matrix scatterplots

To create many different scatterplots on a single graph, one can use the official command **graph matrix**, which produces matrix scatterplot graphs. Users must specify a variable list, which will result in all possible combinations of scatterplots being graphed as a matrix plot. Figure 21.3 below demonstrates.

Creating figure 21.3

```
. * Matrix scatterplot
. graph matrix price mpg weight length displacement
```

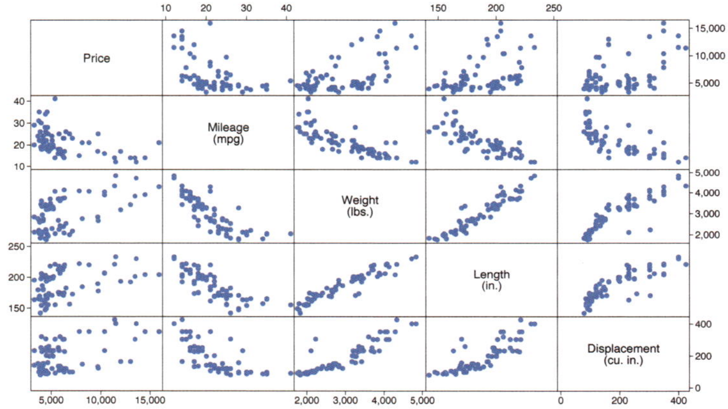

Figure 21.3. Matrix scatterplot

Figure 21.3 presents a basic matrix scatterplot. Each subgraph contains a scatterplot where two of the five possible variables are scattered against one another. It is now possible to delve behind the statistical correlations and explore the underlying data in more detail. For example, the relationship between **weight** and **displacement** appears nonlinear. Other relationships, such as **price** and **weight**, appear to have two groups of data. Accounting for such relationships will be crucial in further multivariate modeling. It would not have been possible to determine such facts from basic correlation coefficients.

The official command **graph matrix** plots both the upper- and the lower-triangular matrices, which are symmetrical. Information is therefore repeated. To plot only the lower-triangular set of scatterplots, the **half** option can be used. In addition, it will often be useful to format the look of markers using the standard **twoway** marker options (see **help marker_options**). Figure 21.4 below demonstrates the use of the **half** option and changes the outline width to very thin (**mlwidth(vvthin)**), the marker fill color to light blue with 20% transparency (**mfcolor(ltblue%80)**), and the marker size to medium–small (**msize(medsmall)**). In addition, the **scale()** option is used to make all text, marker, and line widths smaller.

Creating figure 21.4

```
. * Half-scatter matrix plot with options
. graph matrix price mpg weight length displacement, ///
> mlwidth(vvthin) mfcolor(ltblue%80) msize(medsmall) half scale(0.7)
```

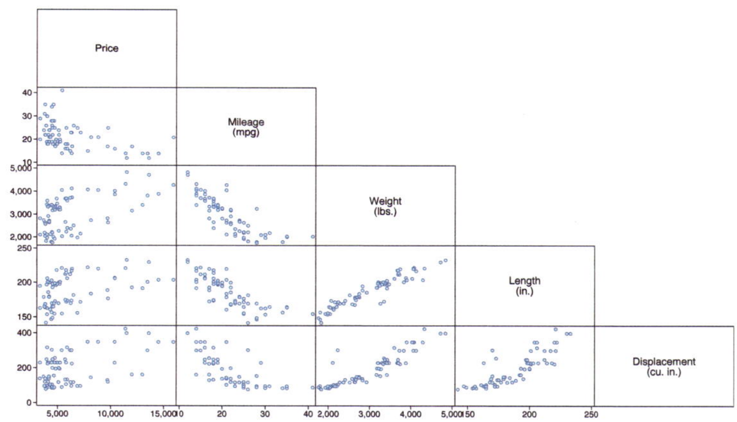

Figure 21.4. Half-scatter matrix plot with custom markers

Figure 21.4 presents a customized scatter matrix plot. The options available in the official command **graph matrix** are generally enough for basic analysis and representation.

21.4 Multiple matrix scatterplots

Many Stata graph commands can make use of the `by()` option to repeat a graph over different subgroups. `graph matrix` is no exception, and figure 21.5 below demonstrates this. The previous matrix scatterplot is repeated over the five subgroups of the variable `foreign`. The `iscale()` option is used to modify the size of symbols and labels within plots.

Creating figure 21.5

```
. * Multiple default matrix plots
. graph matrix price mpg weight length displacement, mlwidth(vvthin) ///
> mfcolor(ltblue%40) msize(medsmall) half iscale(0.6) by(foreign)
```

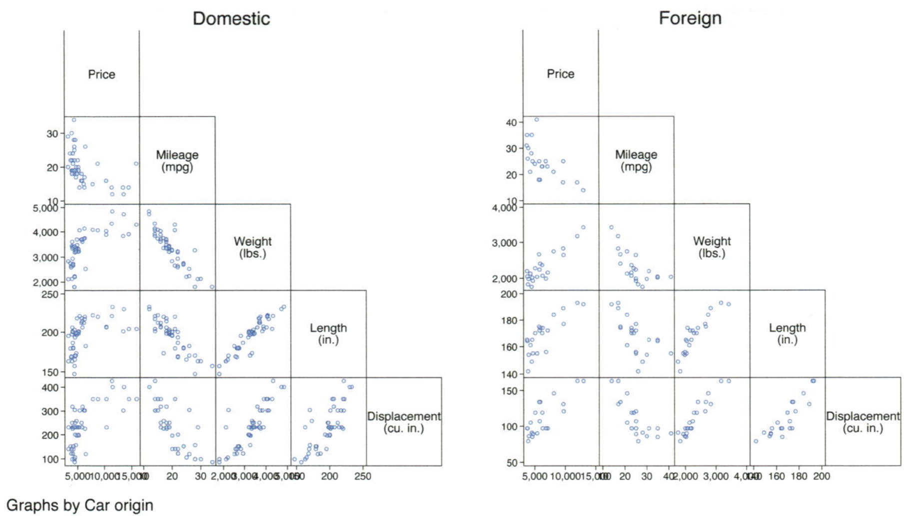

Figure 21.5. Multiple default matrix plots

Figure 21.5 shows how matrix plots can be repeated over different groups. Such graphs do come with a health warning: the number of scatterplots displayed can quickly explode and generate graphs that take a long time to draw or have little visual value. In general, such graphs should be used purely for analysis or data transformation and should not be aimed at external users.

21.5 Cross plots

The `graph matrix` command interacts all variables with each other, creating a large matrix with all possible bivariate combinations of the specified variable list. One who wishes to focus on a subpart of the total matrix plot should use the community-contributed command `crossplot` (Cox 2014a). It takes the form `crossplot (`*yvarlist*`) (`*xvarlist*`)`, and each variable of the *yvarlist* is plotted against each variable of the *xvarlist*. This method allows users more flexibility in defining their array of scatterplots. Figure 21.6 below demonstrates this with four y variables and only one x variable.

Creating figure 21.6

```
. * Cross plots
. crossplot (price mpg weight length) (displacement)
```

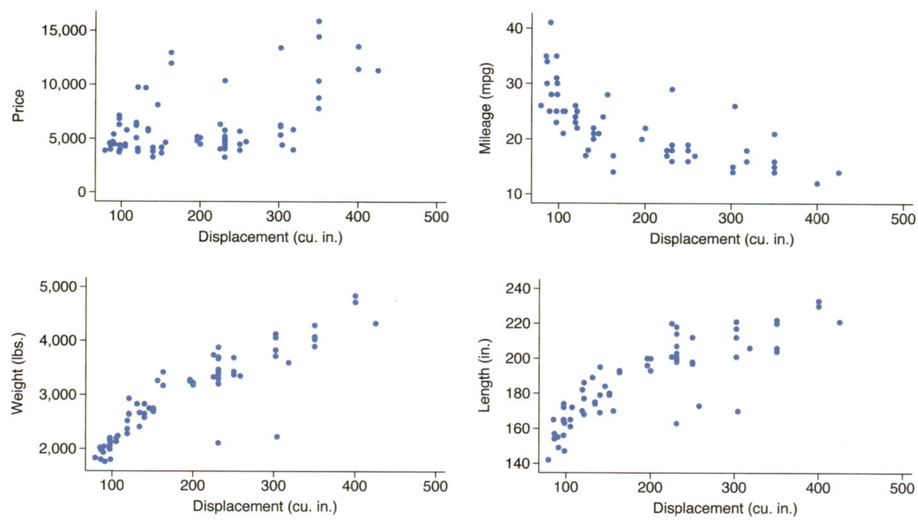

Figure 21.6. Cross plot

Figure 21.6 presents a basic cross plot that contains `displacement` as the single x variable. Only four graphs are shown because only four different y variables (`price`, `mpg`, `weight`, and `length`) were specified. This results in only a portion of the previously shown matrix plot being displayed—in this case, the fifth column of graphs in figure 21.3.

Users who wish to re-create the full matrix plot from figure 21.3 need to specify only two identical variable lists in the *yvarlist* and *xvarlist*. This is demonstrated below in figure 21.7.

Creating figure 21.7

```
. * Cross plots that resemble matrix plots
. crossplot (price mpg weight length displacement)  ///
> (price mpg weight length displacement)
```

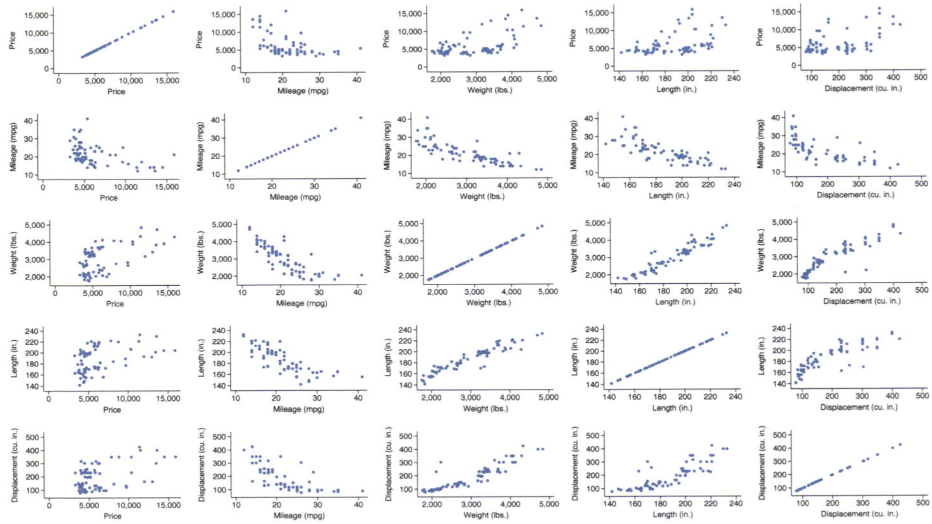

Figure 21.7. Full cross plot

Figure 21.7 demonstrates how the `crossplot` command can be used to re-create the same output as `graph matrix`. The core advantage of `crossplot` over `graph matrix` is its ability to select different subgroups from the entire matrix plot, thereby reducing information overload from unnecessary subgraphs.

21.6 Trellis plots

A useful cousin of the matrix plot is the trellis plot (also known as a lattice plot), which is a type of matrix plot that uses the same scale and axis in each subgraph. This allows users to do a much faster graph-to-graph comparison because each subgraph has identical measurements. Such a plot is useful if one needs to examine the same relationship across many different groups, particularly when one needs to examine bivariate relationships (or even univariate distributions) across two categorical variables, where the two categorical variables create an $n \times k$ matrix.

Official Stata does not have a command for producing trellis plots, but the community-contributed command `trellis` (Mander 2005c) is specifically designed to generate trellis plots. With this command, users generally do not need to specify a variable list; all variables are entered via the options.

A basic example that creates a 5×6 matrix of repeated scatterplots of weight (`weight`) versus blood pressure (`bpsystol`) over the categorical variables health status (`hlthstat`) versus age group (`agegrp`) is demonstrated below using `nhanes2.dta`, which contains information on 10,351 people and their personal and health characteristics. Note that some values in `hlthstat` are recoded to missing values to avoid empty graphs.

```
. * Load data
. webuse nhanes2

. recode hlthstat 8 = .
(14 changes made to hlthstat)

. summarize hlthstat agegrp weight bpsystol

    Variable |        Obs        Mean    Std. dev.       Min        Max
-------------+---------------------------------------------------------
    hlthstat |     10,335    2.586164    1.206196          1          5
      agegrp |     10,351     3.35813    1.735657          1          6
      weight |     10,351    71.89752    15.35642      30.84     175.88
    bpsystol |     10,351    130.8817    23.33265         65        300
```

To operationalize the `trellis` command, two options are required. The first is the `by()` option, which requires two categorical variables that together define an $n \times k$ matrix of plots. The second is the `function()` option, which defines the Stata command that is used to draw each subgraph. In this case, the `twoway scatter` command is used to create a scatterplot of `weight` versus `bpsystol`, and this graph will be repeated in each $n \times k$ subgraph of the trellis plot. The option `msymbol(p)` is added to the scatterplot code to draw each marker as a point because of the large size of the data. Because there are many subgraphs drawn in this example, the `fopt()` option is used to customize the title labels. `fopt()` applies graph options to each graph in the matrix and can be used to modify many graph options; it modifies title labels in this example. To avoid repitition of the variable titles, the individual x- and y-axis titles are switched off via the `xtitle("")` and `ytitle("")` options within the `fopt()` option. This avoids flooding the combined graph with many instances of `Systolic blood pressure` and `Weight (kg)`. Instead, the rarely used options `t1title()` and `l1title()` are used to display single x- and y-axis titles in the combined plot. Note that these options are not specified within the `fopt()` option; they apply only to the combined graph.

Creating figure 21.8

```
. * Trellis plot
. trellis, by(hlthstat agegrp)                                        ///
> function(twoway (scatter weight bpsystol, msymbol(p))) ///
> fopt(xtitle("") ytitle(""))                                         ///
> t1title("Systolic blood pressure") l1title("Weight (kg)")
```

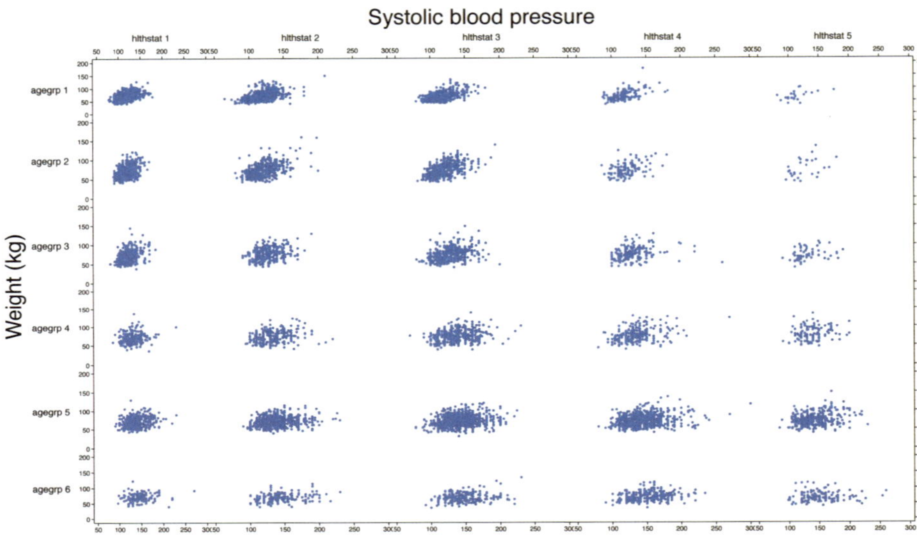

Figure 21.8. Trellis plot with `scatter`

Figure 21.8 is an example of a basic trellis scatterplot. This graph plots the relationship between weight and blood pressure across 30 different combinations of 5 health groups and 6 age groups. A key difference between this plot and the previous plots is that each subgraph has the same x and y axes, making it much easier to compare subgraphs. For example, it is easy to see that the relationship between weight and blood pressure is positive, but in some groups, this relationship appears steeper than in others. Some subgroups also have little data, for example, `agegrp` 1 and `hlthstat` 5, corresponding to 20–29-year-olds with poor health.

It can be hard to identify exact relationships from pure scatterplots alone. Thankfully, the `trellis` command is incredibly flexible, and users can simply extend the set of commands they specify in `function()` to include additional overlay graphs. Figure 21.9 demonstrates how a `qfitci` command is added to the `twoway scatter` command to produce an overlay of quadratic fits to each scatterplot. The `legend(off)` option is specified in the `qfitci` command to avoid drawing a legend in each subgraph. The `fopt()` option remains specified to avoid title label clutter, as described previously.

Creating figure 21.9

```
. * Trellis plot with multiple two-way plots
. trellis, by(hlthstat agegrp)                                    ///
> function(twoway (scatter weight bpsystol, msymbol(p)) ///
> (qfitci weight bpsystol, lcolor(black) legend(off)))  ///
> fopt(xtitle("") ytitle(""))                           ///
> t1title("Systolic blood pressure") l1title("Weight (kg)")
```

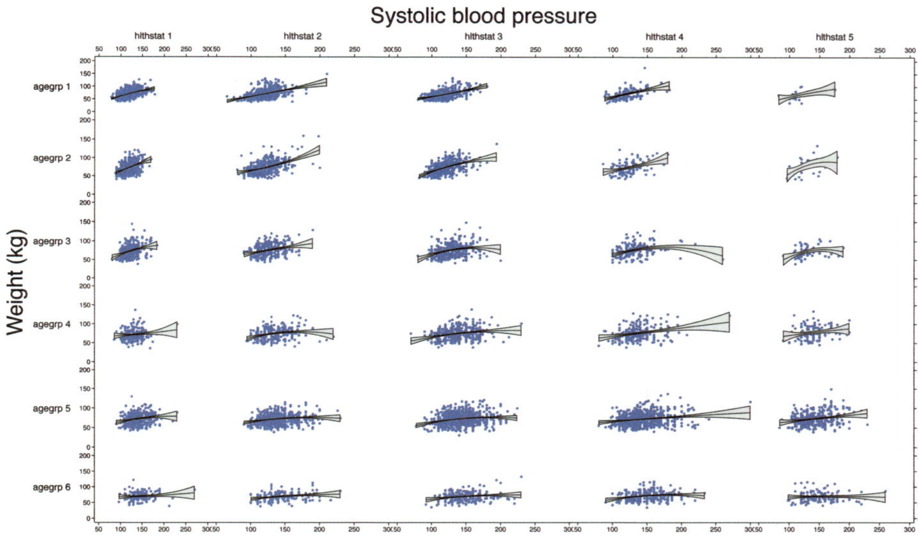

Figure 21.9. Trellis plot with `scatter` and `qfitci`

Figure 21.9 shows how flexible the **trellis** command is by enabling users to examine many different relationships at once over a large set of groups. There seems to be evidence that the relationship between weight and blood pressure is positive for all groups, but in some groups, the gradient is steeper than in others. In particular, younger age groups (for example, `agegrp` 1) appear to have steeper gradients than older age groups.

Trellis plots are not confined to scatterplots or fit plots. The `function()` option allows users to specify any graph command available to their version of Stata—using both official commands and community-contributed commands. Figure 21.10 below demonstrates how box plots are graphed for the 30 different `hlthstat`–`agegrp` groups. In this particular example, the `fopt()` option is also specified with the `over()` option, which results in each individual graph showing a box plot of the variable `weight` computed over three different race groups. Note that because box plots do not have an x-axis title, the `xitle()` option is not specified as in the previous graphs. Users should also note that in some cases, command-specific options should be specified within the relevant command line in the `function()` option (for example, figure 21.8); in other cases, they must be specified in the `fopt()` option (as is the case here).

Creating figure 21.10

```
. * Trellis plot with box plots
. trellis, by(hlthstat agegrp) function(graph box weight) ///
> fopt(over(race) ytitle(""))                              ///
> t1title("Systolic blood pressure") l1title("Weight (kg)")
```

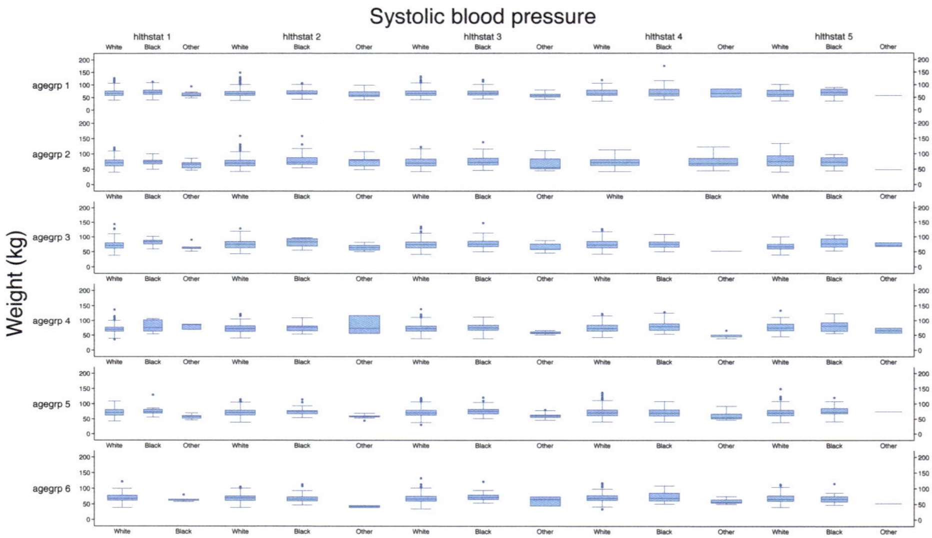

Figure 21.10. Trellis plot with `graph box`

Figure 21.10 demonstrates how trellis plots can be used with nonscatterplot graphs. The core use of `trellis` is to allow users to create many different graphs over a large group of categories. Finally, users can specify any valid installed graph command, including community-contributed commands such as `heatplot`. Figure 21.11 demonstrates.

Creating figure 21.11

```
. * Trellis plot with heat plots
. trellis, by(hlthstat agegrp) function(heatplot height weight) ///
> fopt(legend(off) xtitle("") ytitle(""))                       ///
> t1title("Systolic blood pressure") l1title("Weight (kg)")
```

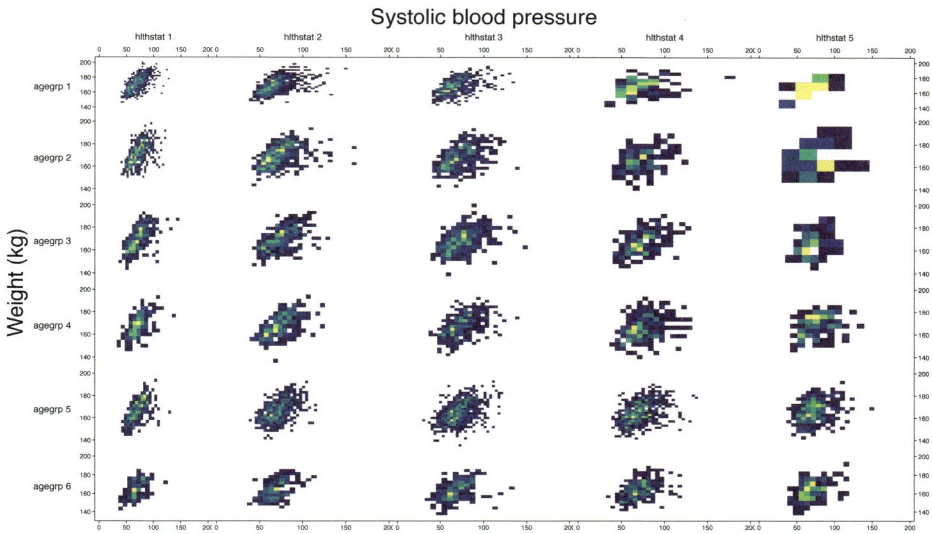

Figure 21.11. Trellis plot with `heatplot`

Figure 21.11 is an example of a trellis heat plot. Many different heat plots are combined and presented in a matrix array that allows end users to quickly and easily compare a large amount of data. The core concept here is that trellis plots are flexible plottypes that can present a wide array of different visualizations.

22 Contour plots

Advantages

1. Can use contour and color lines to provide visualizations of a third dimension (z variable)
2. Resemble maps and are thus easy to read for most people
3. Are most useful when all three variables are continuous
4. Excel when the three-way relationship is smooth and populated with lots of data

Disadvantages

1. Require grid of data; empty spaces can lead to poor interpolation
2. Cannot be used with discrete variables
3. Are affected by color-level and interpolation choices
4. Are not good for exploring random variables

Required community-contributed packages

To install the community-contributed commands that are required to follow along with this chapter completely, type

```
. ssc install palettes
. ssc install colrspace
```

22.1 Introduction

A contour plot is a type of pseudo-3D plot that attempts to provide a visualization of the relationship between three different variables. They do this by plotting two variables in the standard x and y dimensions and then using contours to represent the third variable in the third dimension. Contours are constant "slices" of the third variable (z axis) plotted across the x and y variables (axes). At their core, they allow users to answer the question of how a variable z changes as a function of both x and y.

Contour plots are often used in topological mapping, where the contours represent height. For example, mountains are often represented by ever-smaller concentric rings. A contour plot's similarity to a map is one of its core advantages in data visualization. Most users will understand how contours are related to higher or lower values of the z variable, especially when there is a smooth relationship in the data (as is the case for geographic data). Contour plots are often combined with colors to make it easier to identify numeric values, and overall, this kind of visualization is a user-friendly method to examine three-way relationships.

However, contour plots have some limits. They are best used when all three variables are continuous, although the z variable could be an ordered categorical variable if there are sufficient categories. The y and x variables need to be continuous. Moreover, the contour levels are similar to bin sizes in histograms or bandwidths in kernel density plots; more or fewer levels can significantly alter the plot. Users will need to experiment with different contour levels to suit their analysis type.

Another serious problem is that contour plots need complete or near-complete information across the data grid. Empty spaces in the 3D data cube that make up the x, y, and z coordinates lead to interpolation. While some interpolation is fine, when large data sections are missing, the resulting plot will look messy and convey little information. In such cases, one should use heat plots, which are close cousins of contour plots, to convert the data into a usable form (this is demonstrated at the end of this chapter).

Finally, even with complete data between three variables, a lack of smooth relationships between the x, y, and z variables can lead to messy plots. Again, a heat plot or a method to convert the data into a smoother form are appropriate responses in such cases.

22.2 A basic contour plot

The basic question that contour plots seek to answer is how three variables are jointly related. A classic example of contour plots is geographic data that include y and x coordinates and a third variable representing height. The subsea elevation of Lamont sandstone in an area of Ohio dataset (`sandstone.dta`) contains such data and represents depth characteristics of an area in Ohio, United States. A summary of the dataset is provided below.

```
. * Load data
. sysuse sandstone
(Subsea elevation of Lamont sandstone in an area of Ohio)
. summarize
```

Variable	Obs	Mean	Std. dev.	Min	Max
depth	6,400	7900.934	82.07205	7556	8061
northing	6,400	74991.9	8762.577	60840	89250
easting	6,400	39619.03	5721.486	29240	49610
collection	6,400	2.985469	.1401273	1	3

```
. list depth northing easting in 1/20, separator(0)
```

	depth	northing	easting
1.	7620	60,840	29,240
2.	7627	60,840	29,840
3.	7633	60,840	30,200
4.	7636	60,840	30,350
5.	7642	60,840	30,650
6.	7645	60,840	30,800
7.	7657	60,840	31,280
8.	7666	60,840	31,640
9.	7669	60,840	31,730
10.	7677	60,840	32,000
11.	7685	60,840	32,270
12.	7698	60,840	32,720
13.	7710	60,840	33,080
14.	7723	60,840	33,500
15.	7724	60,840	33,530
16.	7732	60,840	33,770
17.	7733	60,840	33,800
18.	7746	60,840	34,190
19.	7756	60,840	34,520
20.	7768	60,840	34,880

`sandstone.dta` consists of 6,400 observations, of which the first 20 are shown above. The variables `northing` and `easting` identify geographic coordinates, and the variable `depth` identifies depth. A contour plot can provide a visualization of the observations of all three variables jointly.

The official command `twoway contour` produces contour plots. The command is part of the `twoway` command and thus allows multiple overlays of other graphs, though in practice, it is rare to overlay other plottypes over a contour plot. The `twoway contourline` command is sometimes used to give better definition to contour boundaries; this will be demonstrated later. Unlike other `twoway` plottypes (such as `scatter`), `twoway contour` requires the `twoway` command to be specified to operate; it is not possible to call only `contour`. `twoway contour` requires a z, a y, and an x variable. A basic example is shown in figure 22.1.

Creating figure 22.1

```
. * Basic contour plot
. twoway (contour depth northing easting)
```

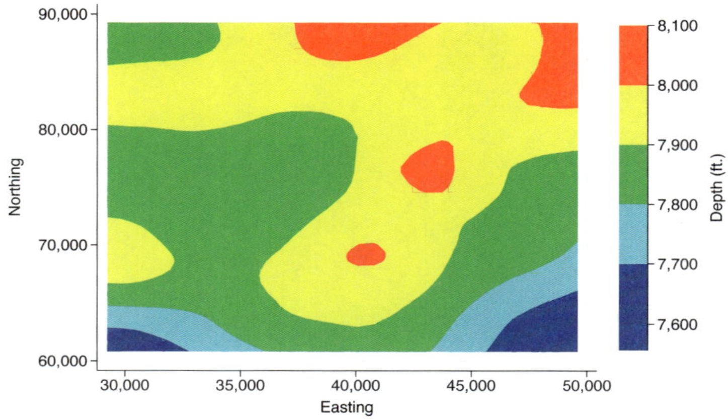

Figure 22.1. A basic contour plot

Figure 22.1 presents a basic contour plot that illustrates the subsea elevation of sandstone in an area in the United States. It is a map, which is how most people will have some familiarity with contour plots. The y and x variables represent a set of coordinates; together, they form an area of land. The z variable measures depth in feet. A legend identifies which color values and contours represent which areas of depth. This is the third dimension, or z axis. Some areas of land are shaded in red and have high depth, while other areas are shaded in blue and have low depth. Overall, a user gets a good understanding of the geography of this area.

22.2.1 Custom contour levels

Contour plots are subject to various choices, including how many contour lines should be computed. Various options allow users to specify different contours. The `ccuts()` option allows users to specify custom contour values, while the `levels()` option allows users to specify how many contour levels should be included. The options are mutually exclusive, and figure 22.2 below presents an example with the `levels()` option where the default of 5 levels is changed to 10. In addition, the `minmax` option is specified, which includes minimum and maximum values of the z variable in the legend. Users can always specify this option for greater clarity.

Creating figure 22.2

```
. * Contour plot with 10 levels and minimum and maximum legend labeling
. twoway (contour depth northing easting, levels(10) minmax)
```

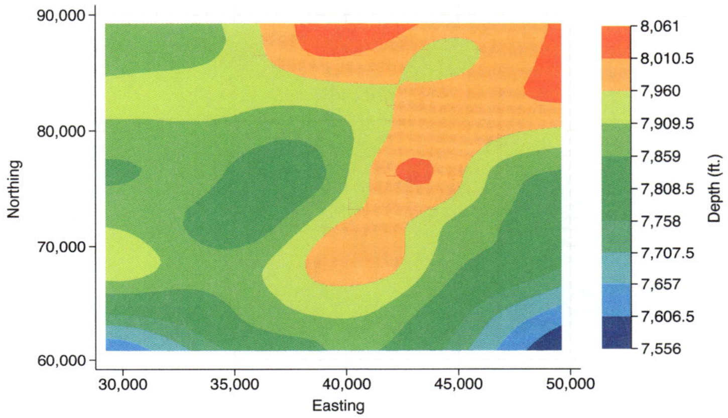

Figure 22.2. Contour plot with custom levels

Figure 22.2 offers more detail than the last contour plot. Multiple mountains and valleys emerge in the landscape, and more detail is shown. The choice of contour levels is important. It often makes sense to increase the number of levels from the default five, which can be somewhat restrictive.

22.3 Contour-line plots

The `twoway contourline` command is similar to the `twoway contour` command, with many of the same options. This command creates only contour lines and removes the coloring of the depicted contours. A basic example is shown below in figure 22.3.

Creating figure 22.3

```
. * Basic contour-line plot
. twoway (contourline depth northing easting)
```

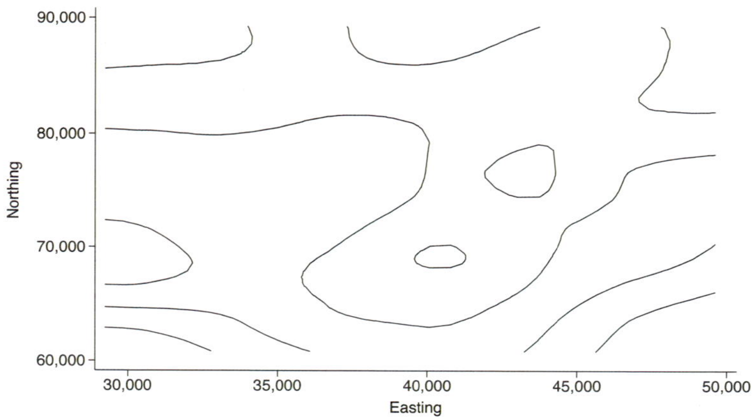

Figure 22.3. Contour-line plot

Figure 22.3 is a basic contour-line plot that shows only the contour boundaries from figure 22.1. By itself, the `contourline` command has little use because readers cannot identify which lines correspond to which boundaries of depth. It is possible to invoke the `colorlines` option, which is unique to `twoway contourlines`. It displays contour lines in different colors, so users can thus identify different levels of the z variable. An example with 20 different contour lines is shown below in figure 22.4.

Creating figure 22.4

```
. * Basic contour-line plot with 20 levels and colored lines
. twoway (contourline depth northing easting, colorlines levels(20))
```

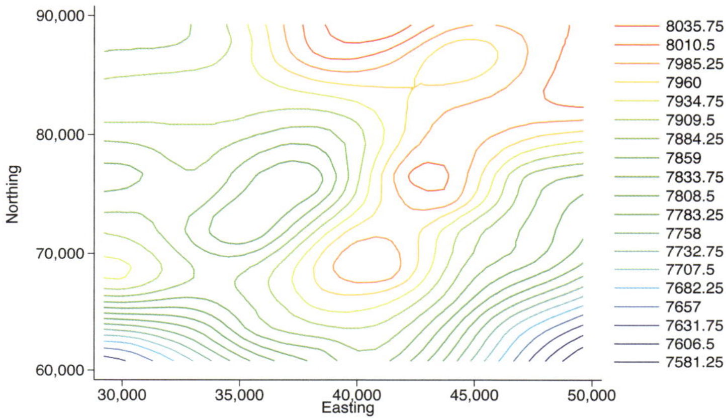

Figure 22.4. Contour-line plot with colored lines

Figure 22.4 presents a colored contour-line plot of `sandstone.dta`. A legend identifies the relevant color values. Standalone, this visualization is more useful than the default monochrome plot; however, compared with a basic contour plot, it offers few advantages. The visual narrative provided by standard contour plots is more striking, and it is easier to identify different values of the z variable.

22.4 Contour plots with custom colors

Color choices matter when using contour plots. They can greatly impact the way end users perceive the graph and data. Both the `twoway contour` and the `twoway contourline` commands contain various color options that allow users to specify custom colors. The simplest option is to use `ccolors()`, which allows users to specify a set of custom colors for each level. A more complex trio of options is `scolor()`, `ecolor()`, and `crule()`. The options `scolor()` and `ecolor()` define start and end colors for a contour plot. Both commands will then automatically interpolate all remaining color levels between these start and end colors. The color interpolation rule can be adjusted via the `crule()` option.

The heavy emphasis of color in contour plots suggests that users may want flexibility in creating their own custom color schemes. Therefore, users may want to consider the command `colorpalette` (part of the `palletes` package [Jann 2017]) to create custom colors. The key to using `colorpalette` with `twoway contour` or `twoway contourline` is the stored results in local macros that are left behind by `colorpalette`. These locals contain all the color RGB values of whatever scheme was defined in `colorpalette`. Importantly, the local `r(p)`, which can be accessed by typing `` `r(p)' ``, contains all color values in one syntax line. This is demonstrated below with the color scheme `viridis`, which has five equally spaced color levels achieved with the option `ipolate(5)`. The option `reverse` reverses the colors, and `nograph` suppresses the graphical output.

```
. * Custom colors
. colorpalette viridis, ipolate(5) reverse nograph
. return list
scalars:
                  r(n) =  5

macros:
              r(ptype) : "color"
              r(pname) : "viridis"
              r(pnote) : "viridis colormap from matplotlib.org"
            r(psource) : "https://github.com/matplotlib/matplotlib/blob/maste.."
             r(pclass) : "sequential"
                  r(p) : ""253 231 37" "94 200 99" "33 144 140" "59 82 138" ".."
            r(p5info) : "Jab 19 18 -16"
                r(p5) : "68 1 84"
            r(p4info) : "Jab 37 -4 -22"
                r(p4) : "59 82 138"
            r(p3info) : "Jab 55 -21 -6"
                r(p3) : "33 144 140"
            r(p2info) : "Jab 74 -23 19"
                r(p2) : "94 200 99"
            r(p1info) : "Jab 92 -5 33"
                r(p1) : "253 231 37"
```

Now that a local with all custom colors has been established, it can be directly inputted into the `twoway contour` (and `contourlines`) command via the `ccolors()` option. An example is shown below in figure 22.5. Note that the `ipolate()` value is the same as the `levels()` value. Also note that because the local is forgotten after execution, the below code reruns the `colorpalette` command in quiet mode via the additional prefix command `quietly`.

Creating figure 22.5

```
. * Custom color contour plot
. quietly colorpalette viridis, ipolate(5) reverse nograph
. twoway (contour depth northing easting, ccolors(`r(p)´) levels(5))
```

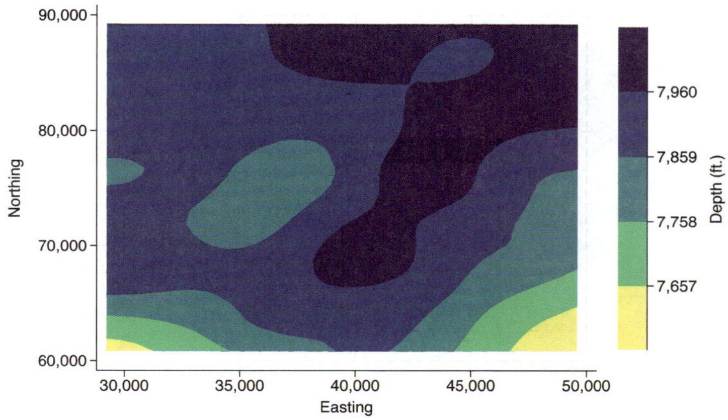

Figure 22.5. Contour plot with custom colors

Figure 22.5 presents a contour plot with custom color levels defined by the command `colorpalette`. Many color combinations and options are available, and users should examine the associated help file (see `help colorpalette`).

A second example is shown below in figure 22.6. In this example, 20 different color levels are defined using the `HCL terrain2` color scheme. In addition, a `twoway contourline` plot is overlaid on the original contour plot. The start and end colors are both white, resulting in only white lines that produce a clear differentiation of boundaries.

Creating figure 22.6

```
. * Custom colors and overlaid plots
. quietly colorpalette HCL terrain2, ipolate(20) reverse nograph
. twoway (contour depth northing easting, ccolors(`r(p)') levels(20))  ///
> (contourline depth northing easting, plegend(off) scolor(white)      ///
> ecolor(white) crule(linear) colorlines levels(20))
```

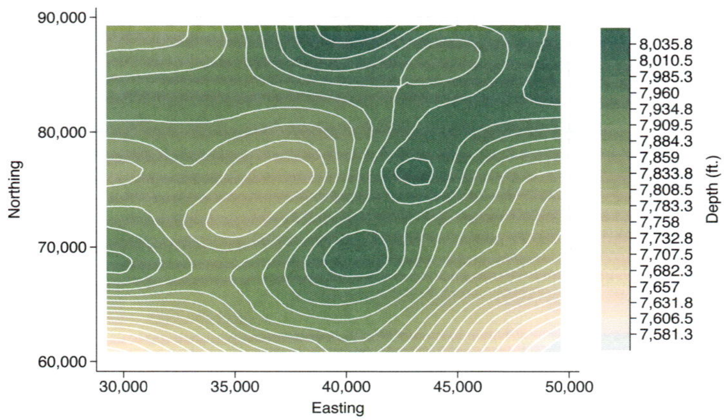

Figure 22.6. Contour plot with custom colors and custom contour lines

Figure 22.6 shows how `contour` plots can be combined with `contourlines` to create one contour map. In many applied cases, this is the most used overlay for `twoway contour` because other plottypes such as scatterplots or histograms would add little value.

22.5 Contour plots with small and incomplete datasets

All previous examples were created with `sandstone.dta`, which nicely exemplifies contour plots. This dataset is unique in that it has no missing values, and there are data on depth across the entire x and y grids (although the variable `collection` makes it clear that even in this dataset, over 99% of data points are estimated or interpolated). In many datasets, this would not be the case, because missing values or lack of observations will lead to large empty areas in a contour plot. Moreover, relationships between the x, y, and z variables may not be smooth.

In the following example, `auto.dta` is used to demonstrate why contour plots may be less useful when faced with incomplete data.

```
. * Using contour when data does not fill regular grid
. sysuse auto
(1978 automobile data)

. summarize
```

Variable	Obs	Mean	Std. dev.	Min	Max
make	0				
price	74	6165.257	2949.496	3291	15906
mpg	74	21.2973	5.785503	12	41
rep78	69	3.405797	.9899323	1	5
headroom	74	2.993243	.8459948	1.5	5
trunk	74	13.75676	4.277404	5	23
weight	74	3019.459	777.1936	1760	4840
length	74	187.9324	22.26634	142	233
turn	74	39.64865	4.399354	31	51
displacement	74	197.2973	91.83722	79	425
gear_ratio	74	3.014865	.4562871	2.19	3.89
foreign	74	.2972973	.4601885	0	1

The dataset contains only 74 observations. A user who is interested in visualizing the relationship between miles per gallon (`mpg`), car weight (`weight`), and car length (`length`) may initially attempt to create a contour plot as shown in figure 22.7.

Creating figure 22.7

```
. * Contour plot with little data
. twoway (contour mpg weight length, levels(5))
```

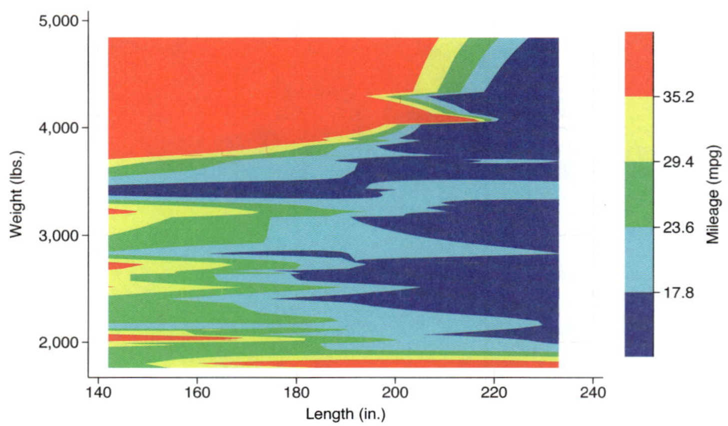

Figure 22.7. A poor contour plot

Figure 22.7 clearly shows that a contour plot applied to a small dataset results in a poor plot that has little applied value. There is simply a lack of observations.

The solution in this situation is to resort to parametric modeling and then visualize the parametric model. In this example, the relationship between `mpg`, `weight`, and `length` is first modeled via a basic regression analysis using the `regress` command, where `weight` and `length` are fully interacted with each other via the factor interaction operator `##`. The `c.` prefixes identify `weight` and `length` as continuous variables. The results from this model are then used to predict the values of `mpg` at different levels of `weight` and `length`. This is done with the `margins` command, with its output suppressed with the `quietly` command. Importantly, the `at()` option allows users to specify at what values of the x variables miles per gallon should be predicted. In this case, the values used are 1,760 to 4,840 in increments of 100 for `weight` and values of 142 to 233 in increments of 5 for `length`. Predicting more values leads to better-defined contour plots, although there is a tradeoff between computational length and graph creation time versus graph definition. These predicted values are saved to another dataset called `predictions.dta` using the `saving()` option. This new dataset is then loaded.

```
. * Model relationship
. regress mpg c.weight##c.length

      Source |       SS           df       MS      Number of obs   =        74
-------------+----------------------------------   F(3, 70)        =     48.17
       Model |  1646.04758         3  548.682526   Prob > F        =    0.0000
    Residual |   797.41188        70  11.3915983   R-squared       =    0.6737
-------------+----------------------------------   Adj R-squared   =    0.6597
       Total |  2443.45946        73  33.4720474   Root MSE        =    3.3751

------------------------------------------------------------------------------
         mpg | Coefficient  Std. err.      t    P>|t|     [95% conf. interval]
-------------+----------------------------------------------------------------
      weight |  -.0113212   .0048651    -2.33   0.023    -.0210244    -.001618
      length |  -.1805868   .0829033    -2.18   0.033    -.3459323   -.0152413
             |
   c.weight#|
    c.length |   .0000376   .0000232     1.62   0.109    -8.65e-06    .0000839
             |
       _cons |   67.45002   13.48129     5.00   0.000     40.56243    94.33762
------------------------------------------------------------------------------

. * Predict values across grid
. quietly margins, at(weight=(1760(100)4840) length=(142(5)233))  ///
> saving(predictions, replace)

. * Load prediction
. use predictions
(Created by command margins; also see char list)
```

The regression model suggests that for each pound increase of car weight, miles per gallon decreases by -0.011, while for each inch of car length, miles per gallon decreases by -0.18. However, as weight and length increase together, miles per gallon increases; this is represented by the positive sign on the interaction term. We do not find evidence that the coefficient on the interaction term is different from zero, but that is ignored in this analysis.

Once the new predictions are loaded, a contour plot on the predicted values can be created. This is shown below in figure 22.8. The variable `_margin` represents the predictions of `mpg`, the variable `_at1` represents weight, and the variable `_at2` represents length.

Creating figure 22.8

```
. * New contour plot
. twoway (contour _margin _at1 _at2, levels(20))
```

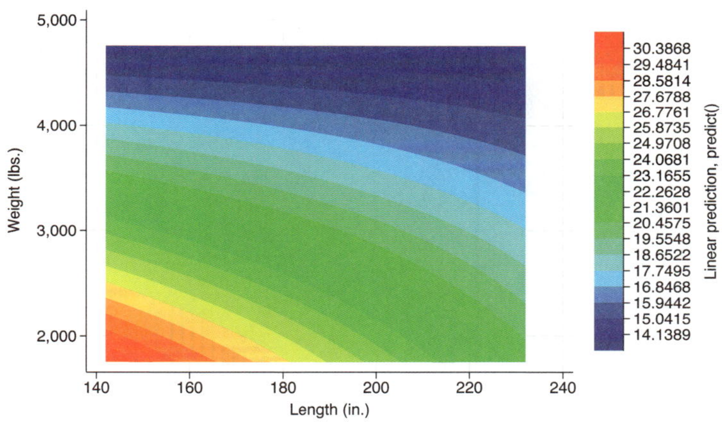

Figure 22.8. A predicted contour plot

Figure 22.8 presents a much smoother visualization of the relationship between `mpg`, `weight`, and `length`. Assuming the underlying regression model is correct, the contour plot suggests that low values of weight or length are associated with high mileage, while increasing weight or length is associated with lower values of car mileage. However, the relationship is not linear; higher weight values lead to more rapid mileage drops than higher length values.

Of course, whether the underlying parametric model is correct is another debate. The primary point of figure 22.8 is to demonstrate how contour plots can be used on predicted data from modeling. Even a small dataset with only 74 observations can support a contour plot if the underlying modeling is correct and many predictions are generated.

23 Trivariate heat plots

Advantages

1. Can provide visualizations of the third dimension via colors
2. Are useful for large datasets (overplotting)
3. Can also be applied to discrete variables

Disadvantages

1. Provide less statistical information
2. Are affected by bin-size and color choices
3. Require further accessibility considerations (for example, color disabilities)

Required community-contributed packages

To install the community-contributed commands that are required to follow along with this chapter completely, type

```
. ssc install palettes
. ssc install colrspace
. ssc install heatplot
```

23.1 Introduction

Heat plots are graphs that use color gradients to represent additional statistical information within standard 2D plots. The basic two-variable use of heat plots has already been discussed in chapter 13, and readers should first become familiar with the operations outlined there. This chapter will focus only on the three-variable case of heat plots and omit lengthy discussion around custom bin sizes and custom color levels and how such choices shape the output of a heat plot.

A critical difference between two- and three-variable heat plots is that in the two-variable case, color is used to represent numerical information about the number of observations in each bin, which can be in proportions, percent, or frequencies. The three-variable case emphasizes the statistical values of the third variable (z). This could be means, medians, standard deviations, or other statistical concepts.

Moreover, the types of variable heat plots use are not restricted only to continuous variables. Discrete variables can be introduced to morph the visualization of a heat plot toward a table plot but with colors as the key statistical indicator instead of bars or other symbols. This makes heat plots very versatile when examining trivariate distributions among continuous and discrete variables.

Users should remember that all the standard heat plot caveats apply. Bin and color choices matter greatly, especially when continuous variables are used. Moreover, color accessibility should be given some weight because end users may have problems with particular color sets (for example, red or green). Furthermore, color shading is relatively poor at conveying detailed numeric information.

23.2 A basic trivariate heat plot

The basic problem that three-way heat plots attempt to solve is the visualization of three variables. The example below uses `nhanes2.dta`, which contains 10,351 indivdual observations and 58 variables. Only five variables are summarized below.

```
. * Load data
. webuse nhanes2
. summarize agegrp sex bpsystol weight height
    Variable |        Obs        Mean    Std. dev.       Min        Max
-------------+---------------------------------------------------------
      agegrp |     10,351     3.35813    1.735657          1          6
         sex |     10,351    1.525167    .4993904          1          2
    bpsystol |     10,351    130.8817    23.33265         65        300
      weight |     10,351    71.89752    15.35642      30.84     175.88
      height |     10,351    167.6509    9.655916      135.5        200
```

```
. list agegrp sex bpsystol weight height in 1/20
```

	agegrp	sex	bpsystol	weight	height
1.	50-59	Male	106	62.48	174.598
2.	40-49	Female	108	48.76	152.297
3.	20-29	Female	98	67.25	164.098
4.	60-69	Female	180	94.46	162.598
5.	60-69	Female	120	74.28	163.098
6.	60-69	Female	180	66	147.098
7.	60-69	Female	100	54.55	153.898
8.	50-59	Female	154	58.97	160
9.	60-69	Female	120	68.95	164
10.	60-69	Male	104	65.43	176.598
11.	70+	Female	186	76.77	156.199
12.	60-69	Female	90	58.4	170.098
13.	30-39	Female	104	65.77	151.797
14.	20-29	Female	100	48.54	154.199
15.	30-39	Male	110	73.03	171.098
16.	60-69	Male	140	68.72	169.297
17.	60-69	Female	156	88.11	151.398
18.	70+	Female	154	68.04	162
19.	70+	Female	130	61.58	171.797
20.	60-69	Male	136	52.5	170.098

The first 20 observations of **nhanes2.dta** are presented. This medical dataset contains information on medical outcomes (for example, blood pressure) and individual characteristics (such as age, female, weight, and height). A researcher's key interest might be how the **weight** and **height** variables relate to other variables (such as **agegrp**, **sex**, or **bpsystol**). These variables are a mix of continuous and discrete variables.

To create trivariate heat plots, one must use the community-contributed command **heatplot** (Jann 2019). **heatplot** must include y and x variables to function (this form has been explored in chapter 13), but it can also include a z variable if desired. The command's syntax is **heatplot** $\left[z\right]$ y x, where z is an optional variable that can be binary or continuous. Specifying a z variable leads to a color legend that identifies the mean values of z in each bin. Figure 23.1 below demonstrates an example with three continuous variables: **bpsystol**, **weight**, and **height**.

Creating figure 23.1

```
. * Trivariate heat plot with three continuous variables
. heatplot bpsystol weight height
```

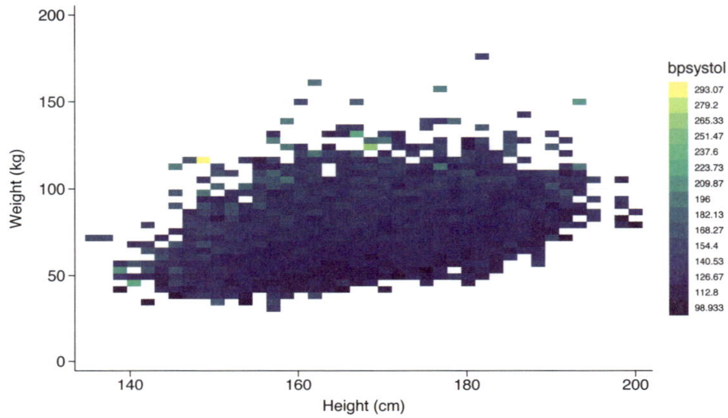

Figure 23.1. A basic trivariate heat plot

Figure 23.1 is an example of a trivariate heat plot where the x and y axes define 2D bins (from the variables `weight` and `height`) and the color gradient inside the bins reflects the mean values of a third variable (`bpsystol`). The graph indicates that there is some trend in the three-way relationship. High blood pressure values are generally related to those with higher weights, but larger height values moderate this. In other words, higher height reduces average blood pressure values for those with higher weight.

Color choices present a problem in figure 23.1. Some extreme values, which occur rarely, lead to a graph that is primarily blue in color gradient, making it hard to identify a clear pattern. In the case of blood pressure, there are standard medical values that define normal, elevated, hypertension 1, hypertension 2, and hypertensive crisis statuses. The `cuts()` option can be used to define such custom levels, and an example of this is shown below in figure 23.2.

Creating figure 23.2

```
. * Trivariate heat plot with custom color bins
. heatplot bpsystol weight height, cuts(0, 120, 130, 140, 160, 180, 300)
```

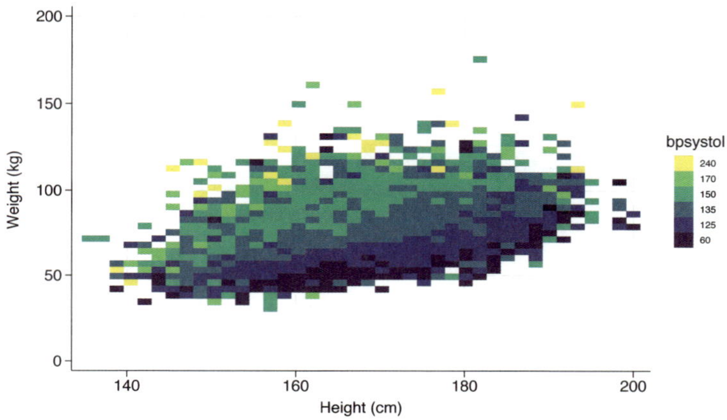

Figure 23.2. Trivariate heat plot with custom color levels

Figure 23.2 presents a more useful version of the previous graph. Bins are now coded by their medical categorical status, and it is easy to identify where certain blood pressure groups are located on the bivariate **weight–height** distribution. It is also easy to identify that critical blood pressures (180–300) are located at the outer edges of the distribution, indicating that extreme **weight–height** values might contribute to severe blood pressures.

23.2.1 Trivariate heat plot with a binary z variable

The z variable in `heatplot` does not need to be continuous. It can also be dichotomous, in which case the color legend will identify the proportion of cases that takes the highest value in each bin. Figure 23.3 below presents an example by using the binary variable $[0,1]$ `female` as the z variable.

Creating figure 23.3

```
. * Trivariate heat plot with two continuous variables and
. * one dichotomous z variable
. heatplot female weight height
```

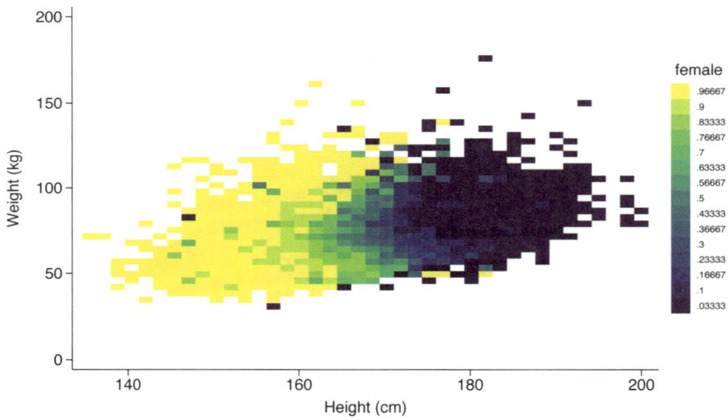

Figure 23.3. Trivariate heat plot with binary z variable

Figure 23.3 replaces the continuous variable `bpsystol` with the binary variable `female`. The z axis now corresponds to the proportion of females in each bin, and it is visually easy to identify two distinct groups in the data. Women are generally of lower height, while men are taller. There is some overlap in the 160–170 cm height region.

23.2.2 Trivariate heat plot scaled color fields

Because the color legend now shows the relevant statistics of the z variable (which can be defined with the `statistic()` option) instead of the underlying frequency values, users may wish to have some visual indication of how many observations are in each bin. This can better help users distinguish which bin areas are important analytically. The option `sizeprop` allows users to scale the size of each bin by relative frequencies. An example is shown in figure 23.4.

Creating figure 23.4

```
. * Trivariate heat plot with scaled color fields
. heatplot bpsystol weight height, cuts(0, 120, 130, 140, 160, 180, 300) ///
> sizeprop
```

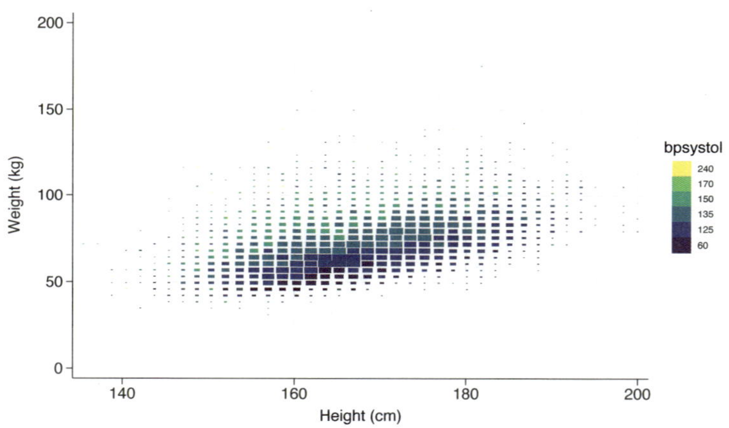

Figure 23.4. Trivariate heat plot with scaled color fields

Figure 23.4 identifies how many relative observations are in each bin in the underlying trivariate distribution. It is now possible to identify that bins outside the central mass have relatively low-frequency populations, while those inside the data mass have high frequencies. Nonetheless, a distinct gradient remains noticeable in the central mass, indicating that blood pressure rises with weight but is somewhat offset by height.

23.2.3 Trivariate heat plot with a categorical x variable

The `heatplot` command is able to integrate discrete variables (with many categories) into its syntax. The options `ydiscrete()` and `xdiscrete()` allow users to treat variables in the core syntax as discrete and omit binning. A value between 0 and 1 can be specified that determines the relative size of the bars presented. Users can choose a value of 0.8 or 0.9 to ensure that individual categories can be easily identified. An example with the binary variable `female` as the x variable is shown below in figure 23.5.

Creating figure 23.5

```
. * Trivariate discrete heat plot
. heatplot weight height female, xdiscrete(0.9)
```

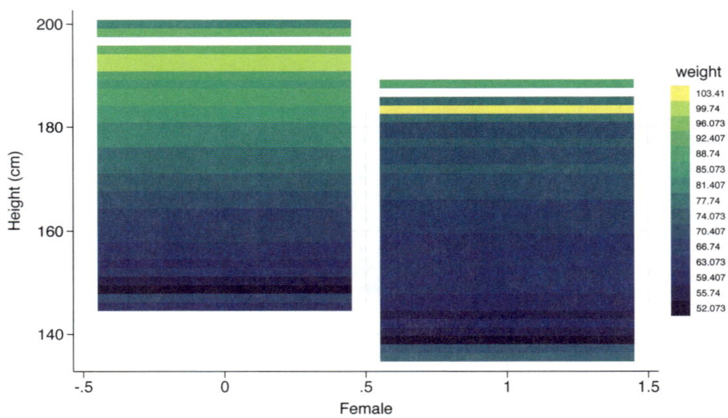

Figure 23.5. Trivariate heat plot with binary x variable

Figure 23.5 presents a heat plot where the x variable is discrete. Height is banded into one-dimensional bins plotted over male and female categories. The individual bin colors represent numerical values of weight. For example, it appears that height is associated with increased weight for both men and women. The advantage of this kind of visualization is that it leads to very fast inference across groups. A scatterplot of `weight–height` by both genders would require users to compare marker trends between two different groups; here this comparison is reduced to an almost qualitative color comparison.

Users are not limited only to simple binary variables. Categorical variables with many categories can be used with the `i.` prefix, which indicates to Stata that variables should be expanded. An example with the variable `agegrp` is shown in figure 23.6. Here the `sizeprop` option is used to indicate relative bin-size frequencies, and the `colors()` option is used to specify an alternative color scheme from the community-contributed command `colorpalette`.

Creating figure 23.6

```
. * Trivariate discrete heat plot with relative frequencies
. heatplot weight height i.agegrp, xdiscrete(0.9) sizeprop colors(twilight)
```

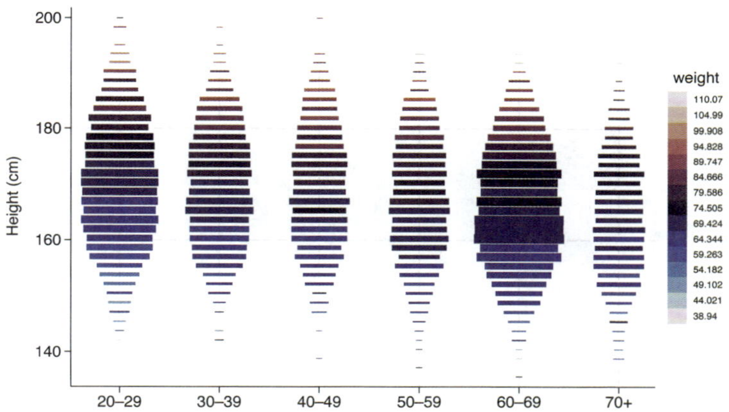

Figure 23.6. Trivariate heat plot with discrete x variable

Figure 23.6 presents a categorical heat map that extends the example from figure 23.5 to multiple categories. It is clear from this graph that weight and height continue to be positively related in each age group, although the sample sizes in each bin vary. There is also evidence that median height is smaller for higher age groups. A key advantage of this visualization is its simplicity and the ability to mix discrete and continuous variables in a trivariate analysis.

23.2.4 Trivariate heat plot with categorical y and x variables

The use of categorical variables is not limited to one per `heatplot` command. The x, y, and z variables can be categorical (although the z variable should be normally be dichotomous because summary statistics are computed over this variable). Using two categorical variables in the x- and y-variable positions will result in a plot that depicts a numeric two-way table via cell colors instead of cell numbers. An example is shown in figure 23.7, which uses the two categorical variables `hlthstat` and `agegrp` as y and x variables, respectively. To avoid showing missing cells, an `if` qualifier is used.

Creating figure 23.7

```
. * Trivariate heat plot with categorical x and y variables
. heatplot weight i.hlthstat i.agegrp if hlthstat != 8, ydiscrete(0.9)   ///
> xdiscrete(0.9)
```

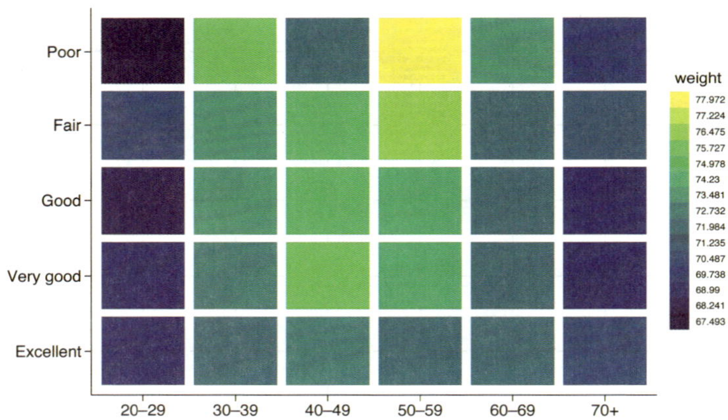

Figure 23.7. Trivariate heat plot with discrete y and x variables

Figure 23.7 presents a matrix-like tabular visualization that contains categorical values in the x and y planes. Each square corresponds to a unique bivariate category, and cell color values indicate the size of the measured z statistic, in this case mean weight. Results suggest that those aged 20–29 are generally of low weight no matter their health status, while those aged 50–59 with poor health have the heaviest average weight.

Similarly to previous graphs, the **sizeprop** option can be used to evaluate individual cells with respect to frequency. This is demonstrated in figure 23.8.

Creating figure 23.8

```
. * Trivariate heat plot with categorical x and y variables
. heatplot weight i.hlthstat i.agegrp if hlthstat != 8, ydiscrete(0.9)  ///
> xdiscrete(0.9) sizeprop
```

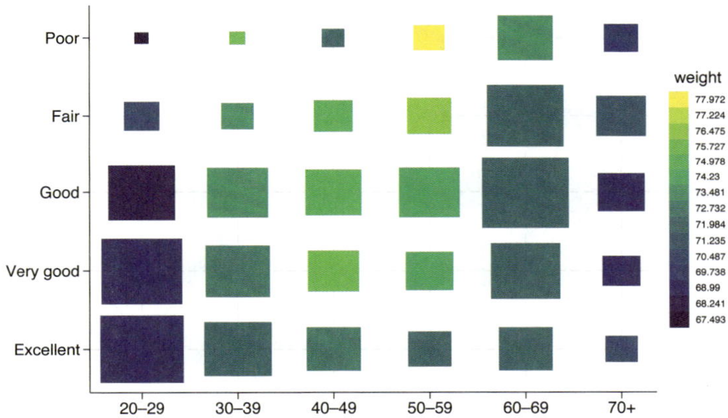

Figure 23.8. Trivariate heat plot with discrete y and x variables and relative bin sizes

Figure 23.8 identifies how many observations underpin each cell. It should not be surprising that few people aged 20–29 have poor health status, while those in the 60–69 age category have the highest share of poor health. Using heat plots in this way offers a novel alternative to standard two-way tables and allows users to convey information on three different variables in a 2D graphic.

24 Bubble plots

Advantages

1. Can provide visualizations of the third dimension without a fake z axis or color variation
2. Are easy to understand for most people
3. Excel when all three variables are continuous

Disadvantages

1. Have less numerically accurate z-values
2. Make it difficult to compare exact areas
3. Are prone to overplotting even with small samples

Required community-contributed package

To install the community-contributed command that is required to follow along with this chapter completely, type

```
. ssc install sepscatter
```

24.1 Introduction

Bubble plots are modified scatterplots that use varying marker sizes to represent a third dimension (variable). They are easy ways to graph a third variable without having to fake a third dimension graphically or resorting to color differentiation. A typical application of bubble plots is when users face three continuous variables that they wish to analyze jointly. The first two variables are scattered normally, while the third, a z variable, inflates the markers accordingly. Bubble plots can also be used when the z variable is discrete, although having only a few categories will lead to poorer plots.

Because bubble plots are based on scatterplots, they are easy to understand for laypeople. In contrast to graphs that use color as the third numerical identifier (for example, heat plots), most end users can intuitively understand marker sizes. Large markers signify large z-variable values; small markers signify small z-variable values.

However, some caution is advised. Marker sizes represent numerical values by their surface areas and not by their diameters. This means that the visual relationship between markers of different sizes is squared and not linear. In other words, markers that have twice the data will not be twice the diameter but twice the surface area. That can make distinguishing numeric differences for large and very large values difficult because markers will look approximately similar. Ultimately, the numerical accuracy of the z variable is compromised in bubble plots. Moreover, different marker shapes can lead to problems of comparison.

Another problem that bubble plots often experience is overplotting. Expanding marker sizes can quickly lead to visually full scatterplots that contain many overlapping markers. Therefore, bubble plots are primarily valuable for smaller datasets, and one should use color transparency and color differentiation when bubble plots become crowded. However, this introduces color shading to the bubble plot, which can present its own problems of color overlap and differentiation.

24.2 A basic bubble plot

A basic case for bubble plots is outlined below with the 1980 Census data by state dataset (`census.dta`), which contains 50 observations, one for each US state. The dataset also contains various demographic characteristics such as death rates (`death`) and median ages (`medage`). A summary and overview of the first 10 observations is shown below.

```
. * Load data
. sysuse census
(1980 Census data by state)
. summarize
```

Variable	Obs	Mean	Std. dev.	Min	Max
state	0				
state2	0				
region	50	2.66	1.061574	1	4
pop	50	4518149	4715038	401851	2.37e+07
poplt5	50	326277.8	331585.1	35998	1708400
pop5_17	50	945951.6	959372.8	91796	4680558
pop18p	50	3245920	3430531	271106	1.73e+07
pop65p	50	509502.8	538932.4	11547	2414250
popurban	50	3328253	4090178	172735	2.16e+07
medage	50	29.54	1.693445	24.2	34.7
death	50	39474.26	41742.35	1604	186428
marriage	50	47701.4	45130.42	4437	210864
divorce	50	23679.44	25094.01	2142	133541

```
. list medage death state pop in 1/10, separator(0)

     +----------------------------------------------+
     | medage     death   state                 pop |
     |----------------------------------------------|
  1. |  29.30    35,305   Alabama         3,893,888 |
  2. |  26.10     1,604   Alaska            401,851 |
  3. |  29.20    21,226   Arizona         2,718,215 |
  4. |  30.60    22,676   Arkansas        2,286,435 |
  5. |  29.90   186,428   California     23,667,902 |
  6. |  28.60    18,925   Colorado        2,889,964 |
  7. |  32.00    26,005   Connecticut     3,107,576 |
  8. |  29.80     5,123   Delaware          594,338 |
  9. |  34.70   104,190   Florida         9,746,324 |
 10. |  28.70    44,230   Georgia         5,463,105 |
     +----------------------------------------------+
```

Any analysis of age and deaths across these geographic areas must be contextualized within the overall size of the state. The variable `pop` measures state population, and any graphical representation of these data will often require users to be able to account for total population size. This is what bubble plots excel in.

Bubble plots are created via the standard `twoway scatter` command. The third variable is entered as a weight variable using Stata's [*weight*] modifier. The [*weight*] modifier is available to many official Stata commands; if weights are specified in the syntax of `twoway scatter`, the size of the scatterplot's markers are scaled according to the size of the specified weight variable. `twoway scatter` can use analytical weights, frequency weights, and probability weights. For more information on the different kinds of weights, users should explore the help file on weights (see `help weight`).

A basic example is shown below in figure 24.1. The variables `medage` and `death` populate the x and y axes, respectively. The third variable `pop` is used as a [*weight*] modifier. In this dataset, `pop` is a frequency variable, so the `[fweight]` modifier is specified.

Creating figure 24.1

```
. * A basic bubble plot
. twoway (scatter medage death [fweight = pop])
```

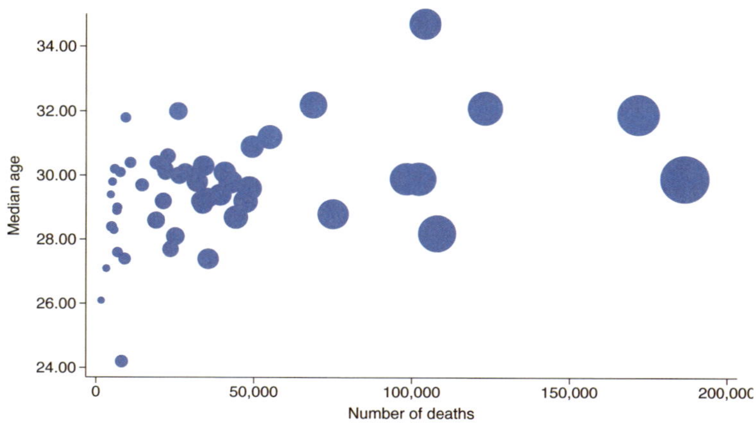

Figure 24.1. A basic bubble plot

Figure 24.1 presents a basic bubble plot. Each marker represents one US state, and the size of the marker is a function of the state's population. Users can now examine how the three variables `medage`, `death`, and `pop` are jointly related. Overall, this plot helps users identify how much importance they should place on each individual marker.

A problem with figure 24.1 is that it contains a lot of overlapping markers. Some individual markers are completely hidden, and others overlap too much to make much sense of. Therefore, users can modify the markers via the options available in the marker options help file (see `help scatter##marker_options`). Core options that customize markers are `mlcolor()`, which defines marker line colors and opacity; `mlwidth()`, which defines marker line widths; and `mfcolor()`, which defines marker fill colors and opacity. Figure 24.1 presents an example where marker outlines are distinct and marker fill colors are transparent. Line thickness is made very thin.

Creating figure 24.2

```
. * Bubble plot formatted
. twoway (scatter medage death [fweight = pop], mlcolor(black) ///
> mlwidth(vthin) mfcolor(ltblue%70))
```

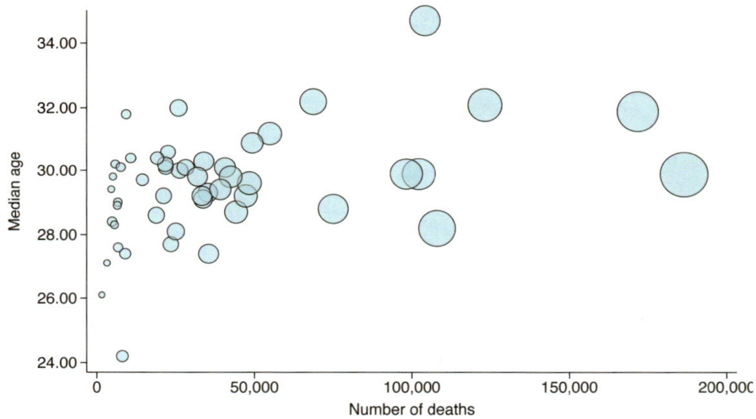

Figure 24.2. Bubble plot with transparent bubbles

Figure 24.2 is a much more manageable plot than figure 24.1. It is now easy to identify individual markers and determine their appropriate sizes. It is obvious from this graph that the number of deaths in a state is strongly related to the overall population size because markers become progressively larger along the x axis.

24.2.1 Bubble plot with marker labels

Bubble plots are often used in smaller datasets because the size increase of markers quickly fills the available chart area, and too many markers overlap into a "giant blob". In many cases, when datasets are small, adding individual marker identifiers to each marker may be helpful so that users can gain a qualitative overview of the plotted area. This can easily be done with the `mlabel()` option, which allows users to specify a variable with marker names (or values).

Users should note that entering the `mlabel()` option directly into the original `twoway scatter` graph nullifies the weight inflation of markers. An example is shown below in figure 24.3.

Creating figure 24.3

```
. * Bubble plot with wrong marker labels
. twoway (scatter medage death [fweight = pop], mlcolor(black)  ///
> mlwidth(vthin) mfcolor(ltblue%70) mlabel(state2))
```

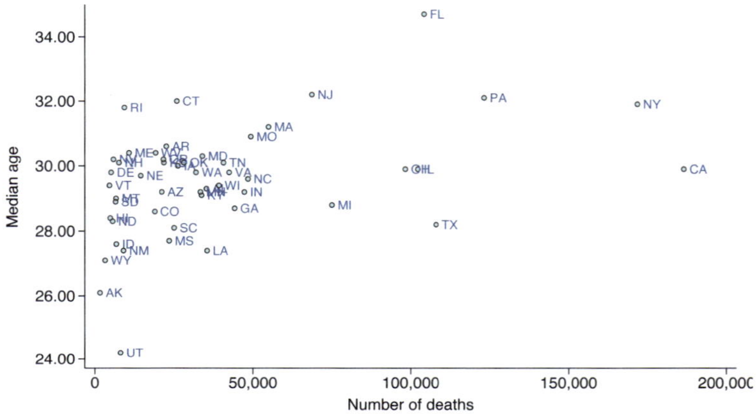

Figure 24.3. Bubble plot with "wrong" markers

Figure 24.3 is not the desired graph. While each marker is now labeled with a state identifier, the population information that was depicted via different marker sizes has now disappeared.

The solution is to insert a second scatter plot via the `twoway` command. The second scatterplot repeats the original plot without weights and uses the `msymbol(i)` option to hide a second set of markers. At the same time, the second `scatter` command uses the `mlabel()` option to add markers. It also invokes additional marker label options to change the marker label size (`mlabsize()`), change marker label color (`mlabcolor()`), and change the position of each marker label to the center of the marker (`mlabposition()`). The result is a second overlay containing no additional markers but all the necessary labeling information. The combined result is figure 24.4.

Creating figure 24.4

```
. * Bubble plot with correct marker labels
. twoway (scatter medage death [fweight = pop], mlcolor(black)              ///
> mlwidth(vthin) mfcolor(ltblue%70)) (scatter medage death, msymbol(i)  ///
> mlabel(state2) mlabsize(*0.7) mlabcolor(black) mlabposition(0))
```

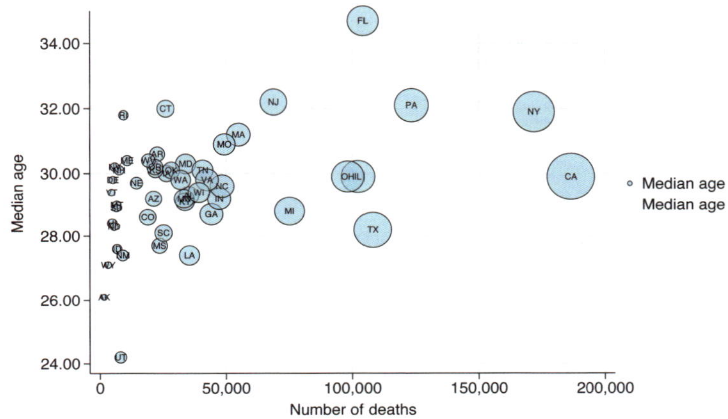

Figure 24.4. Bubble plot with "correct" markers

Figure 24.4 demonstrates how marker labels can be added to bubble plots to allow users to identify individual data markers. California, New York, and Texas appear to be the most populous US states. Of course, users are not limited to inserting marker identifiers; they can also insert numerical information, such as the actual population size, if they wish to do so.

24.2.2 Bubble plot with different symbols

Bubble plots are so named because most graphs depict differently sized markers as circles, hence the analogy to bubbles. But there is no reason why users should be limited only to circles. Other options, such as squares or diamonds, are easily available. Stata can generate a wide variety of symbols in its graphs, and the command `palette symbolpalette` creates a new graph depicting all the available symbols and their associated *symbolstyle*. Figure 24.5 presents an example.

Creating figure 24.5

```
. * Stata marker symbols
. palette symbolpalette
```

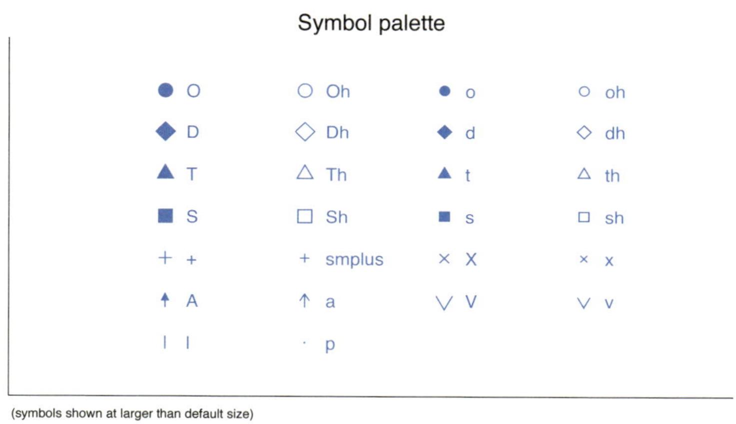

Figure 24.5. Stata's symbol palette

Figure 24.5 presents the available *symbolstyle* choices that official Stata users can access. Any of these symbols can be used to create a bubble plot. The only consideration users should have is whether the desired symbol is more useful than a traditional circle. For example, if users want to compute the difference between marker sizes, then square markers may lead to a more straightforward computational process than round markers.

An example of a bubble plot with diamond markers is shown below in figure 24.6. The `msymbol()` option changes the marker symbols.

Creating figure 24.6

```
. * Bubble plot with diamond markers
. twoway (scatter medage death [fweight = pop], mlcolor(black)  ///
> mlwidth(vthin) mfcolor(ltblue%70) msymbol(D))
```

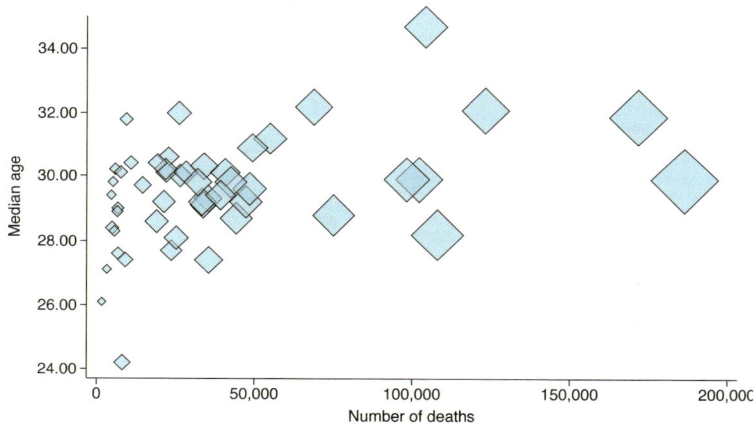

Figure 24.6. Bubble plot with diamond markers

Figure 24.6 presents a modified bubble plot with different markers. Users should think carefully about whether a different marker symbology adds value to a plot because most end users will be familiar with circular markers.

24.2.3 Bubble plot over multiple groups

Differential marker symbols can add great value to a bubble plot in the case of multiple categories. If users wish to visualize how the x, y, and z variables are distributed over a fourth categorical variable, then symbol and color differentiations become useful tools.

The default method to accommodate a fourth categorical variable is via the `if` qualifier, which limits a scatterplot to a particular category. Multiple scatterplots within the `twoway` framework can then be overlaid to build up a graph that contains different symbols and colors. An example is presented in figure 24.7, which overlays four different `scatter` commands, each representing one region by using the `if` qualifier. All contain the same x, y, and z variables. The `msymbol()` and `mfcolor()` options are used to modify individual scatterplot markers. In addition, each marker is made 70% opaque. The `legend()` option is used to place the legend text below the graph in one row via the suboptions `position()` and `row()`. In addition, legend labels are used via the `legend()` suboption `label()` to label the numerous parts of the plot.

Creating figure 24.7

```
. * Bubble plot over multiple groups
. twoway (scatter medage death [fweight = pop] if region == 1,              ///
> mlcolor(black) mlwidth(vthin) msymbol(O) mfcolor(ltblue%70))              ///
> (scatter medage death [fweight = pop] if region == 2,                     ///
> mlcolor(black) mlwidth(vthin) msymbol(D) mfcolor(erose%70))               ///
> (scatter medage death [fweight = pop] if region == 3, mlcolor(black)      ///
> mlwidth(vthin) msymbol(T) mfcolor(eltgreen%70))                           ///
> (scatter medage death [fweight = pop] if region == 4, mlcolor(black)      ///
> mlwidth(vthin) msymbol(S) mfcolor(edkblue%70)                             ///
> legend(position(6) row(1) label(1 "NE") label(2 "N Cntrl")                ///
> label(3 "South") label(4 "West")))
```

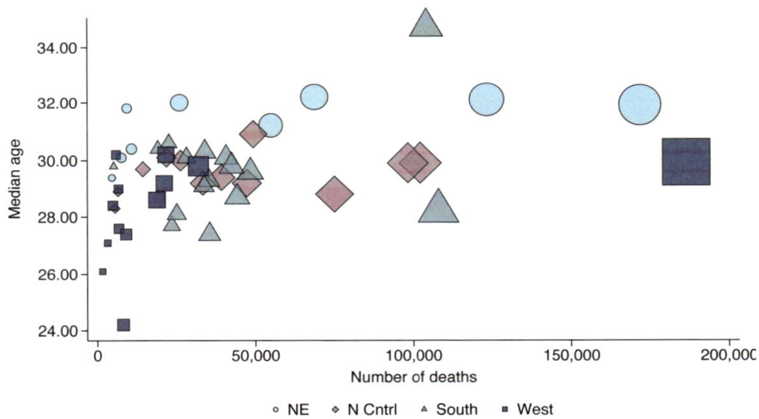

Figure 24.7. Bubble plot over multiple groups

Figure 24.7 demonstrates how different marker symbologies and colors can be used to identify different categories in the data. For example, regions in the West are predominantly small and have few deaths except for California, which is the largest.

Admittedly, the code for figure 24.7 is relatively long. When faced with lengthy graph code, one should try to identify parts of the code that can be repeatedly reused using the copy-and-paste function with only minor modifications. For example, the code `(scatter medage death [fweight = pop] if region == 1, mlcolor(black) mlwidth(vthin) msymbol(O) mfcolor(ltblue%70)))` requires changing only the region number and the `mfcolor()` type within the code block. An even faster option is the community-contributed command `sepscatter` (Cox 2014b), which automatically creates different marker symbols and colors different categories using the `separate()`

option. Like `scatter`, it allows the use of [*weight*] and can thus be easily extended to create bubble plots. An example is shown in figure 24.8.

Creating figure 24.8

```
. * Bubble plot over multiple groups with sepscatter
. sepscatter medage death [fweight = pop], separate(region)
```

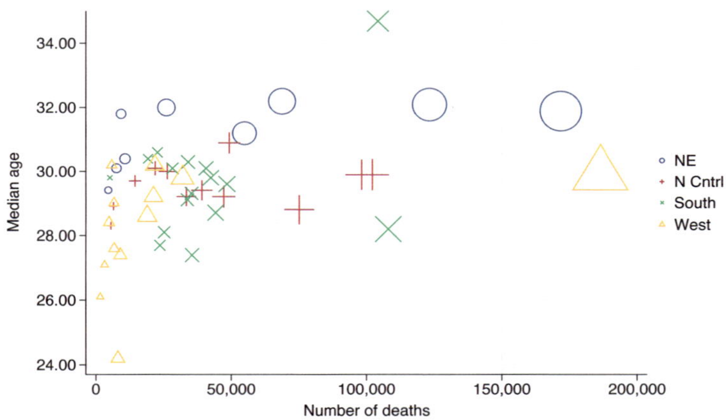

Figure 24.8. Bubble plot over multiple groups with `sepscatter`

Figure 24.8 presents a faster way to achieve marker and color differentiation in a bubble plot.

25 Chernoff faces

Advantages

1. Can visualize numeric data from many variables
2. Are intuitive to examine
3. Are great for qualitative comparisons of observations

Disadvantages

1. Are inferior for accurate statistical information
2. Excel only with small datasets with few observations
3. Do not clarify which shapes, curves, and colors are associated with high or low values
4. Are not true multivariate
5. Can be subjective in interpretation

Required community-contributed package

To install the community-contributed command that is required to follow along with this chapter completely, type

```
. net install https://www.stata-journal.com/software/sj9-3/gr0038
```

25.1 Introduction

Chernoff faces can be used to visualize multivariate data across a small(ish) set of observations. The core idea is that individual observations are represented via human faces and that face characteristics such as hair color, eyebrow size, mouth shapes, and nose shapes all represent different quantities of data. Because humans are naturally trained to identify faces, it can be easier to spot outliers or similar data patterns. When two faces look identical, the underlying numeric data values will be similar. When two faces look different, the underlying numerical values will be different.

Like sparkline plots, Chernoff faces attempt to reduce the complexity of data and present a simple set of visuals that do not rely on axis labeling. Chernoff faces are very qualitative graphs that deemphasize numeric detail. They should not be used for visualization that requires detailed statistical information to be transmitted or analyzed. However, when users search for generic patterns in a set of multivariate data, Chernoff faces excel because similar and dissimilar faces will quickly stand out. Comparison of facial features is much faster than extensive tabular examination.

An important note is that Chernoff faces work on observations, each of which is drawn into a face. This limits the realistic use of Chernoff faces to samples of approximately 50 or less. In addition, because numeric data ranges are presented via shapes, it is not immediately clear which shapes, shading, or colors are associated with high or low values. Chernoff faces specialize in identifying similar or dissimilar observations and should be used in such contexts.

Finally, Chernoff faces are not true multivariate visualizations in the context of examining many variables jointly. They are multivariate in the sense that they can illustrate many variables independently across a range of observations, but they do not allow users to examine joint relationships between three or more variables.

25.2 Basic Chernoff faces

The basic problem that Chernoff faces solve is the following: a user has a small dataset with multiple variables and wishes to identify which observations are similar or dissimilar across a range of variables. An example is shown below with `auto.dta`, which contains multiple variables on car characteristics; however, this dataset is too large at 74 observations, so a random sample of 40% is taken with the `sample` command. The `set seed` command ensures replication.

```
. * Load data
. sysuse auto
(1978 automobile data)
. set seed 1234
. * Sample data
. sample 40
(44 observations deleted)
```

```
. * Examine data
. list make price mpg headroom trunk weight length, separator(0)
```

	make	price	mpg	headroom	trunk	weight	length
1.	Merc. Monarch	4,516	18	3.0	15	3,370	198
2.	Audi 5000	9,690	17	3.0	15	2,830	189
3.	Linc. Versailles	13,466	14	3.5	15	3,830	201
4.	Chev. Monte Carlo	5,104	22	2.0	16	3,220	200
5.	AMC Pacer	4,749	17	3.0	11	3,350	173
6.	Pont. Firebird	4,934	18	1.5	7	3,470	198
7.	VW Dasher	7,140	23	2.5	12	2,160	172
8.	Chev. Monza	3,667	24	2.0	7	2,750	179
9.	Subaru	3,798	35	2.5	11	2,050	164
10.	Ford Mustang	4,187	21	2.0	10	2,650	179
11.	Buick Riviera	10,372	16	3.5	17	3,880	207
12.	Datsun 510	5,079	24	2.5	8	2,280	170
13.	Cad. Eldorado	14,500	14	3.5	16	3,900	204
14.	Chev. Impala	5,705	16	4.0	20	3,690	212
15.	BMW 320i	9,735	25	2.5	12	2,650	177
16.	Buick Electra	7,827	15	4.0	20	4,080	222
17.	Dodge Diplomat	4,010	18	4.0	17	3,600	206
18.	VW Scirocco	6,850	25	2.0	16	1,990	156
19.	Datsun 810	8,129	21	2.5	8	2,750	184
20.	Merc. Bobcat	3,829	22	3.0	9	2,580	169
21.	Chev. Nova	3,955	19	3.5	13	3,430	197
22.	Olds Toronado	10,371	16	3.5	17	4,030	206
23.	Datsun 200	6,229	23	1.5	6	2,370	170
24.	VW Rabbit	4,697	25	3.0	15	1,930	155
25.	Dodge St. Regis	6,342	17	4.5	21	3,740	220
26.	Volvo 260	11,995	17	2.5	14	3,170	193
27.	Datsun 210	4,589	35	2.0	8	2,020	165
28.	Pont. Catalina	5,798	18	4.0	20	3,700	214
29.	Plym. Sapporo	6,486	26	1.5	8	2,520	182
30.	Plym. Volare	4,060	18	5.0	16	3,330	201

Each car in the remaining data has various characteristics, including price, miles per gallon, headroom in inches, trunk size in cubic feet, weight in pounds, and length in inches. Glancing at this table briefly most likely would not reveal any particular or important data structures. One could spend a lot of time examining this table and comparing each cell value, but a better way is to represent these multivariate data quickly via Chernoff faces.

Chernoff faces can be created with the community-contributed command `chernoff` (Raciborski 2009). The command requires that all variables be specified in the options, not in the primary syntax. Each variable must be assigned to a particular facial characteristic. A list of 18 possible options is shown below in table 25.1:

Table 25.1. `chernoff` facial options

Option	Facial feature
`isize()`	Eye size
`iangle()`	Eye angle
`ihor()`	Eye horizontal position
`ivert()`	Eye vertical position
`psize()`	Pupil size
`ppos()`	Pupil position
`bcurv()`	Brow curvature
`bdens()`	Brow density
`bhor()`	Brow horizontal position
`bvert()`	Brow vertical position
`fline()`	Face line
`hupper()`	Hair upper line
`hlower()`	Hair lower line
`hdark()`	Hair darkness
`hslant()`	Hair shading slant
`nose()`	Nose line
`msize()`	Mouth size
`mcurv()`	Mouth curvature

Each relevant variable needs to be placed into one of the above-listed options. Figure 25.1 demonstrates how 6 variables are used to form a facial representation using the facial options. When users select fewer than 18 variables, they should choose primary facial features, such as eye size, brow curvature, face line, hair darkness, and mouth size, because they are easier to identify visually. The `ititle()` option with the variable `make` is used to identify each individual Chernoff face.

Creating figure 25.1

```
. * Chernoff faces
. chernoff, isize(price) psize(mpg) bdens(headroom) fline(trunk)  ///
> hdark(weight) msize(length) ititle(make)
```

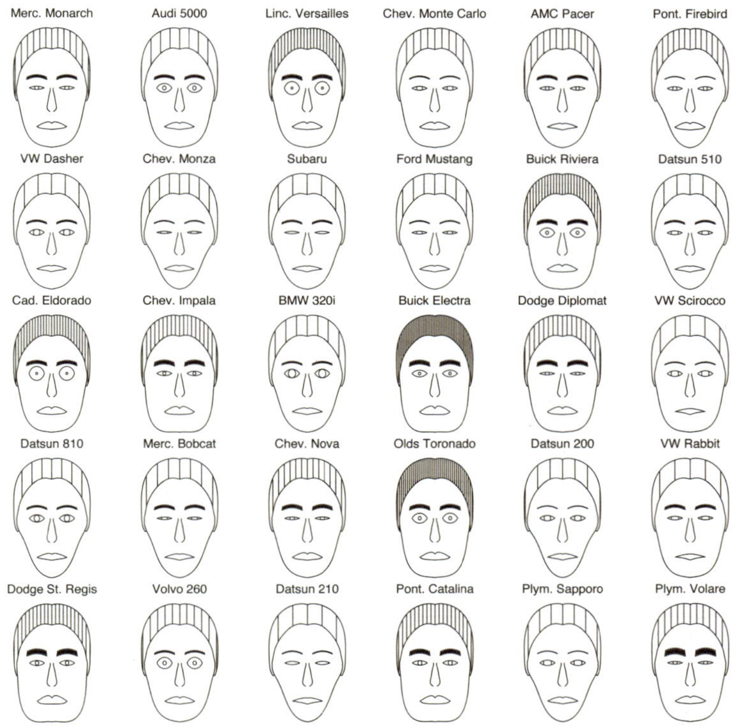

Figure 25.1. Basic Chernoff faces

Figure 25.1 presents a multivariate analysis of `auto.dta` using Chernoff faces. Each facial characteristic in this graph is associated with a different variable. Higher values of one variable will lead to more curvature in the mouth, darker eyebrows, or darker hair. Even without understanding what each facial characteristic represents and what quantities are being measured, it is still possible to make some basic qualitative comparisons across these 30 observations.

For example, it looks like the Buick Electra is similar to the Olds Toronado in terms of car characteristics. In addition, the Ford Mustang appears similar to the Datsun 510, but both are quite different from the previous couple. The ultimate aim of Chernoff faces is to allow users to quickly and intuitively determine which observations are very different and which are similar across a wide range of variables.

To help users evaluate Chernoff faces in more detail, it makes sense to add a legend to the bottom of the graph that identifies which facial characteristics measure which variables. This can be done with the `legend()` option, where a number identifies how many columns the legend should be split over. Figure 25.2 demonstrates.

Creating figure 25.2

```
. * With legend()
. chernoff, isize(price) psize(mpg) bdens(headroom) fline(trunk)  ///
> hdark(weight) msize(length) ititle(make) legend(2)
```

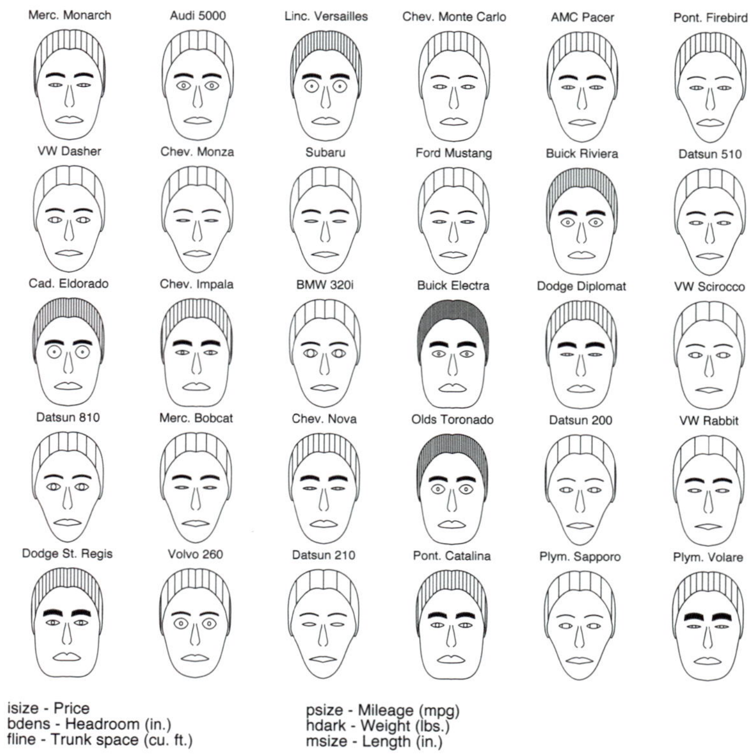

Figure 25.2. Chernoff faces with legend

Figure 25.2 adds important statistical detail to the graph by adding a legend identifying which variables represent which facial characteristic. It is up to end users to determine which variable matters more or less in their analysis. Note that some facial characteristics carry much more pictorial real estate than others (for example, hair shading versus pupil size), which may falsely lead to the impression that some variables matter more than others. This is a general disadvantage of Chernoff faces, and users should take care not to associate the size of facial features with importance.

Creators should also carefully consider that labeling is a function of what they want to convey. Should graphs provide users with more detail so that they can make up their own minds, or should graphs limit their access to information to keep discussion of the data more qualitative?

25.2.1 Chernoff faces with custom ordering

Often, one specific variable dominates the analysis of multivariate data. Specific focus might be on car weight, for example, and how cars of different weights compare with each other. The `chernoff` command includes the option `order()`, which draws faces sorted by a variable list. An example with the variable `weight` is shown in figure 25.3.

Creating figure 25.3

```
. * With custom ordering
. chernoff, isize(price) psize(mpg) bdens(headroom) fline(trunk)    ///
> hdark(weight) msize(length) ititle(make) legend(2) order(weight)
```

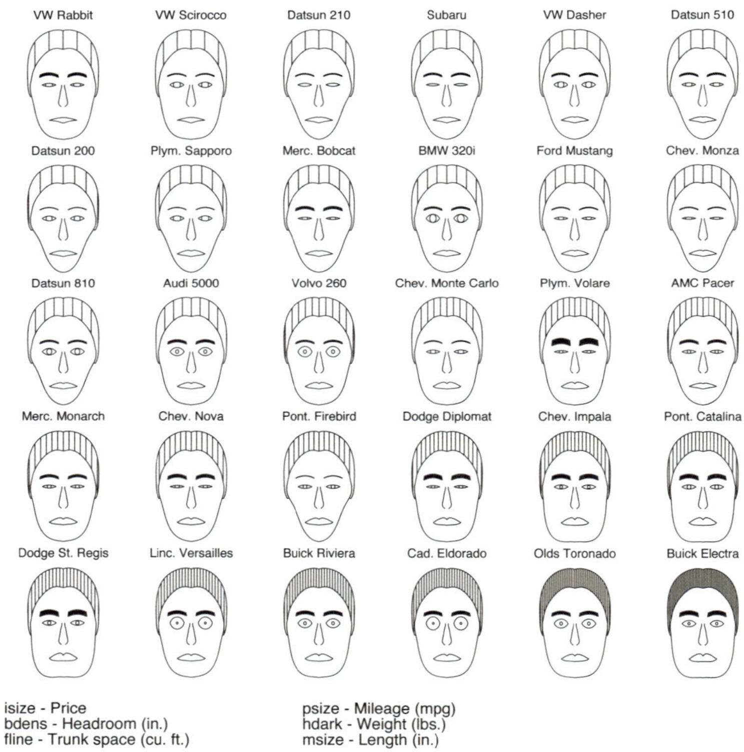

Figure 25.3. Chernoff faces with custom ordering

Figure 25.3 shows how ordering faces can help identify similar or dissimilar observations. Multiple variables can be specified in the `order()` option, allowing users to build custom order profiles that allow for rapid facial comparison based on one or multiple key variables.

25.2.2 Chernoff half faces

Faces are symmetrical, which means they repeat information. When users need to draw many faces on a graph, they may want to consider half faces that retain all statistical information but use only half of the available graph area. This saves space but comes with a loss of visual intuition. The `lhalf` and `rhalf` options can be used to draw left- and right-half faces. One can also use the `hspace()` option to reduce the distance between faces. The minimum value is 0.5, and the maximum is 1.0 (the default is `hspace(0.75)`). An example that draws all 30 faces on two rows is shown below in figure 25.4.

Creating figure 25.4

```
. * With half faces and custom rows
. chernoff, isize(price) psize(mpg) bdens(headroom) fline(trunk)     ///
> hdark(weight) msize(length) ititle(make) legend(2) order(weight)   ///
> rhalf hspace(0.5) rows(2)
```

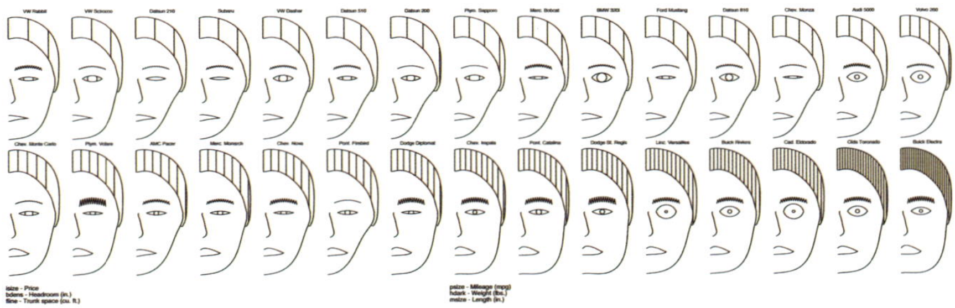

Figure 25.4. Chernoff half faces

Figure 25.4 presents how Chernoff half faces can be used to draw many faces on a single plot. When faced with larger datasets, this can be a useful way to retain Chernoff face analysis, especially when combined with custom ordering of faces. Note that a graph will become much harder to read if the number of faces exceeds 50.

26 Triplots

Advantages

1. Can provide 2D visualizations of trivariate distributions
2. Show relative proportions effectively
3. Are useful for compositional analysis

Disadvantages

1. Are limited to variables that add up to a constant sum (proportion or percent)
2. Feature triangular coordinates that can be hard to identify
3. Are not commonly used and may require explaining

Required community-contributed package

To install the community-contributed command that is required to follow along with this chapter completely, type

```
. ssc install triplot
```

26.1 Introduction

Triplots (also called ternary plots or de Finetti plots) are helpful when users are faced with three variables that add up to a constant sum. Tabulating or graphing three variables jointly can be tricky because tables and graphics must be extended into 3D space. However, in the unique case where three variables add up to a constant sum (such as proportion), the data points are constrained to a triangular subset of the higher dimensional plane. This implies that all data points can be plotted on a 2D graph without loss of information.

Example data that add up to a constant sum might be voting intentions between three political parties, the three-way chemical composition of compounds, or what shares of gross domestic product are determined by industry, services, and agriculture. Importantly, these measures add up to a constant such as 1 (fraction) or 100 (percent). All such data can be plotted on triplots.

The key to understanding triplots is that they do not use Cartesian coordinates $[x, y]$ but barycentric coordinates $[x, y, z]$. Each data point must be interpreted within three axes to understand the composition of that data point. This makes triplots relatively complex to understand, especially for laypeople not used to reading three separate axes in two dimensions.

Triplots excel at depicting trivariate data that add up to a constant sum. Users can immediately identify the general proportions of data points and their locations. Repeating triplots over groups allows rapid group-level analysis of three proportional variables that cannot be achieved with standard scatterplots.

26.2 A basic triplot

The basic problem is outlined below. Data from `nlsw88.dta` are reconfigured to create three proportional categories. In this case, the variable `education` is taken as a use-case example. The original `grade` variable is coded into three `education` categories representing low, medium, and high education levels. The data are then collapsed (`collapse`) over the variable `occupation` so that the resulting dataset contains the proportion of low, medium, and high education levels for each occupation category. The final collapsed data are listed below.

```
. * Load data
. sysuse nlsw88
(NLSW, 1988 extract)

. * Create 3 education groups
. egen education = cut(grade), group(3)
(2 missing values generated)

. tabulate education, generate(education)

  education |      Freq.     Percent        Cum.
------------+-----------------------------------
          0 |        334       14.88       14.88
          1 |      1,119       49.87       64.75
          2 |        791       35.25      100.00
------------+-----------------------------------
      Total |      2,244      100.00

. * Recode occupation
. drop if occupation == .
(9 observations deleted)

. recode occupation 9/12 = 13
(28 changes made to occupation)

. * Collapse data
. collapse (mean) education1 education2 education3, by(occupation)

. * Label education variables
. label variable education1 "Low"

. label variable education2 "Medium"

. label variable education3 "High"
```

```
. * List data
. list, nolabel separator(9)

     +-------------------------------------------+
     | occupa~n   educat~1   educat~2   educat~3 |
     |-------------------------------------------|
  1. |        1    .037855    .242902    .719243 |
  2. |        2     .08365     .51711     .39924 |
  3. |        3    .081267    .668044    .250689 |
  4. |        4    .127451    .568627    .303922 |
  5. |        5    .207547    .603774    .188679 |
  6. |        6     .37551    .546939    .077551 |
  7. |        7    .571429    .392857    .035714 |
  8. |        8    .321678    .562937    .115385 |
  9. |       13    .069767    .093023    .837209 |
     +-------------------------------------------+
```

Illustrating these proportions jointly is the primary problem. While a table displaying all the data will likely suffice for small datasets like this, a basic tabular approach may not work when sample size increases or when multiple subgroups are required for the analysis. Triplots, however, can provide effective visualizations of such data.

To create triplots, one must use the community-contributed command **triplot** (Cox 1998b). The command requires that three variables are specified in the form **triplot** *leftvar rightvar botvar*. An example using the educational variables from **nlsw88.dta** is shown below in figure 26.1. Note that users with Stata 18 or later may need to add the **xlabel(, nogrid)** option to their code to remove the light-gray grid lines that come from the default graph scheme. However, because the default grid lines conveniently fall on important numerical values on the third axis, they can also be retained. The below graph retains the default gridlines, while other graphs in this chapter suppress the gridlines.

Creating figure 26.1

```
. * Basic triplot
. triplot education1 education2 education3
```

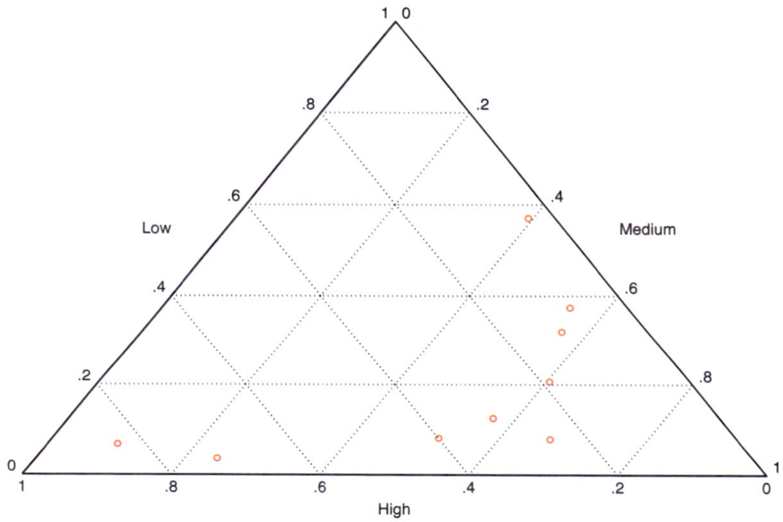

Figure 26.1. A basic triplot

Figure 26.1 presents a triplot of how three educational proportions differ across nine occupations. The data markers represent individual occupations, and the coordinates of each data marker consist of three values $[x, y, z]$. The three edges of the graph represent fractional or proportional values between 0 and 1. There are multiple ways to read the values of a data point in a triplot; the grid method is likely the most intuitive for users. For each data point, the x axis (*leftvar*) is read by following a horizontal line to the edge of the triangle on the left. The y axis (*rightvar*) is read by following a diagonal line upward to the right side of the triangle. The z axis (*botvar*) is read by following the diagonal (not vertical!) line toward the bottom right of the triangle.

As an example, the leftmost data point in this graph has a low value for low education (0.07), a low value for medium education (0.09), and a high value for high education (0.84). To aid interpretation, one may wish to use the `mlabel()` function to label each data point. To create individual coordinates, the `egen` command with the `concat()` function can be used to create $[x, y, z]$ coordinates for each observation. In the example below, the `format()` and `punct()` suboptions are also used to format each coordinate to two decimal places and to separate each value by a comma. In addition, the marker symbols are turned into solid circles using the `msymbol(O)` option, and the marker label color is made black for better identification via the `mlabcolor(black)` option.

```
. * Create coordinates
. egen coord = concat(education1 education2 education3), format(%9.2f)  ///
> punct(,)
```

Creating figure 26.2

```
. * Basic triplot with coordinates
. triplot education1 education2 education3, msymbol(O) mlabcolor(black)  ///
> mlabel(coord) xlabel(, nogrid)
```

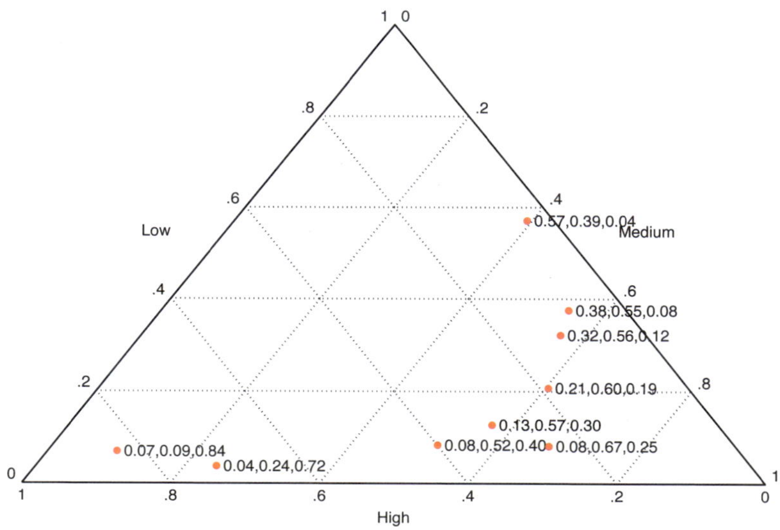

Figure 26.2. Triplot with coordinates

Figure 26.2 repeats the triplot graph, this time with marker coordinates allowing users to identify individual values better. Note that coordinate labels cause some clutter in the graph. Such approaches should be used primarily to identify individual data points or for learning purposes.

26.2.1 Important triplot options

The command `triplot` contains several useful options that can greatly enhance its analytical purpose. The options `msymbol()`, `mlabel()`, `mlabcolor()`, and `mlabsize()` allow users to modify the shape and size of markers and marker labels. For example, instead of plotting coordinates, observation identifiers could be plotted next to markers to signify which occupation belongs to each data point. The `max()` option indicates the upper limit of each variable and the sum of all three variables. The default is `max(1)`, which works for proportional data, but when users are faced with percentage data, they will want to consider using the `max()` option with a value of 100. The `label()` option allows users to specify a list of numeric labels to be shown on each side of the triangle that, in turn, defines the grid reference lines. The `y` option specifies that a "Y" is to be drawn into the graph that divides the triangle into regions in which each variable is greater than the other two. This aids in the rapid discovery of which variable dominates each data point. Figure 26.3 presents an example of all the aforementioned options.

Creating figure 26.3

```
. * Basic triplot with custom markers
. triplot education1 education2 education3, msymbol(O) max(1) y       ///
> label(0(0.1)1) mlabel(occupation) mlabcolor(black) mlabsize(*0.9)   ///
> xlabel(, nogrid)
```

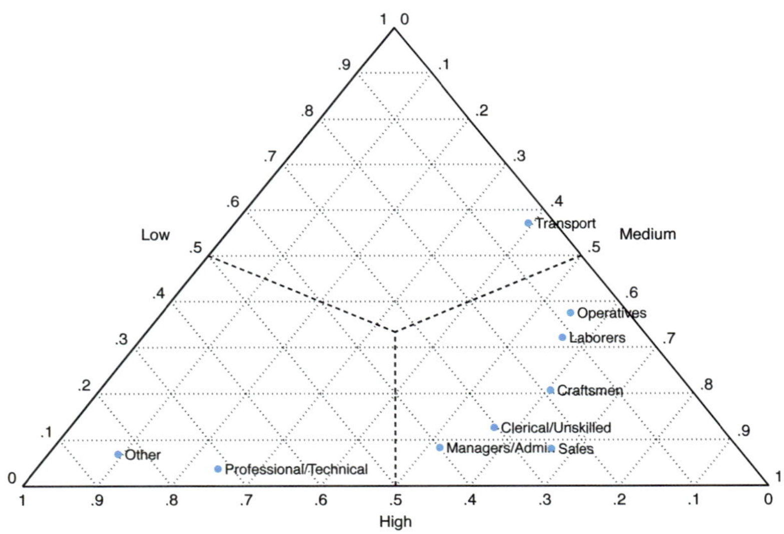

Figure 26.3. Triplot with custom markers

Figure 26.3 displays a triplot that labels individual marker values by their occupational category. Combined with the numerical information of education, it is now easy to identify that "Other" and "Professional/Technical" occupations have the highest share of high-education workers. Moreover, the "Y" lines that divide the triangle into three separate areas quickly allow users to identify which education variable dominates each occupation. The "Transport" occupation is the only one dominated by low-education workers.

26.2.2 Multiple triplots

Users may wish to create multiple triplots. This can be done in two distinct ways. The `separate()` option indicates to divide observations, based on a given categorical variable, into classes. This allows users to quickly identify whether there are group differences in the trivariate distribution of the relevant variables. The code below reloads `nlsw88.dta` and collapses the education groups by `occupation` and the dummy variable `south`, which then separates the resulting triplot.

Note that to ensure both sets of data markers are labeled, two occupational variables are included in the `mlabel()` option. The second variable, `occupation2`, is simply the clone of the first occupation variable. The command `clonevar` retains all labels in the creation of the new variable. Other label options repeat their specifications so that both sets of markers look similar.

```
. * Load data
. sysuse nlsw88, clear
(NLSW, 1988 extract)

. * Create 3 education groups
. egen education = cut(grade), group(3)
(2 missing values generated)

. tabulate education, generate(education)

  education |      Freq.     Percent        Cum.
------------+-----------------------------------
          0 |        334       14.88       14.88
          1 |      1,119       49.87       64.75
          2 |        791       35.25      100.00
------------+-----------------------------------
      Total |      2,244      100.00
. * Recode occupation
. drop if occupation == .
(9 observations deleted)

. recode occupation 9/12 = 13
(28 changes made to occupation)

. * Collapse data
. collapse (mean) education1 education2 education3, by(occupation south)

. * Label education variables
. label variable education1 "Low"

. label variable education2 "Medium"

. label variable education3 "High"

. clonevar occupation2 = occupation
```

```
. * List data
. list occupation south education*, nolabel separator(2)
```

	occupa~n	south	educat~1	educat~2	educat~3
1.	1	0	.025126	.226131	.748744
2.	1	1	.059322	.271186	.669492
3.	2	0	.071856	.526946	.401198
4.	2	1	.104167	.5	.395833
5.	3	0	.07033	.674725	.254945
6.	3	1	.099631	.656827	.243542
7.	4	0	.107692	.553846	.338462
8.	4	1	.162162	.594595	.243243
9.	5	0	.172414	.62069	.206897
10.	5	1	.25	.583333	.166667
11.	6	0	.34188	.564103	.094017
12.	6	1	.40625	.53125	.0625
13.	7	0	.454545	.454545	.090909
14.	7	1	.647059	.352941	0
15.	8	0	.255034	.597315	.147651
16.	8	1	.394161	.525547	.080292
17.	13	0	.037383	.046729	.915888
18.	13	1	.101852	.138889	.759259

Creating figure 26.4

```
. * Basic triplot over multiple groups
. triplot education1 education2 education3, separate(south)             ///
> mlabel(occupation occupation2) msymbol(O) max(1) y label(0(0.1)1)  ///
> mlabcolor(black black) mlabsize(*0.9 *0.9) xlabel(, nogrid)
```

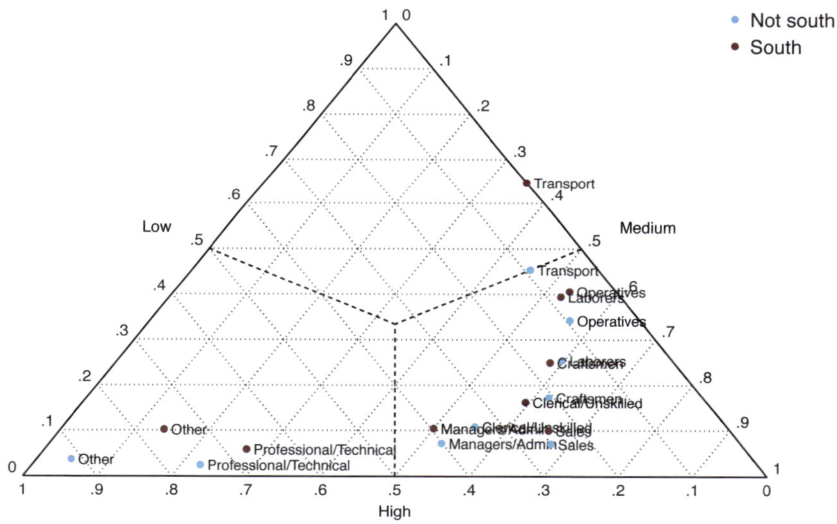

Figure 26.4. Multiple triplots with `separate()`

Figure 26.4 demonstrates how groups can be displayed in triplots. For example, it appears that Southern transport workers have substantially lower education than non-Southern transport workers. Moreover, there are no high-education Southern transport workers, while about 9% of non-Southern transport workers are high education.

Triplots can also be repeated by graph with the `by()` option. Figure 26.5 presents an example where the original data are reloaded and then collapsed over the variables `occupation`, `south`, and `c_city`. Both `south` and `c_city` are dichotomous variables and included in the `by()` option. This results in four different graphs being drawn.

```
. * Load data
. sysuse nlsw88, clear
(NLSW, 1988 extract)

. * Create 3 education groups
. egen education = cut(grade), group(3)
(2 missing values generated)

. tabulate education, generate(education)

  education |      Freq.     Percent        Cum.
------------+-----------------------------------
          0 |        334       14.88       14.88
          1 |      1,119       49.87       64.75
          2 |        791       35.25      100.00
------------+-----------------------------------
      Total |      2,244      100.00
. * Recode occupation
. drop if occupation == .
(9 observations deleted)

. recode occupation 9/12 = 13
(28 changes made to occupation)

. * Collapse data
. collapse (mean) education1 education2 education3, by(occupation south c_city)

. * Label education variables
. label variable education1 "Low"

. label variable education2 "Medium"

. label variable education3 "High"
```

```
. list occupation south c_city education*, nolabel separator(4)

     +-----------------------------------------------------------------+
     | occupa~n   south   c_city   educat~1   educat~2   educat~3 |
     |-----------------------------------------------------------------|
  1. |        1       0        0    .027397    .219178    .753425 |
  2. |        1       0        1    .018868    .245283    .735849 |
  3. |        1       1        0    .089744    .320513    .589744 |
  4. |        1       1        1          0       .175       .825 |
     |-----------------------------------------------------------------|
  5. |        2       0        0    .045802    .564885    .389313 |
  6. |        2       0        1    .166667    .388889    .444444 |
  7. |        2       1        0    .131148     .57377    .295082 |
  8. |        2       1        1    .057143    .371429    .571429 |
     |-----------------------------------------------------------------|
  9. |        3       0        0    .069277    .689759    .240964 |
 10. |        3       0        1    .073171    .634146    .292683 |
 11. |        3       1        0    .076923    .704142    .218935 |
 12. |        3       1        1    .137255    .578431    .284314 |
     |-----------------------------------------------------------------|
 13. |        4       0        0    .065217    .652174    .282609 |
 14. |        4       0        1    .210526    .315789    .473684 |
 15. |        4       1        0    .181818    .681818    .136364 |
 16. |        4       1        1    .133333    .466667         .4 |
     |-----------------------------------------------------------------|
 17. |        5       0        0    .190476    .619048    .190476 |
 18. |        5       0        1       .125       .625        .25 |
 19. |        5       1        0    .263158    .578947    .157895 |
 20. |        5       1        1         .2         .6         .2 |
     |-----------------------------------------------------------------|
 21. |        6       0        0    .316456    .594937    .088608 |
 22. |        6       0        1    .394737         .5    .105263 |
 23. |        6       1        0    .418367    .520408    .061224 |
 24. |        6       1        1    .366667    .566667    .066667 |
     |-----------------------------------------------------------------|
 25. |        7       0        0    .444444    .444444    .111111 |
 26. |        7       0        1         .5         .5          0 |
 27. |        7       1        0    .692308    .307692          0 |
 28. |        7       1        1         .5         .5          0 |
     |-----------------------------------------------------------------|
 29. |        8       0        0    .238938    .628319    .132743 |
 30. |        8       0        1    .305556         .5    .194444 |
 31. |        8       1        0    .376344    .537634    .086022 |
 32. |        8       1        1    .431818         .5    .068182 |
     |-----------------------------------------------------------------|
 33. |       13       0        0    .039474    .052632    .907895 |
 34. |       13       0        1    .032258    .032258    .935484 |
 35. |       13       1        0    .139241    .126582    .734177 |
 36. |       13       1        1          0    .172414    .827586 |
     +-----------------------------------------------------------------+
```

Creating figure 26.5

```
. * Triplots by groups
. triplot education1 education2 education3, by(south c_city)    ///
> mlabel(occupation) msymbol(O) max(1) y label(0(0.1)1)          ///
> mlabcolor(black) mlabsize(*0.6) scale(0.6) xlabel(, nogrid)
```

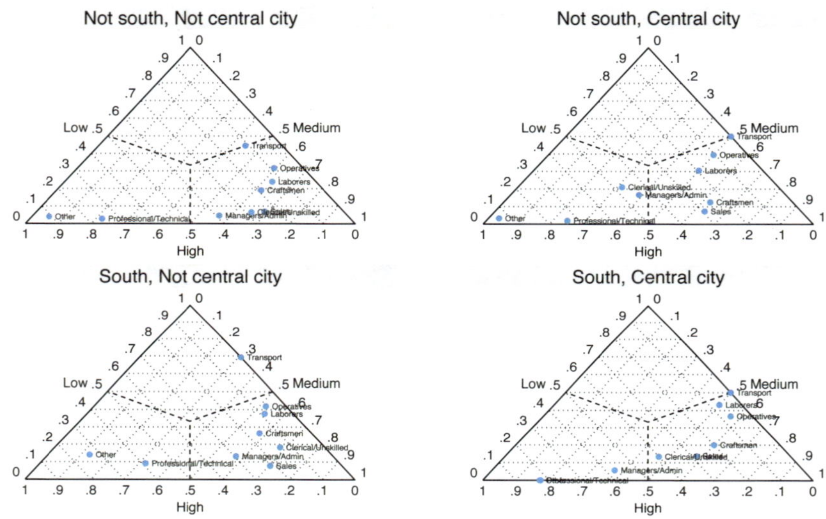

Figure 26.5. Multiple triplots with `by()`

Figure 26.5 exemplifies the `by()` option of `triplot`. The `by()` option allows users to group data over several graphs and examine multiple triplots in one graph. `triplot`, with both the `by()` and the `separate()` options, therefore offers a substantial range of visualization methods for grouped data.

Triplots are unique graphs that apply to very particular data setups. They are unlikely to be used often in practice, but when three variables add up to a constant sum, triplots provide powerful visualizations of trivariate distributions without loss of information and without having to resort to z-axis trickery.

27 3D scatterplots

Advantages

1. Can provide visualizations of the trivariate relationship of variables
2. Are good for investigating nonlinearities in small datasets
3. Are intuitive to understand

Disadvantages

1. Require rotation for full information
2. Lose precise values because users cannot easily trace markers to axes

Required community-contributed packages

To install the community-contributed commands that are required to follow along with this chapter completely, type

```
. ssc install scat3
. ssc install graph3d
. ssc install surface
```

27.1 Introduction

3D plots are natural extensions to 2D plots often seen in data visualization. Living in a 3D world means we intuitively understand three different axes. Thus, it makes sense to attempt to include a third dimension when graphing data to present complex relationships better and understand how multiple variables are distributed and related to each other. Unfortunately, the 2D nature of paper or the computer screen means that 3D plots will always be pseudo-3D plots, where a third dimension is "faked" into the graph. This can lead to problems of clear identification and measurement of values. 3D plots should be used sparingly and only in certain circumstances.

A third dimension is often frowned upon in data visualization, especially when it adds no extra value, such as in 3D pie charts or bar charts, where the third dimension is simply an artistic feature. However, in 3D scatterplots, a third dimension can add extra value because it allows users to incorporate an additional variable into their analysis. This results in the ability to examine trivariate relationships. The key to using such graphs is the ability to rotate the 3D plot, because only rotation will allow users to visualize all angles of the scatterplot and thereby identify important data features.

Unfortunately, Stata does not include an official command for creating 3D plots, so users will rely on community-contributed commands to generate such graphs.

Note that 3D scatterplots should be limited to smaller datasets because large samples will quickly overwhelm the plot. Some leeway is available via smaller or transparent markers, but 3D scatterplots are generally more prone to overplotting than 2D scatterplots.

27.2 A basic 3D scatterplot

The basic problem is that a user wishes to examine how three variables are jointly (trivariate) correlated. Using `auto.dta`, a user may wish to focus on how the variables `price`, `mpg`, and `weight` are all related to each other. In particular, as weight increases, car mileage is expected to decrease. The relationships between weight and price and between mileage and price, however, are more uncertain. A summary of the dataset is shown below.

```
. * Load data
. sysuse auto
(1978 automobile data)

. summarize

    Variable |        Obs        Mean    Std. dev.       Min        Max
-------------+---------------------------------------------------------
        make |          0
       price |         74    6165.257    2949.496       3291      15906
         mpg |         74     21.2973    5.785503         12         41
       rep78 |         69    3.405797    .9899323          1          5
    headroom |         74    2.993243    .8459948        1.5          5
-------------+---------------------------------------------------------
       trunk |         74    13.75676    4.277404          5         23
      weight |         74    3019.459    777.1936       1760       4840
      length |         74    187.9324    22.26634        142        233
        turn |         74    39.64865    4.399354         31         51
displacement |         74    197.2973    91.83722         79        425
-------------+---------------------------------------------------------
  gear_ratio |         74    3.014865    .4562871       2.19       3.89
     foreign |         74    .2972973    .4601885          0          1
```

While some graphs, such as contour plots or heat plots, use color to illustrate a third dimension, 3D plots allow users to visualize the exact coordinates of each data point. In other words, each value of `price`, `mpg`, and `weight` for an observation can be displayed in its raw form, and no data-reduction process is needed.

The community-contributed command `scat3` (Cox 2004) allows Stata users to create basic 3D scatterplots. It requires that three variables are specified in the order `scat3` *xvar yvar zvar*. By default, a 3D scatterplot features data markers, spikes extending from data markers toward the x–y plane, and x, y, and z axes, including titles. The rotation and elevation angle of the plot are set to 45 degrees. Figure 27.1 demonstrates a basic use case for the continuous variables `price`, `mpg`, and `weight`.

Creating figure 27.1

```
. * A basic 3D scatterplot
. scat3 price mpg weight
```

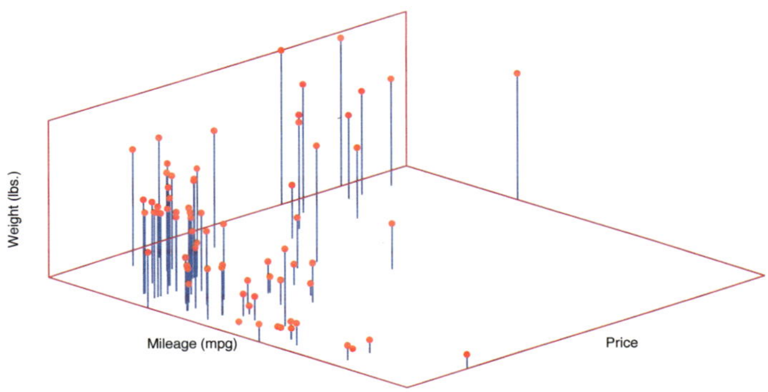

Figure 27.1. A basic 3D scatterplot

Figure 27.1 presents a basic 3D scatterplot in Stata. The plot presents markers that symbolize data values in three dimensions, allowing users to infer the kind of relationship that the three variables have with each other. In this case, it appears that the relationship between weight and mileage is negative, the relationship between weight and price is somewhat positive, and the relationship between price and mileage is negative.

Three important points must be considered when viewing graphs from `scat3`. The first point is that `scat3` autoscales the axes (and thus variables) so that the resulting graph presents a reasonable cuboid. Had `scat3` not scaled the axes, then the mileage axis would be incredibly small because `weight` and `price` are measured in thousands but `mpg` in tens. In other words, the true underlying graph would appear more like a 2D slice. This autoscaling makes `scat3` particularly useful to investigate basic relationships and nonlinearities between three variables. However, it also implies that the axes cannot be labeled properly, because they do not resemble their true values. It is not possible to add value labels to axes from `scat3`.

The second point is that the y and z planes are different from normal graphs, where y is the vertical and z the depth plane. As seen above, `price` is on the traditional z plane but is specified as a y variable. `weight` is on the traditional y plane but is specified as a z variable. Users must be aware of how `scat3` operationalizes their x, y, and z specifications into the resulting graph.

27.2.1 Rotation

The third point is the rotation of the 3D scatterplot. The default aspect of **scat3** is generally functional, but any applied use of a 3D graph will ultimately require rotation. Unfortunately, Stata cannot rotate graphs on the fly via the Data Editor; this must be done via the graph options and multiple graphs. In **scat3**, the options **rotate()** and **elevate()** specify the rotation angle and elevation angle in degrees, respectively. Users should experiment with different angles; doing so can greatly impact the structure of the data revealed. Figure 27.2 below displays four graphs that are rotated, stored, and then merged into a combined graph using the **graph combine** command.

Creating figure 27.2

```
. * Rotating 3D scatterplots
. scat3 price mpg weight, rotate(30)  elevate(45) name(g1)
. scat3 price mpg weight, rotate(120) elevate(45) name(g2)
. scat3 price mpg weight, rotate(210) elevate(45) name(g3)
. scat3 price mpg weight, rotate(300) elevate(45) name(g4)
. graph combine g1 g2 g3 g4
```

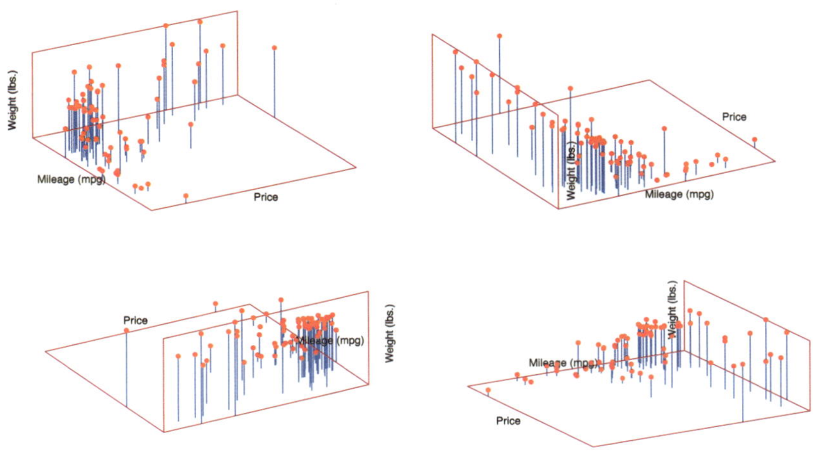

Figure 27.2. Rotated 3D scatterplots

Figure 27.2 presents four different 3D scatterplots, each rotated by 90 degrees. All aspects of the data are now well presented, and users can identify parts of the plot they are most interested in. Including at least two rotated versions of a 3D scatterplot is recommended when presenting these graphs to end users; however, four is best.

27.2.2 Modifying markers

`scat3` can call upon various `twoway` options to modify marker labels. This allows users to customize their 3D scatterplots better. Figure 27.3 presents an example with diamond markers via the `msymbol()` option and olive-colored markers that are transparent via the `mcolor()` option. The `spikes()` option allows users to change the representation of the spikes by allowing users to call options from the command `twoway spike`. In this case, the line pattern is dashed via the `lpattern()` option, and the line width is 0.5 of the default width via the `lwidth()` option.

Creating figure 27.3

```
. * Customized 3D scatterplot
. scat3 price mpg weight, msymbol(D) mcolor(olive%50)     ///
> spikes(lpattern(dash) lwidth(*.5))
```

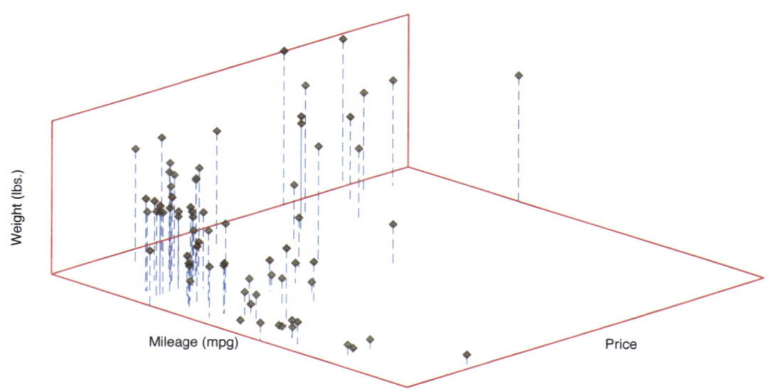

Figure 27.3. Custom 3D scatterplot

Figure 27.3 presents a customized 3D scatterplot. The use of transparent markers makes it easier to identify individual observations in dense clusters of data, while the dashed lines to the x–y plane now attract less visual attention but continue to signify height on the z plane.

27.2.3 3D scatterplots over groups

A useful option of `scat3` is `separate()`, which divides observations of a categorical variable into classes. The variable `foreign` is a dichotomous variable in `auto.dta`. An example of markers being divided over these two groups is shown below in figure 27.4.

Creating figure 27.4

```
. * 3D scatterplot over groups
. scat3 price mpg weight, separate(foreign) spikes(lcolor(gs10))
```

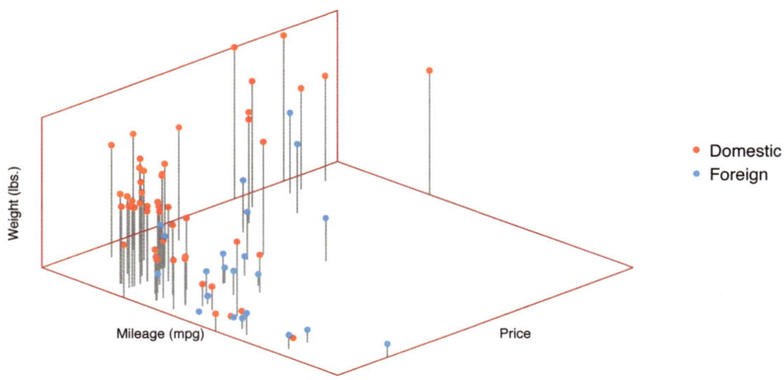

Figure 27.4. 3D scatterplot over group variable

Figure 27.4 demonstrates how `scat3` can be used to plot three variables over a fourth categorical variable. This gives users additional flexibility to examine data and identify differences in the trivariate relationship over different groups.

27.2.4 A note on large datasets

3D plots are generally not suited for large datasets. Plotting three coordinates in a 2D graph space is too complex to accommodate many observations or different groups. Figure 27.5 below demonstrates such an example using `nlsw88.dta`, which contains 2,246 observations on middle-aged female workers in the United States from 1988. A user may be interested in how wages (`wages`), working hours (`hours`), and ages (`age`) are trivariately related. To ensure that the graph is not immediately overplotted, marker sizes are reduced using the `msize()` option. Two `msize()` values are specified because the example below also uses the `separate()` option to divide the plot over the two groups of `collgrad`, an indicator of higher education.

```
. * Load data
. sysuse nlsw88, clear
(NLSW, 1988 extract)
. summarize wage hours age
```

Variable	Obs	Mean	Std. dev.	Min	Max
wage	2,246	7.766949	5.755523	1.004952	40.74659
hours	2,242	37.21811	10.50914	1	80
age	2,246	39.15316	3.060002	34	46

Creating figure 27.5

```
. * 3D scatterplot with many observations
scat3 wage hours age, separate(collgrad) msize(*0.2 *0.2) ///
> spikes(lpattern(dash) lwidth(*.5 *.5) lcolor(gs10 gs10))
```

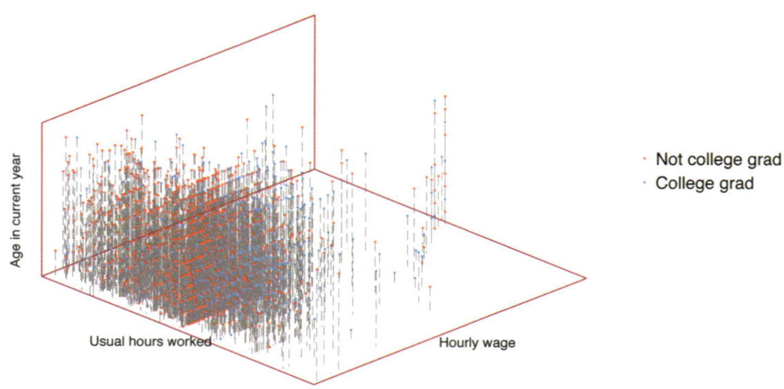

Figure 27.5. An overplotted 3D scatterplot

Figure 27.5 demonstrates that even with careful marker size adjustment, it is difficult to identify what is happening in this plot. There is a large mass of data, and no clear relationship is identifiable. The only takeaway is that there are few high-wage observations, which can easily be identified by examining the wage distribution via univariate methods. Users should avoid such plots where possible.

27.3 3D scatterplot without autoscaling

A primary feature of the `scat3` command is its autoscaling feature, which presents scaled cuboids to make relationship identification easier. Users who wish to avoid this feature or present the data in their raw forms can make use of the community-contributed command `graph3d` (Rostam-Afschar and Robin 2014). This command works similarly to `scat3` in that it requires three variables—x, y, and z—to be specified. A basic example is shown below in figure 27.6. To avoid overplotting, `auto.dta` is reloaded.

```
. * Load data
. sysuse auto, clear
(1978 automobile data)
```

Creating figure 27.6

```
. * graph3d
. graph3d price mpg weight
```

Figure 27.6. 3D scatterplot without autoscaling

Figure 27.6 shows a basic use case of `graph3d` without any options. This default graph clearly has little applied use. The options `cuboid` and `innergrid` are minimal options that users should specify to better visualize the 3D setting of this plot. `cuboid` will draw the edges of a cuboid, allowing users to identify the 3D box shape in which the scatterplot sits. innergrid draws additional grid lines in the cuboid. Figure 27.7 presents these options.

Creating figure 27.7

```
. * graph3d with grid
. graph3d price mpg weight, cuboid innergrid
```

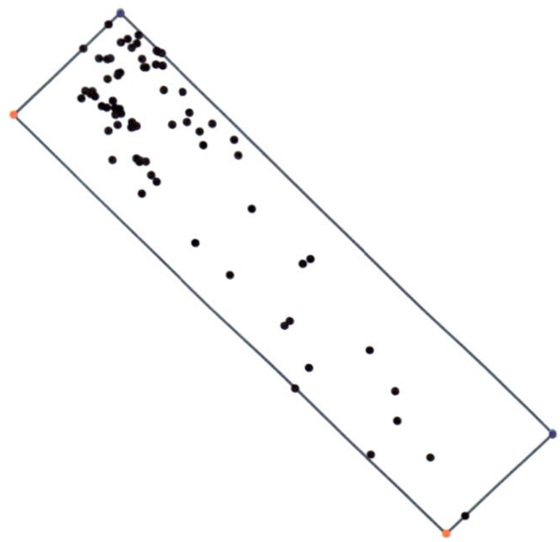

Figure 27.7. 3D scatterplot without autoscaling with grid

Figure 27.7 looks marginally better than figure 27.6, but it still appears to be 2D; there remains little evidence of a 3D plot. This is not due to poorly specified options but rather due to the scaling difference between the three variables. The variable mpg is measured from 12 to 41, while the variables price and weight are measured in thousands. This difference results in an extremely small y axis because no automatic scaling is applied. This, in turn, results in a 3D plot that looks like a standard 2D plot.

The solution is to manually rescale the mpg variable, for example, by multiplying its values by 100.

```
. * Rescaling mpg
. summarize price mpg weight
    Variable |        Obs        Mean    Std. dev.       Min        Max
-------------+---------------------------------------------------------
       price |         74    6165.257    2949.496       3291      15906
         mpg |         74     21.2973    5.785503         12         41
      weight |         74    3019.459    777.1936       1760       4840
. replace mpg = mpg*100
(74 real changes made)
```

```
. summarize price mpg weight

    Variable |        Obs        Mean    Std. dev.       Min        Max
-------------+---------------------------------------------------------
       price |         74    6165.257    2949.496       3291      15906
         mpg |         74     2129.73    578.5503       1200       4100
      weight |         74    3019.459    777.1936       1760       4840
```

After the data transformation, the same code results in the following plot:

Creating figure 27.8

```
. * graph3d with grid and rescaled mpg
. graph3d price mpg weight, cuboid innergrid
```

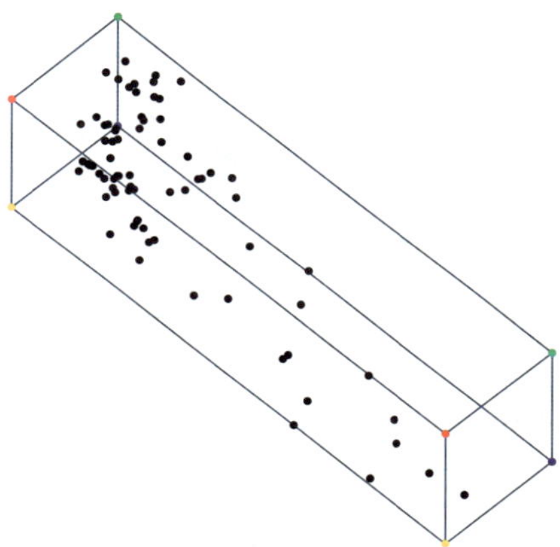

Figure 27.8. 3D scatterplot with custom scaling

Figures 27.7 and 27.8 show how important scaling is in creating 3D graphs. The rescaled `mpg` variable made a big difference in being able to identify and visualize the relationship between all three variables.

`graph3d` has many options that users can specify to customize their plots. Users should read the associated help file carefully (see `help graph3d`) because users often must significantly modify the original plot. An example is shown below in figure 27.9. Here the `xangle()`, `yangle()`, and `zangle()` options are used to specify the rotation angle about different axes. The `xlabel()`, `ylabel()`, and `zlabel()` options label the respective axes. The `xlangle()`, `ylangle()`, and `zlangle()` options modify the angle of the axis labels. This allows for easier identification of the axes. The `markeroptions()` option allows users to call upon the `twoway` marker options (see `help marker_options`). The `xlpos()`, `ylpos()`, and `zlpos()` options specify the compass direction of the labels.

Finally, the `coordinates()` option specifies that the coordinates of the cuboid vertices are shown. Up to eight can be selected.

Creating figure 27.9

```
. * graph3d with angle, label, and marker options
. graph3d price mpg weight, cuboid innergrid xangle(30) yangle(45)       ///
> zangle(0) xlabel(price) ylabel(mpg100) zlabel(weight) xlangle(-40)     ///
> ylangle(40) zlangle(90) markeroptions(mcolor(%50)) xlpos(10) ylpos(11) ///
> zlpos(6) coordinates(2 5)
```

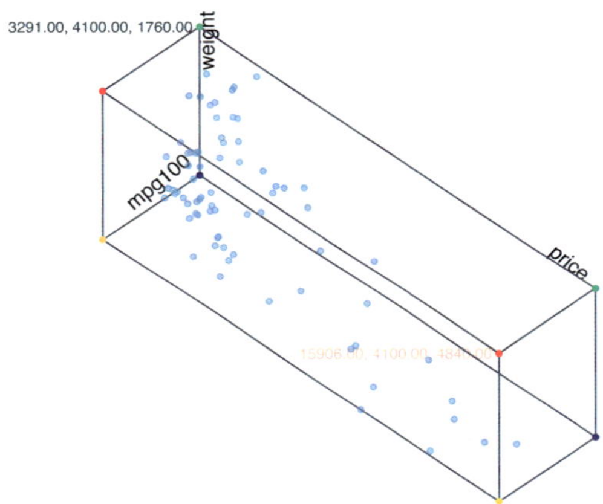

Figure 27.9. 3D scatterplot with custom scaling and markers

Figure 27.9 presents an improved 3D scatterplot using the command **graph3d**, but it required the use of many options. In general, **scat3** is a more user-friendly command; therefore, in exploratory data analysis, users are often better served with **scat3**.

One advantage of **graph3d** over **scat3** is its ability to include a fourth variable in the variable list. This fourth variable is then colored to represent different values. This allows users to create a pseudo-4D multivariate scatterplot in three dimensions. The fourth variable can be a binary, categorical, or continuous variable. Figure 27.10 demonstrates a simple use case with the binary variable `foreign`. The `colorscheme()` option must be specified. In this case, the `cr` scheme (cyan and red) is used because the variable `foreign` contains only two values.

Creating figure 27.10

```
. * graph3d over fourth binary variable
. graph3d price mpg weight foreign, cuboid innergrid colorscheme(cr)
```

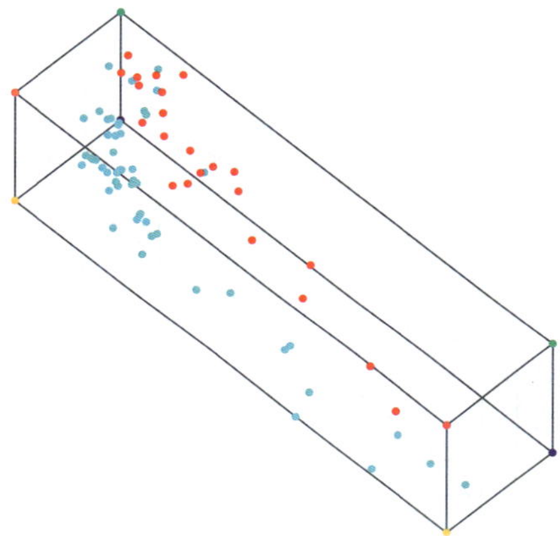

Figure 27.10. Pseudo-4D scatterplot (binary)

Figure 27.10 demonstrates how the `graph3d` command can be used to visualize binary groups in a 3D scatterplot. This is very similar to `scat3`'s ability to plot multiple groups via its `separate()` option. `graph3d` can also specify a continuous fourth variable. Users must simply insert the continuous variable as the fourth variable in the command syntax and specify the gray color scheme using the `colorscheme(fadetogrey)` option. An example is shown below in figure 27.11.

Creating figure 27.11

```
. * graph3d over fourth continuous variable
. graph3d price mpg weight length, cuboid innergrid colorscheme(fadetogrey)
```

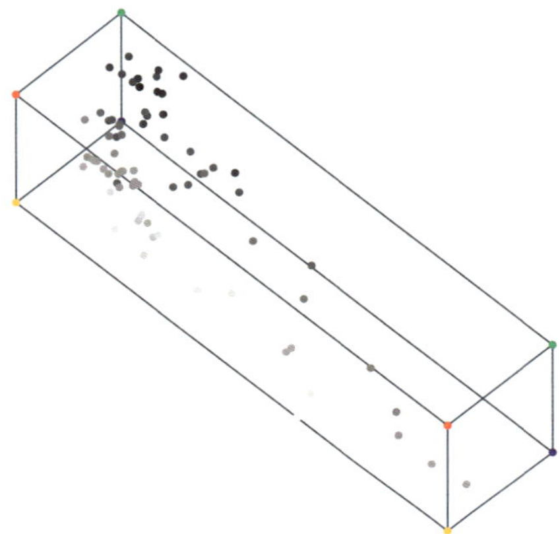

Figure 27.11. Pseudo-4D scatterplot (continuous)

Figure 27.11 is an example of a 3D scatterplot where a fourth continuous variable is represented via shades of gray. There is evidence in this plot that higher weight values are associated with higher length values, but additional rotation of this plot is required to be certain.

27.4 3D surface plots

Users can create 3D wireframe surface plots with the community-contributed command `surface` (Mander 2005b). Three variables—x, y, and z—must be specified. Similarly to a line plot, this visualization aims to connect 3D markers via 3D lines. However, in many circumstances, using this plot on unsorted data will lead to a "spaghettification" of the visualization because data points are connected seemingly randomly. Figure 27.12 below demonstrates this using `auto.dta` with the previously used variables `price`, `mpg`, and `weight`.

```
. * Load data
. sysuse auto, clear
(1978 automobile data)
```

Creating figure 27.12

```
. * 3D surface plot
. surface price mpg weight
```

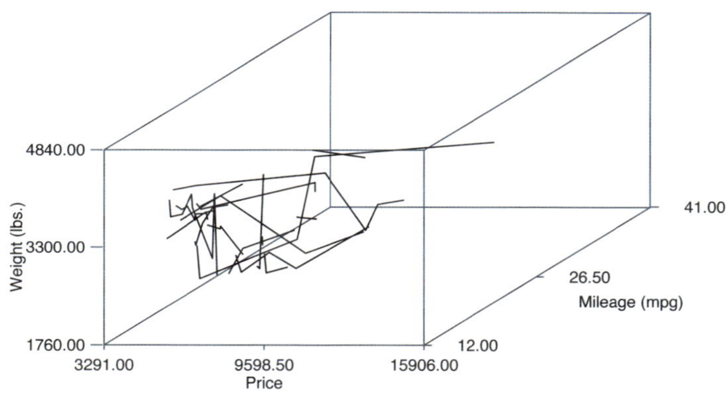

Figure 27.12. 3D surface plot

Figure 27.12 clearly indicates that applying a 3D wireframe surface plot to unorganized data results in a graph that has little use. In most cases where data are clustered randomly in a 3D space, 3D surface plots offer little value, but they excel at illustrating smooth and clean relationships between three different variables. A particularly good use case is graphing interaction terms in applied regression analysis. Interaction terms among continuous variables can be problematic to understand fully. The example below regresses car prices on car mileage, car weight, and their interaction. At first glance, the negative interaction coefficient seems at odds with the positive mileage and weight coefficients.

```
. * Parametric model with predictions
. regress price c.mpg##c.weight
      Source |       SS           df       MS      Number of obs   =        74
-------------+----------------------------------   F(3, 70)        =     13.11
       Model |  228430463         3  76143487.7   Prob > F        =    0.0000
    Residual |  406634933        70  5809070.47   R-squared       =    0.3597
-------------+----------------------------------   Adj R-squared   =    0.3323
       Total |  635065396        73  8699525.97   Root MSE        =    2410.2

------------------------------------------------------------------------------
       price | Coefficient  Std. err.      t    P>|t|     [95% conf. interval]
-------------+----------------------------------------------------------------
         mpg |   396.7844   185.2023     2.14   0.036     27.41003    766.1587
      weight |   5.067008   1.378057     3.68   0.000      2.31856    7.815455
             |
      c.mpg#|
    c.weight |  -.1916795   .0711936    -2.69   0.009    -.3336706   -.0496885
             |
       _cons |  -5944.881   4525.706    -1.31   0.193    -14971.12    3081.356
------------------------------------------------------------------------------

. quietly margins, at(mpg=(12(1)41) weight=(1760(100)4840))  ///
> saving(predictions, replace)
. use predictions, clear
(Created by command margins; also see char list)
. summarize
    Variable |        Obs        Mean    Std. dev.       Min        Max
-------------+---------------------------------------------------------
      _deriv |          0
       _term |        930           1           0          1          1
    _predict |          0
         _at |        930       465.5    268.6122          1        930
      _atopt |        930           1           0          1          1
-------------+---------------------------------------------------------
     _margin |        930    4529.154    2471.089  -2965.945   11986.75
  _se_margin |        930    1313.682      947.39   341.1097   4896.445
  _statistic |        930    6.225093    5.038142  -.6057344   20.43537
     _pvalue |        930    .1227156    .2588751   3.21e-31   .9980601
      _ci_lb |        930    1909.098    4125.765   -12731.6   9774.726
-------------+---------------------------------------------------------
      _ci_ub |        930    7149.211      1526.8   5317.936   14198.78
        _at1 |        930        26.5    8.660099         12         41
        _at2 |        930        3260    894.9085       1760       4760
```

A 3D surface plot can help. The command **margins** is used to make car price predictions from the regression model for incremental values of mileage and weight. The **quietly** prefix command suppresses a long stream of output, and the predictions of price at different values of **mpg** and **weight** are stored using the **saving()** option. The saved predictions are then loaded, at which point the **surface** command can be called. The variables **_margin**, **_at1**, and **_at2**, which represent the price predictions, values of **mpg**, and values of **weight**, respectively, are specified in the command as x, y, and z variables, respectively. The result is shown in figure 27.13.

Creating figure 27.13

```
. * 3D surface plot
. surface _margin _at1 _at2
```

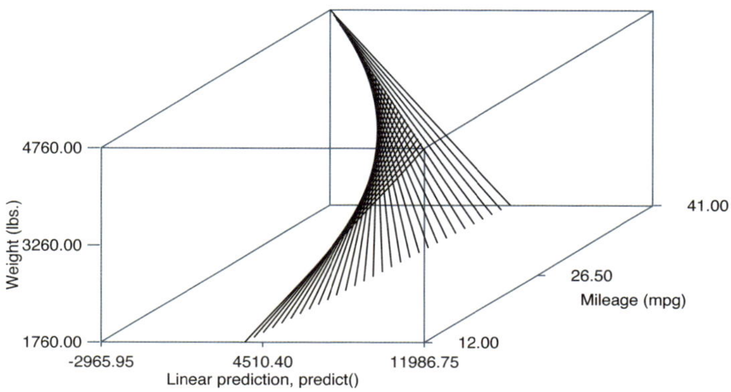

Figure 27.13. 3D surface plot from predicted parametric model

Figure 27.13 demonstrates how 3D wireframe surface plots could be used. The plot shows that the relationship between car weight and car price is positive when car mileage values are low but then inverts into a negative relationship when mileage values are high. This is a clear and visual interpretation of the three coefficients from the underlying regression model. Moreover, it would be relatively simple to extend this model to include nonlinear functional forms.

3D plots should be used carefully. There will often be better ways to graph data, but in unique circumstances, they offer unmatched visual and analytical abilities.

Part IV

Special graphs

Not all graphs can be easily placed in univariate, bivariate, and multivariate hierarchies. This part of the book focuses on special graphs and techniques to give users access to a broader set of capabilities that is not purely focused on data relationships. Animated graphs, for example, offer amazing opportunities to add an additional layer of data that moves and shifts over time. While such graphs have little use in textbooks, the modern medium of data communication has moved far beyond pen and paper. There are many ways in which moving graphs can be used, for example, on websites, social media feeds, asynchronous learning videos, or building projections. Alternatively, the core trade tool for many data analysts is often regression analysis. Why always present such results in overplotted and dense statistical tables? Visualizing such results can leave a significant impression on end users and help data analysts effectively communicate complex findings.

This part of the book deals with such problems. It presents various graphing possibilities that will enhance users' abilities to communicate and present data to others and thus achieve more effective impact and engagement. The graphs and ideas chosen are but a small subset of the myriad of possibilities that Stata offers, but hopefully they will offer powerful additional insights into the capabilities that Stata users have.

28 Animated graphs

Advantages

1. Can communicate time, providing an important additional dimension to visualization
2. Can be powerful learning and communication aids
3. Draw attention
4. Can reveal trends and patterns that static graphs may miss

Disadvantages

1. Cannot be printed on paper
2. Require complex looping code
3. Must be custom-specified each time
4. Require external programs
5. Depend on viewing platform's capabilities

Required community-contributed package

To install the community-contributed command that is required to follow along with this chapter completely, type

```
. ssc install coefplot
```

Required external software

The program `ffmpeg` is required to stitch graphs together. It can be obtained at https://ffmpeg.org.

28.1 Introduction

Animated graphs can be used when users want to highlight transition states from a start graph to an end graph. Often, the transitions between these two time points are cut into tens or hundreds of tiny graph movements that resemble animations where data are slowly shuffled across the graph, distributions change their shapes, or other visual elements are changed.

Core to the idea of creating graph animations is that graphs are repeated many times. Therefore, any situation that requires many graphs to be displayed may be suitable for animations. Ultimately, animation condenses many graphs into one video of changing graphs. The advantage of this is that animation is a powerful visual technique to learn or communicate complex ideas. It draws attention and can elevate a user's ability to transfer statistical knowledge. A major disadvantage of animated graphs is the medium. Computer and video screens are required to display moving graphs; they cannot be printed in books or articles.

Stata can be used to create animated graphs using some creativity, loops, and a free software package called `ffmpeg`. This means such graphs are always bespoke, which generally requires lengthier coding and lots of trial and error. External programs such as video editors or compilers are required to stitch individual graphs together to create an animated video or movie. This takes users outside the Stata environment and into different software packages.

28.2 A basic animated graph

The core ideas required to create animated graphs in Stata are the following:

- A graph for each video frame. Videos often contain 24–30 frames per second.
- A loop code that generates many graphs quickly.
- A way to splice graphs together into a video.

The code and output below demonstrate all of these principles. To understand the complex code below, it is best to start in the middle, where data are generated. The first `generate` commands create variables containing random numbers that are normally distributed by using the `rnormal()` function. Variable `e` represents an error term, and variables `x1`, `x2`, and `x3` represent observed covariates. One particular variable, `x3`, is correlated with `e`. All variables are then used to create a final `y` variable, which represents a dependent variable. Parameters define the relationship between y and the x variables (in this example, they are all set to 1).

A regression (`regress`) is then used to estimate the parameters of `x1`, `x2`, and `x3`. In this case, `x3` is an endogenous variable, and the strength of the endogeneity is determined by the local `` `i' ``. Higher values of `` `i' `` will lead to more bias in the resulting output. In this case, `` `i' `` is part of the `forvalues` loop, which counts from 0 to 1 in increments of 0.02. The output from the regression is visualized via the community-contributed command `coefplot` (Jann 2013). The output is then saved as a `.png` file. Note that the saved file has a counter that starts at 1 and increments by 1 unit for each loop. This will lead to saved graphs that end in `1.png`, `2.png`, `3.png`, etc. Finally, to avoid lots of noise and graphs popping up, the `quietly` prefix command is used before the loop and the graphs are disabled via `set graphics off`. They are reenabled at the end.

The output of all of this code is 50 different `coefplot` graphs that are stored in the (Mac) directory `/users/buschaf`. These 50 graphs are then stitched together using an external command called `ffmpeg`. External commands can be executed within Stata via the `shell` or `winexec` commands (Mac or Windows, respectively). `ffmpeg` is free software that allows users to compile many different pictures into a video. A pathway is specified to the `ffmpeg` command, and the option `-y` tells `ffmpeg` to overwrite current files. The option `-i` asks `ffmpeg` to read from an arbitrary number of input files and write to an arbitrary number of output files. A pathway to the input files is specified, and the `%d` option acts as a wildcard that will pick up all graphs in numeric order starting at 1. The `-b:v` option specifies the video's bitrate, in this case 512k. The result is a video file that plays all 50 graphs in 2 seconds. A picture of the video files is shown below in figure 28.1.

Creating figure 28.1

```
. * Disable graphics
. set graphics off
.
. * Set directory
. cd /users/buschaf
.
. * A quiet loop
. quietly forvalues i = 0(0.02)1 {
.
.       * Create an incremental count local
.       local j = `j´ + 1
.
.       * Clear
.       clear
.
.       * Set observations and seed
.       set obs 1000
.       set seed 1234
.
.       generate e = rnormal()
.       generate x1 = rnormal()
.       generate x2 = rnormal()
.       generate x3 = rnormal() + `i´*e
.
.       generate y = 1 + 1*x1 + 1*x2 + 1*x3 + 1*e
.
.       * Regress data
.       regress y x1 x2 x3
.
.       * Plot estimates
.       coefplot, xline(1) xlabel(0(1)2)
.
.       * Save graphs as .png
.       graph export graph`j´.png, as(png) replace
.
. }
. * Turn graphics back on
. set graphics on
.
. * Use ffmpeg to merge individual graphs into video
. shell /users/buschaf/ffmpeg -y -i /users/buschaf/graph%d.png   ///
> -b:v 512k graph.mpg
```

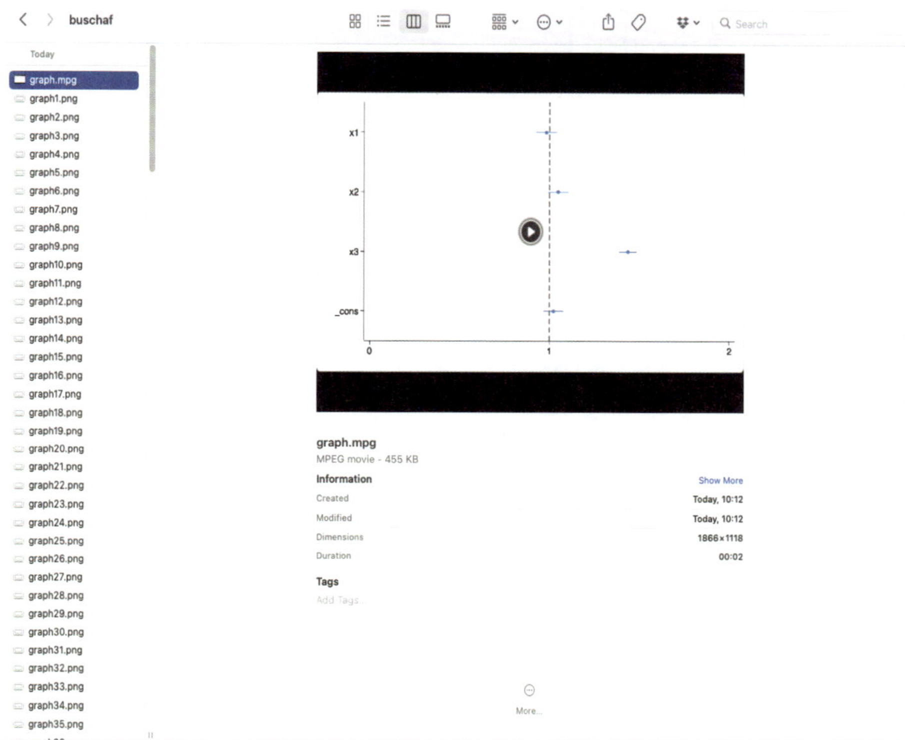

Figure 28.1. An animated Stata graph video

Figure 28.1 above shows a screenshot of the video that was created by `ffmpeg`. It is two seconds long and stored as a `.mpg` file. The video shows how the estimate of `x3` starts at the value of 1 and then slowly increases as the correlation between `x3` and the error term, `e`, increases. Ultimately, the video is an animated example of endogeneity in action and what kind of bias ordinary least-squares regression introduces when data do not meet the estimator assumptions.

There are many different possibilities in varying the above. Users can visualize how changing distributional parameters affect distributions, how regression slopes converge to lines of best fit, how scatterplots vary according to a third variable, etc. The core concept will always be the same: to create many graphs that each minutely differ from the preceding graph, which are then stitched together into a video animation.

29 Rainbow plots

Advantages

1. Are great for making sense of overplotted graphs
2. Allow replacement of legends in crowded graphs
3. Can be used in many circumstances
4. Enhance visual appeal and engagement

Disadvantages

1. Require potentially lengthy coding
2. Require ordering of data
3. Can misrepresent colors as statistical values
4. May not be accessible to colorblind viewers

Required community-contributed packages

To install the community-contributed commands that are required to follow along with this chapter completely, type

```
. ssc install palettes
. ssc install colrspace
```

29.1 Introduction

Rainbow plots are plottypes that emphasize color gradation across a range of ordered data. They are particularly useful when there are a lot of overlapping data, such as with time-series plots, or if many ordered categories need to be displayed. Moreover, color can be used to replace overcrowded legends, making it easier to identify subtrends.

Rainbow plots often use rainbow colors because doing so encompasses the full spectrum of visible colors that the human eye can determine, thereby giving end users access to the broadest possible range of ordering that is represented by color gradation. However, rainbow plots do not necessarily need to be in rainbow colors; they can be in any color. The key emphasis is the gradual change of colors across the ordered space. This makes rainbow plots very similar to heat plots, although they have a wider field of use across different graph types. Importantly, in rainbow plots, color represents data ordering and not data density.

A disadvantage of rainbow plots is that they are not helpful for end users with color vision impairment, and they may require significant fine-tuning in code because there is no standard plottype, although time-series plots tend to be a stereotypical example. Another disadvantage is that it can be difficult to identify small or very specific subgroups because color hues can vary only minutely between different data series or values. It may not be possible to successfully identify a yellow data group compared with another slightly yellower group. This means comparative detail is lost. At the same time, the loss of detail is specific to rainbow plots because they identify central masses of the ordered groups, such as quartiles or medians.

29.2 A basic rainbow plot

The basic problem for rainbow plots can come in multiple forms; the key to all of these is overplotted data, or too many groups of data on a graph. A basic example can be created with `uslifeexp.dta`, which contains time-series information on US life expectancy for various groups in the 20th century. A summary of the dataset is shown below.

```
. * Load data
. sysuse uslifeexp
(US life expectancy, 1900-1999)
. summarize
```

Variable	Obs	Mean	Std. dev.	Min	Max
year	100	1949.5	29.01149	1900	1999
le	100	64.829	9.158628	39.1	76.7
le_male	100	62.302	8.436369	36.6	73.9
le_female	100	67.51	9.834987	42.2	79.5
le_w	100	65.688	9.171269	39.8	77.3
le_wmale	100	63.143	8.503954	37.1	74.6
le_wfemale	100	68.434	9.797167	43.2	80
le_b	100	56.033	12.48937	30.8	71.4
le_bmale	100	53.589	11.4569	29.1	67.8
le_bfemale	100	58.567	13.5409	32.5	74.8

The `le*` variables identify life expectancy time-series variables for different groups. A plot of all of them on a graph using the `twoway line` command is shown below in figure 29.1. `line` can be used as a standalone command; it does not need to be preceded by `twoway`, and it can take multiple y variables. The final variable is treated as the x variable, in this case `year`.

Creating figure 29.1

```
. * Many line plots
. line le_wfemale le_female le_w le le_wmale le_male le_bfemale le_b ///
> le_bmale year
```

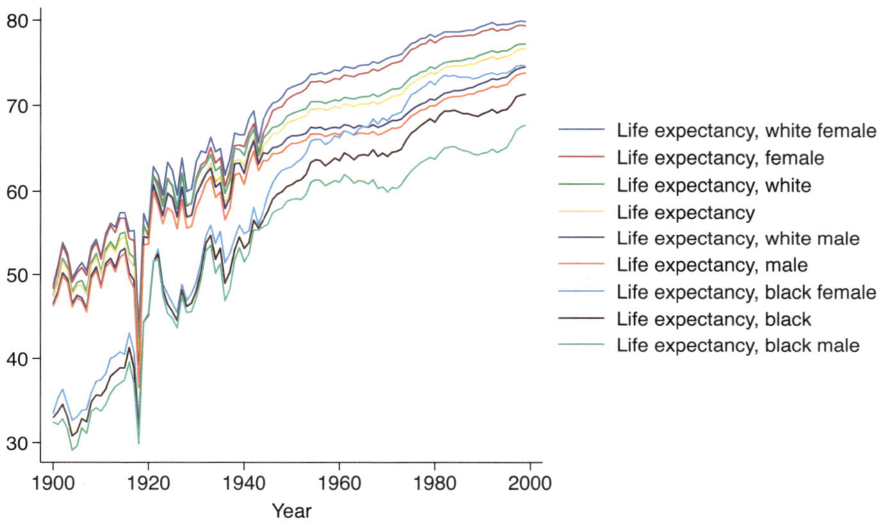

Figure 29.1. Multiple time-series plots

Figure 29.1 presents multiple line plots of life expectancy in the United States over the 20th century. This dataset has nine different groups, which are all plotted on the graph. This makes the graph relatively full of information, and while the default coloring of Stata does its best to produce a visually effective graph, it remains hard to read. A rainbow plot can help make this graph more clear.

There is no official or community-contributed rainbow plot command in Stata. The key to creating rainbow plots in Stata is to modify individual graphs via the color options that come with each graph. That means rainbow coloring can be applied to any graph, although different graphs will have different ways of coding such options. It would be impossible to go through all possible variations in this chapter, but the key to coding appropriate colors into each graph command is the `colorpalette` command, which is part of the `palette` (Jann 2017) package.

The command `colorpalette` allows users to view and display color values from various color palettes. The important option `n()` defines the size of the palette (for example, the number of colors). Below is a basic example that uses the `spmap` color collection. There are many collections, and each contains one or many palettes. Users should examine the associated help file (see `help colorpalette`). In this case, the palette `rainbow` is selected from the `spmap` collection and nine different color values are specified. Figure 29.2 presents the results.

Creating figure 29.2

```
. * Rainbow color palette with nine values
. colorpalette spmap rainbow, n(9)
```

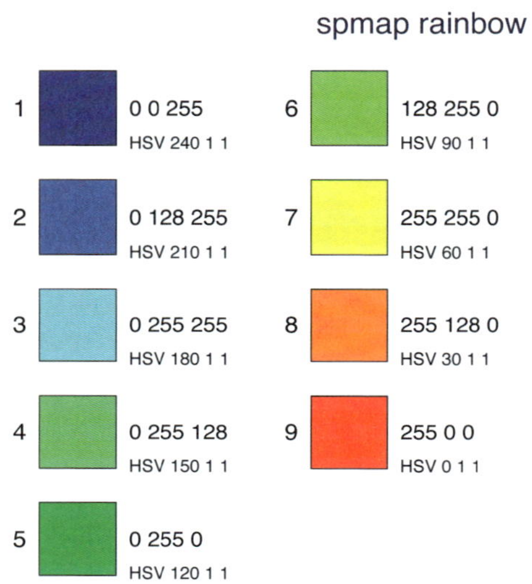

Figure 29.2. Color values using `colorpalette`

Figure 29.2 presents a rainbow color palette and the associated color values. In this case, the start and end colors are dark blue and bright red, respectively. The remaining colors are created so that nine unique and equidistant colors are presented between the start and end colors. An important component of this process is that the associated red, green, and blue (RGB) values are displayed next to each color. These numeric values can now be inserted into a Stata graph command to re-create the same colors.

However, manually entering each color value into the graph command is unlikely to be productive. A better option is to access the RGB values from the local macros stored as part of the `colorpalette` command. These locals can then be accessed by any other Stata command. Users can also access these locals interactively via the `return list` command, as demonstrated below.

```
. * Return list
. return list

scalars:
                  r(n) =  9

macros:
              r(ptype) : "color"
              r(pname) : "spmap rainbow"
              r(pnote) : "rainbow color scheme from the spmap package by Pis.."
            r(psource) : "spmap_color.ado from https://ideas.repec.org/c/boc.."
             r(pclass) : "sequential"
                 r(p) : ""0 0 255" "0 128 255" "0 255 255" "0 255 128" "0 2.."
             r(p9info) : "HSV 0 1 1"
                 r(p9) : "255 0 0"
             r(p8info) : "HSV 30 1 1"
                 r(p8) : "255 128 0"
             r(p7info) : "HSV 60 1 1"
                 r(p7) : "255 255 0"
             r(p6info) : "HSV 90 1 1"
                 r(p6) : "128 255 0"
             r(p5info) : "HSV 120 1 1"
                 r(p5) : "0 255 0"
             r(p4info) : "HSV 150 1 1"
                 r(p4) : "0 255 128"
             r(p3info) : "HSV 180 1 1"
                 r(p3) : "0 255 255"
             r(p2info) : "HSV 210 1 1"
                 r(p2) : "0 128 255"
             r(p1info) : "HSV 240 1 1"
                 r(p1) : "0 0 255"
```

`colorpalette` leaves individual color values for each color in multiple `r(p#)` macros, and it also leaves a list of all colors in the `r(p)` macro. Users can now insert these local macros into their Stata code. This is demonstrated in figure 29.3, which uses the `lcolor()` option of the `line` command to insert the `r(p)` macro. In other words, all nine color values are automatically inserted into the `lcolor()` option.

Note that users must execute the `colorpalette` command together with the `line` command in their do-file because the macros are local macros that disappear after execution of all highlighted code.

Creating figure 29.3

```
. * Line plot with rainbow colors
. colorpalette spmap rainbow, n(9) nograph
. line le_wfemale le_female le_w le le_wmale le_male le_bfemale le_b ///
> le_bmale year, lcolor(`r(p)')
```

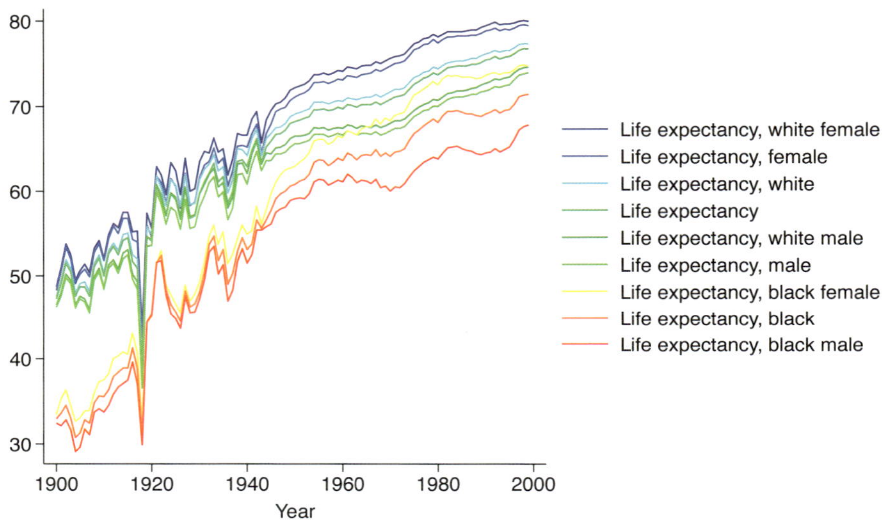

Figure 29.3. A basic rainbow line plot

Figure 29.3 presents a basic rainbow plot. Each data series is given a slightly different color from the rainbow, the important aspect being that the colors gradate from violet to red, as per a rainbow. Red and yellow colors represent data series with low average values, and blue and violet colors represent data series with high average values. The coloring is not related to any statistical value; it simply denotes the initial ordering of the data. Thanks to the unique coloring, it is easier to identify, for example, that the yellow line does not remain parallel to the other lines but intersects with the green lines. It is also easier to identify that many other line series remain parallel to each other.

29.3 A complex rainbow line plot

The previous example was a relatively basic example of a rainbow plot that most likely did not need rainbow coloring. Standard legend colors and different line patterns would have sufficed to identify data series. A more interesting case is when there are many more time series on a plot.

This is shown using the longitudinal analysis of pig weights dataset (`pig.dta`), which contains information on 48 different animals (`id`) and their weight (`weight`) evolution over time (`week`). The dataset is longitudinal in nature, so the `xtset` command is used to set the data to panel data.

```
. * Load data
. webuse pig
(Longitudinal analysis of pig weights)
. summarize
    Variable |        Obs        Mean    Std. dev.       Min        Max
-------------+---------------------------------------------------------
          id |        432        24.5    13.86946          1         48
        week |        432           5    2.584983          1          9
      weight |        432    50.40509    16.64113         20         88
. xtset id week
Panel variable: id (strongly balanced)
 Time variable: week, 1 to 9
         Delta: 1 unit
```

A basic initial analysis of how weight evolved across all individuals in this dataset can be done with the `xtline` command, which plots many line plots, one for each panel. The result is shown below in figure 29.4. The `overlay` option is used to plot all line plots on a single graph, while the `legend(off)` option is used to suppress the 48 individual legends that cluttered the graph.

Creating figure 29.4

```
. * Overlaid line plots
. xtline weight, overlay legend(off)
```

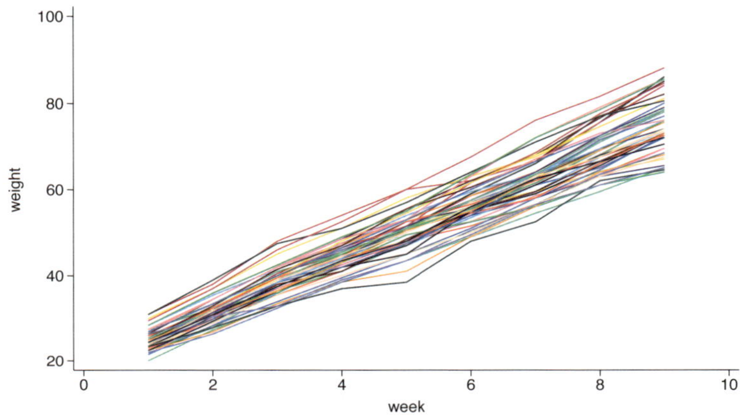

Figure 29.4. Panel line plots

The result is a big mess. This graph has far too many time series to identify individual observations or even examine how individual trends vary over time. Color is used poorly and repeatedly. The solution is to apply a rainbow color scheme that orders the time-series plots.

A natural way to order the data might be by starting weight. This is done in the code below, where a new variable is generated, `mw`, that contains the first `weight` observation of each observation in the panel. The data are then sorted by weight, panel ID, and time. The original panel identifier `id` is dropped, as is the `mw` variable (the data have now been sorted). A new identifier variable is computed that generates sequential numbering in blocks of 9 using the `egen` command.

```
. * Sort data
. bys id: generate mw = weight[1]
. sort mw id week
. drop id mw
. egen id = seq(), block(9)
```

The outcome of this code is a new panel dataset that resembles the one below. Observations have now been ranked by starting weight and given new panel identifiers in accordance with their new ranks.

week[1] 1

	week	weight	id
1	1	20	1
2	2	27.5	1
3	3	33	1
4	4	39	1
5	5	43.5	1
6	6	49	1
7	7	54.5	1
8	8	59.5	1
9	9	65	1
10	1	21.5	2
11	2	30.5	2
12	3	37	2
13	4	42.5	2
14	5	48	2
15	6	52.5	2
16	7	58.5	2
17	8	63	2
18	9	69.5	2
19	1	21.5	3
20	2	28.5	3
21	3	34	3
22	4	39.5	3
23	5	45	3

Figure 29.5. Restructured `pig.dta`

The next step is to create a line graph for each panel observation and apply incremental rainbow colors to each unique time series. There are 48 panels, which implies that 48 different colors are needed. These RGB color values can be created via the `colorpalette` command with the `n(48)` option, which will generate 48 unique colors stored in local macros `r(p)` and `r(p1)`–`r(p48)`. Unfortunately, the shortcut of inserting the local macro `r(p)` into one color option of the command `line` (such as `line` *y1 y2 y3 x*`, lcolor("`r(p)'")`) is not available. Only 20 consecutive color values can be specified in a single `lcolor()` option. Moreover, the individual time series are not cast over 48 variables but must be accessed via panel identifier subgroups. This requires using the `if` qualifier when using a basic `line` command.

There are many potential ways to solve the problem of coding 48 different colors into 48 different `line` graphs, but the below example is a basic and straightforward solution. Each panel identifier is plotted as a separate line graph, which is then overlaid via the `twoway` command. The basic setup of such a code block needs multiple instances of `line weight week if id ==` *#*`, lcolor("`r(p`*#*`)'"))`, where # identifies the panel identifier and `lcolor()` defines the custom color for each identifier.

To avoid manually typing out 48 different lines of code, the first piece of graph code can be copied into Excel, where the # values are placed into a separate column. It is then a simple matter of dragging the first row of cells down 48 times to create the 48 individual graph codes. The use of the `///` code tells Stata to attach the next line of code to the previous line, thereby creating one long line of code that is spread over multiple rows. This makes identification of the code easier. It also allows users to edit the code within Stata Do-file Editor by using the *Ctrl+Alt* (PC) or *Option+Command* (Mac) column mode selector. The column selector allows users to select rectangles in their code, which is perfect for lengthy code blocks such as the one below.

(line weight week if id ==	1	, lcolor("`r(p	1	)'"))	///
(line weight week if id ==	2	, lcolor("`r(p	2	)'"))	///
(line weight week if id ==	3	, lcolor("`r(p	3	)'"))	///
(line weight week if id ==	4	, lcolor("`r(p	4	)'"))	///

Figure 29.6. Building a code block in Excel

The ultimate code block result, after Excel surgery, is shown below. Advanced users are likely able to condense this code using a `forvalues` loop, but the primary purpose of the code below is to exemplify how individual graphs can be populated with many different color gradients from `colorpalette`. Figure 29.7 demonstrates the outcome.

Creating figure 29.7

```
. * Complex rainbow line plot
. colorpalette spmap rainbow, n(48) nograph
. twoway                                                          ///
> (line weight week if id == 1, lcolor("`r(p1)´"))               ///
> (line weight week if id == 2, lcolor("`r(p2)´"))               ///
> (line weight week if id == 3, lcolor("`r(p3)´"))               ///
> (line weight week if id == 4, lcolor("`r(p4)´"))               ///
> (line weight week if id == 5, lcolor("`r(p5)´"))               ///
> (line weight week if id == 6, lcolor("`r(p6)´"))               ///
> (line weight week if id == 7, lcolor("`r(p7)´"))               ///
> (line weight week if id == 8, lcolor("`r(p8)´"))               ///
> (line weight week if id == 9, lcolor("`r(p9)´"))               ///
> (line weight week if id == 10, lcolor("`r(p10)´"))             ///
> (line weight week if id == 11, lcolor("`r(p11)´"))             ///
> (line weight week if id == 12, lcolor("`r(p12)´"))             ///
> (line weight week if id == 13, lcolor("`r(p13)´"))             ///
> (line weight week if id == 14, lcolor("`r(p14)´"))             ///
> (line weight week if id == 15, lcolor("`r(p15)´"))             ///
> (line weight week if id == 16, lcolor("`r(p16)´"))             ///
> (line weight week if id == 17, lcolor("`r(p17)´"))             ///
> (line weight week if id == 18, lcolor("`r(p18)´"))             ///
> (line weight week if id == 19, lcolor("`r(p19)´"))             ///
> (line weight week if id == 20, lcolor("`r(p20)´"))             ///
> (line weight week if id == 21, lcolor("`r(p21)´"))             ///
> (line weight week if id == 22, lcolor("`r(p22)´"))             ///
> (line weight week if id == 23, lcolor("`r(p23)´"))             ///
> (line weight week if id == 24, lcolor("`r(p24)´"))             ///
> (line weight week if id == 25, lcolor("`r(p25)´"))             ///
> (line weight week if id == 26, lcolor("`r(p26)´"))             ///
> (line weight week if id == 27, lcolor("`r(p27)´"))             ///
> (line weight week if id == 28, lcolor("`r(p28)´"))             ///
> (line weight week if id == 29, lcolor("`r(p29)´"))             ///
> (line weight week if id == 30, lcolor("`r(p30)´"))             ///
> (line weight week if id == 31, lcolor("`r(p31)´"))             ///
> (line weight week if id == 32, lcolor("`r(p32)´"))             ///
> (line weight week if id == 33, lcolor("`r(p33)´"))             ///
> (line weight week if id == 34, lcolor("`r(p34)´"))             ///
> (line weight week if id == 35, lcolor("`r(p35)´"))             ///
> (line weight week if id == 36, lcolor("`r(p36)´"))             ///
> (line weight week if id == 37, lcolor("`r(p37)´"))             ///
> (line weight week if id == 38, lcolor("`r(p38)´"))             ///
> (line weight week if id == 39, lcolor("`r(p39)´"))             ///
> (line weight week if id == 40, lcolor("`r(p40)´"))             ///
> (line weight week if id == 41, lcolor("`r(p41)´"))             ///
> (line weight week if id == 42, lcolor("`r(p42)´"))             ///
> (line weight week if id == 43, lcolor("`r(p43)´"))             ///
> (line weight week if id == 44, lcolor("`r(p44)´"))             ///
> (line weight week if id == 45, lcolor("`r(p45)´"))             ///
> (line weight week if id == 46, lcolor("`r(p46)´"))             ///
> (line weight week if id == 47, lcolor("`r(p47)´"))             ///
> (line weight week if id == 48, lcolor("`r(p48)´")),            ///
> legend(off)
```

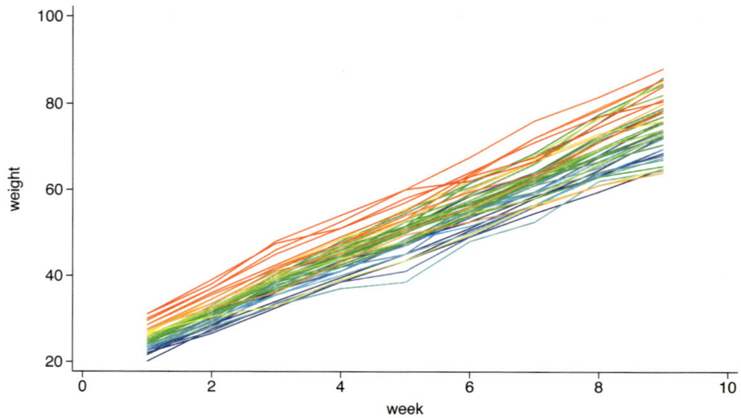

Figure 29.7. Complex rainbow line plot

Figure 29.7 is an example of a rainbow plot. The starting weight of each individual panel observation defines colors from red to violet. By examining the color transition over time, it is possible to gauge how much variation there is in weight gain across the sample. One can observe that by weeks 8 and 9, the initial rainbow pattern has been somewhat disrupted; animals with lower start weights now have higher final weights, and vice versa. This is identified because of the uniform rainbow coloring becoming more mixed.

Users should try different color combinations. Rainbow colors offer multiple color transitions, but two-color transitions are also possible. Users are not required to call upon a specific color scheme in `colorpalette`. Two individual colors can be specified, and `colorpalette` will automatically create a smooth transition of colors between them. An example where the two HTML colors `Gold` and `CadetBlue` are used is shown below in figure 29.8.

Creating figure 29.8

```
. * Complex rainbow line plot v2
. colorpalette Gold CadetBlue, n(48) nograph
. twoway                                                 ///
> (line weight week if id == 1, lcolor("`r(p1)'"))       ///
> (line weight week if id == 2, lcolor("`r(p2)'"))       ///
> (line weight week if id == 3, lcolor("`r(p3)'"))       ///
> (line weight week if id == 4, lcolor("`r(p4)'"))       ///
> (line weight week if id == 5, lcolor("`r(p5)'"))       ///
> (line weight week if id == 6, lcolor("`r(p6)'"))       ///
> (line weight week if id == 7, lcolor("`r(p7)'"))       ///
> (line weight week if id == 8, lcolor("`r(p8)'"))       ///
> (line weight week if id == 9, lcolor("`r(p9)'"))       ///
> (line weight week if id == 10, lcolor("`r(p10)'"))     ///
> (line weight week if id == 11, lcolor("`r(p11)'"))     ///
> (line weight week if id == 12, lcolor("`r(p12)'"))     ///
> (line weight week if id == 13, lcolor("`r(p13)'"))     ///
> (line weight week if id == 14, lcolor("`r(p14)'"))     ///
> (line weight week if id == 15, lcolor("`r(p15)'"))     ///
> (line weight week if id == 16, lcolor("`r(p16)'"))     ///
> (line weight week if id == 17, lcolor("`r(p17)'"))     ///
> (line weight week if id == 18, lcolor("`r(p18)'"))     ///
> (line weight week if id == 19, lcolor("`r(p19)'"))     ///
> (line weight week if id == 20, lcolor("`r(p20)'"))     ///
> (line weight week if id == 21, lcolor("`r(p21)'"))     ///
> (line weight week if id == 22, lcolor("`r(p22)'"))     ///
> (line weight week if id == 23, lcolor("`r(p23)'"))     ///
> (line weight week if id == 24, lcolor("`r(p24)'"))     ///
> (line weight week if id == 25, lcolor("`r(p25)'"))     ///
> (line weight week if id == 26, lcolor("`r(p26)'"))     ///
> (line weight week if id == 27, lcolor("`r(p27)'"))     ///
> (line weight week if id == 28, lcolor("`r(p28)'"))     ///
> (line weight week if id == 29, lcolor("`r(p29)'"))     ///
> (line weight week if id == 30, lcolor("`r(p30)'"))     ///
> (line weight week if id == 31, lcolor("`r(p31)'"))     ///
> (line weight week if id == 32, lcolor("`r(p32)'"))     ///
> (line weight week if id == 33, lcolor("`r(p33)'"))     ///
> (line weight week if id == 34, lcolor("`r(p34)'"))     ///
> (line weight week if id == 35, lcolor("`r(p35)'"))     ///
> (line weight week if id == 36, lcolor("`r(p36)'"))     ///
> (line weight week if id == 37, lcolor("`r(p37)'"))     ///
> (line weight week if id == 38, lcolor("`r(p38)'"))     ///
> (line weight week if id == 39, lcolor("`r(p39)'"))     ///
> (line weight week if id == 40, lcolor("`r(p40)'"))     ///
> (line weight week if id == 41, lcolor("`r(p41)'"))     ///
> (line weight week if id == 42, lcolor("`r(p42)'"))     ///
> (line weight week if id == 43, lcolor("`r(p43)'"))     ///
> (line weight week if id == 44, lcolor("`r(p44)'"))     ///
> (line weight week if id == 45, lcolor("`r(p45)'"))     ///
> (line weight week if id == 46, lcolor("`r(p46)'"))     ///
> (line weight week if id == 47, lcolor("`r(p47)'"))     ///
> (line weight week if id == 48, lcolor("`r(p48)'")),    ///
> legend(off)
```

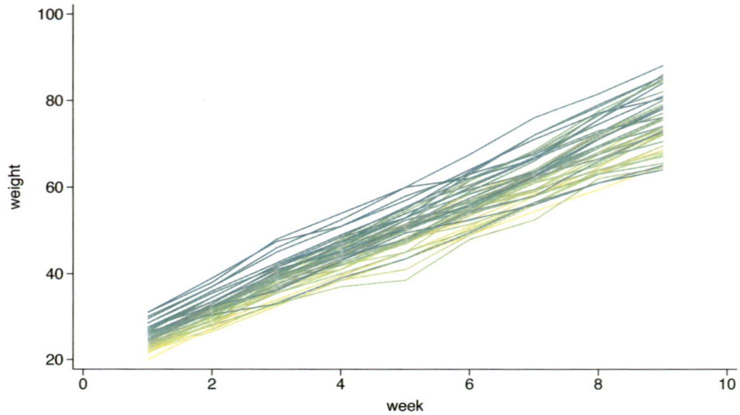

Figure 29.8. Two-color rainbow line plot

Figure 29.8 is an example of a two-color gradient plot. In this particular case, the color scheme also works well and continues to highlight how the aggregate data series evolve over time.

29.4 A complex rainbow box plot

Rainbow coloring, or any gradient coloring, can be applied to many different Stata graphs using the methodology outlined above. The key is finding a repetitive way to insert many of the different color values left behind in `colorpalette`'s `return list` locals.

Figure 29.9 demonstrates an example where multiple box plots are colored in rainbow colors. The `graph box` command is used to draw a box plot of the variable `weight` for each panel member. The `box()` option allows users to specify how individual boxes look. The suboption `color()` is where the local macros from `colorpalette` can be inserted.

Creating figure 29.9

```
. * Complex rainbow box plot
. colorpalette spmap rainbow, n(48) nograph
. graph box weight, over(id) legend(order(1 "Pig 1" 48 "Pig 48"))  ///
> asyvars nooutsides                                                ///
> box(1, color("`r(p1)'"))                                          ///
> box(2, color("`r(p2)'"))                                          ///
> box(3, color("`r(p3)'"))                                          ///
> box(4, color("`r(p4)'"))                                          ///
> box(5, color("`r(p5)'"))                                          ///
> box(6, color("`r(p6)'"))                                          ///
> box(7, color("`r(p7)'"))                                          ///
> box(8, color("`r(p8)'"))                                          ///
> box(9, color("`r(p9)'"))                                          ///
> box(10, color("`r(p10)'"))                                        ///
> box(11, color("`r(p11)'"))                                        ///
> box(12, color("`r(p12)'"))                                        ///
> box(13, color("`r(p13)'"))                                        ///
> box(14, color("`r(p14)'"))                                        ///
> box(15, color("`r(p15)'"))                                        ///
> box(16, color("`r(p16)'"))                                        ///
> box(17, color("`r(p17)'"))                                        ///
> box(18, color("`r(p18)'"))                                        ///
> box(19, color("`r(p19)'"))                                        ///
> box(20, color("`r(p20)'"))                                        ///
> box(21, color("`r(p21)'"))                                        ///
> box(22, color("`r(p22)'"))                                        ///
> box(23, color("`r(p23)'"))                                        ///
> box(24, color("`r(p24)'"))                                        ///
> box(25, color("`r(p25)'"))                                        ///
> box(26, color("`r(p26)'"))                                        ///
> box(27, color("`r(p27)'"))                                        ///
> box(28, color("`r(p28)'"))                                        ///
> box(29, color("`r(p29)'"))                                        ///
> box(30, color("`r(p30)'"))                                        ///
> box(31, color("`r(p31)'"))                                        ///
> box(32, color("`r(p32)'"))                                        ///
> box(33, color("`r(p33)'"))                                        ///
> box(34, color("`r(p34)'"))                                        ///
> box(35, color("`r(p35)'"))                                        ///
> box(36, color("`r(p36)'"))                                        ///
> box(37, color("`r(p37)'"))                                        ///
> box(38, color("`r(p38)'"))                                        ///
> box(39, color("`r(p39)'"))                                        ///
> box(40, color("`r(p40)'"))                                        ///
> box(41, color("`r(p41)'"))                                        ///
> box(42, color("`r(p42)'"))                                        ///
> box(43, color("`r(p43)'"))                                        ///
> box(44, color("`r(p44)'"))                                        ///
> box(45, color("`r(p45)'"))                                        ///
> box(46, color("`r(p46)'"))                                        ///
> box(47, color("`r(p47)'"))                                        ///
> box(48, color("`r(p48)'"))
```

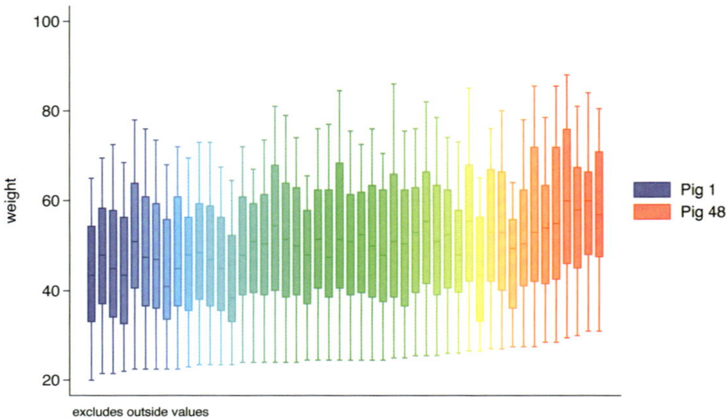

Figure 29.9. Rainbow box plot

Figure 29.9 shows how rainbow colors can create a visual flow across 48 different box plots. Clearly, 48 different legend labels would be impractical, so the legend labels are reduced to only the first and last observations in figure 29.9. The rainbow color gradients act as a replacement legend and help users identify what part of the order distribution a box plot is in.

Rainbow plots are custom plots that require bespoke code and thus advanced coding skills. In addition, imagination about where and how to apply such color schemes is essential. Nonetheless, despite the challenges such plots impose, Stata is versatile and flexible in creating various graphs that use color to convey important information.

30 Plotting regression results

Advantages

1. Makes complex regression tables much more user friendly
2. Facilitates comparison across multiple models
3. Makes identifying trends in group variables easier
4. Highlights confidence intervals visually

Disadvantages

1. May lose precise statistical detail
2. Is more affected by scaling
3. Does not provide diagnostic information
4. May require careful selection of variables to avoid clutter

Required community-contributed package

To install the community-contributed command that is required to follow along with this chapter completely, type

```
. ssc install coefplot
```

30.1 Introduction

Regression analysis is a major methodology in applied data analytics. Modern datasets and regression practice advances have led to more complex models being fit. Results from such models are traditionally presented via tabular means. Some tables, such as Stata's default output, contain a lot of information (much of which is often redundant, such as model diagnostics, or is repeated information such as t statistics, p-values, and confidence intervals). Other regression tables are heavily formatted to reduce the results output to its bare minimum but contain many footnotes, many subnotes, and other fine text that highlight particular aspects of the presented model. Finally, some regression tables are so large that multiple pages are required to present the model.

Displaying these results graphically can often be a more effective approach. Graphing estimates allows users to visually evaluate and identify coefficients and their patterns from a regression table. This is especially useful when results are presented to laypeople or when presentation time is short. Of course, for full analytical and exploratory purposes, one should examine complex tabular results, but even this can become a problem when multiple models must be compared. While Stata can collect regression results and fit them into a custom table (see `help estimates table`), fitting such results into a graphical depiction can often be even more helpful.

30.2 Visualizing coefficients from one regression

The basic problem is outlined below, where a simple regression model is fit on `auto.dta`, which contains information about car characteristics in 1978. A model of car price is fit that uses mileage, headroom, trunk size, car weight, car length, and repair record as explanatory variables. The resulting output is a large table with diagnostic information at the top and with model coefficients and statistical tests at the bottom.

```
. * Load data
. sysuse auto
(1978 automobile data)

. * A normal regression
. regress price mpg headroom trunk weight length i.rep78

      Source |       SS           df       MS      Number of obs   =        69
-------------+----------------------------------   F(9, 59)        =      5.96
       Model |   274615160         9  30512795.5   Prob > F        =    0.0000
    Residual |   302181799        59  5121725.41   R-squared       =    0.4761
-------------+----------------------------------   Adj R-squared   =    0.3962
       Total |   576796959        68  8482308.22   Root MSE        =    2263.1

------------------------------------------------------------------------------
       price | Coefficient  Std. err.      t    P>|t|     [95% conf. interval]
-------------+----------------------------------------------------------------
         mpg |  -119.2102   84.34712    -1.41   0.163    -287.9884      49.568
    headroom |   -683.834   443.1232    -1.54   0.128    -1570.521    202.8536
       trunk |   101.8725   111.2566     0.92   0.364    -120.7515    324.4965
      weight |   5.228857   1.171871     4.46   0.000      2.88395    7.573765
      length |   -125.788   44.65317    -2.82   0.007    -215.1388   -36.43726
             |
       rep78 |
          2  |   1645.096   1899.009     0.87   0.390    -2154.812    5445.004
          3  |   1545.796   1792.391     0.86   0.392    -2040.771    5132.363
          4  |   2591.428   1804.035     1.44   0.156    -1018.436    6201.293
          5  |   4047.228   1870.678     2.16   0.035     304.0092     7790.447
             |
       _cons |   14963.19   6582.287     2.27   0.027     1792.064    28134.32
------------------------------------------------------------------------------
```

While this table remains relatively user-friendly, larger versions will quickly overwhelm users. In addition, diagnostic information such as the sum of squares or the root mean squared error is rarely used in practice. The core focus is generally always on the coefficients that underpin the determinants of the dependent variable.

Visualization of coefficients from a regression model can be achieved with the combined use of the **margins** and **marginsplot** commands. Both commands are incredibly powerful at graphing a large range of predictions from analytical models. Their use is somewhat deemphasized in this chapter because the community-contributed command **coefplot** (Jann 2013) is easier to use when the emphasis is purely on graphing coefficients and not on complex predictions from regression models.

To graph coefficients using the official Stata command, users must run the command **marginsplot**. **marginsplot** requires that the **margins** command is run first because it takes the results from this command and graphs them. This leads to repetition because the regression coefficients must be reestimated via **margins**. Specifically, the **dydx()** option must be used to estimate the marginal effects of the previous model, which, for a linear regression, will result in the same parameters being estimated. Note that this is not the case for nonlinear models, so users should be careful when applying this technique to nonlinear models. Within the **dydx()** option, a * is specified, which signifies that all variable coefficients should be (re)estimated.

Once the **margins** command has been executed, it is a simple matter of running **marginsplot**. It is advisable to recast the graph into a scatterplot using the **recast()** option. This will remove the default line trend between coefficients, which can give a rather odd look. A **yline(0)** option is also added to show where the null is located. The results are shown in figure 30.1.

```
. * Predict marginal values
. margins, dydx(*)

Average marginal effects                                Number of obs = 69
Model VCE: OLS

Expression: Linear prediction, predict()
dy/dx wrt:  mpg headroom trunk weight length 2.rep78 3.rep78 4.rep78 5.rep78

------------------------------------------------------------------------------
             |            Delta-method
             |      dy/dx   std. err.      t    P>|t|     [95% conf. interval]
-------------+----------------------------------------------------------------
         mpg |  -119.2102   84.34712    -1.41   0.163    -287.9884      49.568
    headroom |   -683.834   443.1232    -1.54   0.128    -1570.521    202.8536
       trunk |   101.8725   111.2566     0.92   0.364    -120.7515    324.4965
      weight |   5.228857   1.171871     4.46   0.000      2.88395    7.573765
      length |   -125.788   44.65317    -2.82   0.007    -215.1388   -36.43726
             |
       rep78 |
          2  |   1645.096   1899.009     0.87   0.390    -2154.812    5445.004
          3  |   1545.796   1792.391     0.86   0.392    -2040.771    5132.363
          4  |   2591.428   1804.035     1.44   0.156    -1018.436    6201.293
          5  |   4047.228   1870.678     2.16   0.035     304.0092    7790.447
------------------------------------------------------------------------------
Note: dy/dx for factor levels is the discrete change from the base level.
```

Creating figure 30.1

```
. * Plot predicted values
. marginsplot, recast(scatter) yline(0)
```

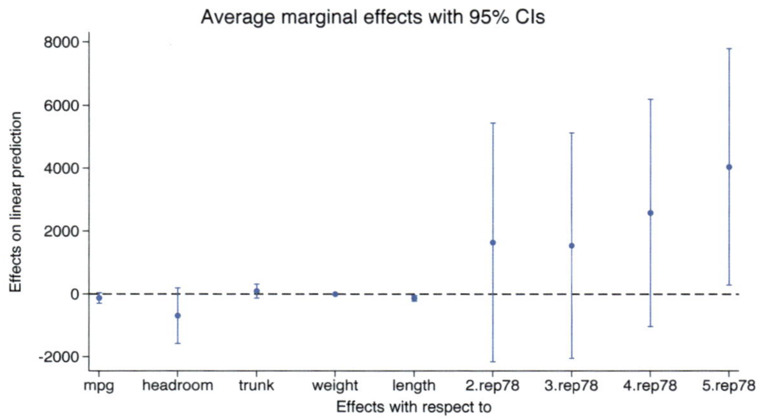

Figure 30.1. Graphing regression coefficients with `marginsplot`

Figure 30.1 presents a graphical depiction of the previous regression results. It is easy to identify that repair status categories have a large effect on car prices, although their confidence intervals mostly overlap with the zero line implying statistical insignificance. Category 5 is statistically significant, though. A key advantage of figure 30.1 is the ease of identification. Note that the coefficient on the constant is not plotted by default when using the `margins` and `marginsplot` commands.

This can be easily addressed by making use of the community-contributed command `coefplot`, which has been specifically designed to graph regression estimates. `coefplot` is a flexible command that can keep or exclude specific variables on the fly and draw results from multiple models into one graph. A basic example is shown in figure 30.2. The command `coefplot` is invoked without any arguments. Only the standard `twoway` option `xline()` is called to add a reference line.

Creating figure 30.2

```
. * Plot all coefficients
. coefplot, xline(0)
```

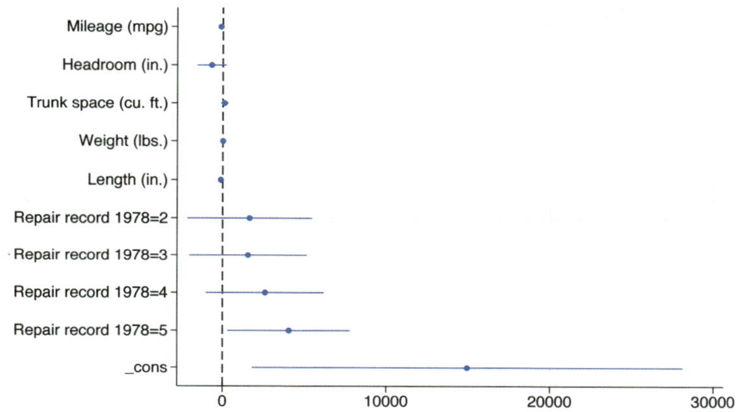

Figure 30.2. Graphing regression coefficients with `coefplot`

Figure 30.2 displays a basic usage of `coefplot`. It presents a similar graph to `marginsplot`, with the critical differences being that 1) the regression constant is included, 2) results are presented horizontally, and 3) the `margins` command did not need to be called. The horizontal aspect is generally advisable when regression models contain many variables or long variable names.

30.2.1 Options for coefplot

There are many options for `coefplot` that give users a lot of customizability when graphing regression estimates. For example, in figures 30.1 and 30.2, some regression coefficients are so small compared with others that they are only shown as dots near the reference line. Excluding variables with large coefficients allows the graph to be rescaled so that smaller coefficients can be evaluated. In `coefplot`, this can be done with the `keep()` and `drop()` options, which keep and drop specific variables, respectively. Wildcards can be used when denoting variables. Figure 30.3 demonstrates this by dropping the constant and the repair record categories from the graphical display. It also rotates the graph using the `vertical` option, adds a reference line using the `yline()` option, and angles the text using the `xlabel(, angle())` option and suboption.

Creating figure 30.3

```
. * Plot coefficients vertically without constant
. coefplot, vertical drop(_cons *rep*) yline(0) xlabel(, angle(45))
```

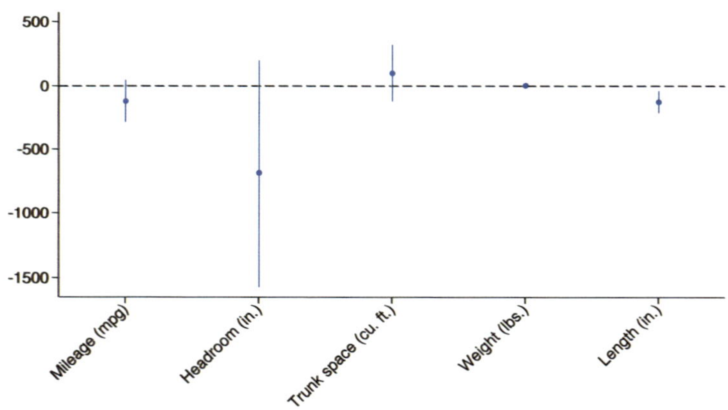

Figure 30.3. Graphing regression coefficients with `coefplot` and dropping selected variables

Figure 30.3 demonstrates how only a subsection of a regression model can be displayed. This allows users to focus only on those variables that matter most to them while ensuring that the coefficients reflect the full fitted model. `coefplot` makes this easy by having `keep()` and `drop()` options and drawing estimates directly from the most recent regression model. No intermediate steps are required.

30.3 Plotting coefficients from two regressions with similar dependent variables

Regression results are seldom presented in isolation. Often, users fit multiple models and would like to compare coefficients across such models. Visualization can add an extra advantage here because different model coefficients can be placed right next to each other for easy comparison.

An example of two different regression models is shown below. The actual models are identical with the same y and x variables; however, the models are fit over different subgroups: domestic and foreign cars. The prefix command `quietly` suppresses regression output (it is not necessary to display all output from these models), and the command `estimates store` stores the regression results in a storage name that can be called upon later by `coefplot`. This setup is demonstrated below.

```
. * Two regressions with same dependent variable
. quietly regress price mpg headroom trunk weight length i.rep78 if foreign == 0
. estimates store foreign
. quietly regress price mpg headroom trunk weight length i.rep78 if foreign == 1
. estimates store domestic
```

After both regression models are fit and their results stored, it is a simple matter of running `coefplot` and inserting the name of the stored estimates. Figure 30.4 presents the result.

Creating figure 30.4

```
. * Plot coefficients of two saved regressions
. coefplot foreign domestic, drop(_cons) xline(0)
```

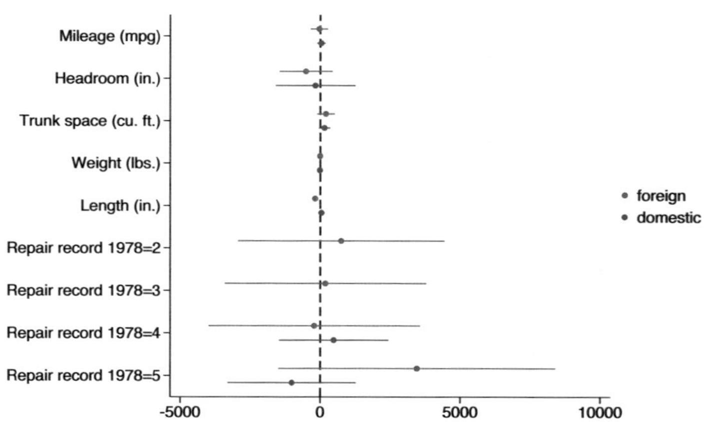

Figure 30.4. Graphing multiple regression coefficients with `coefplot`

Figure 30.4 is a visualization of the coefficients of two regression models. It is now easy and fast to compare how estimated variable coefficients vary across the two regressions. Note that the regression model for foreign cars has missing observations, and thus no coefficients were estimated for the repair record categories 2 and 3.

30.4 Plotting coefficients from four regressions with similar and different dependent variables

The previous example can be extended further to include many regression models, including models that have different dependent variables. The code below demonstrates four different regression estimations: two with `price` as the dependent variable and two with `turn` as the dependent variable.

```
. * Four regressions with two dependent variables
. quietly regress price mpg headroom trunk weight length i.rep78 if foreign == 0
. estimates store foreign
. quietly regress price mpg headroom trunk weight length i.rep78 if foreign == 1
. estimates store domestic
. quietly regress turn mpg headroom trunk weight length i.rep78 if foreign == 0
. estimates store foreign_t
. quietly regress turn mpg headroom trunk weight length i.rep78 if foreign == 1
. estimates store domestic_t
```

To integrate multiple different dependent variables into a visual representation, users should place different dependent variables into distinct subgraphs. These can be created by using the `||` expression in `coefplot`. The general code `coefplot` *subgraph* `||` *subgraph* `||` *subgraph* `||`... allows users to place as many different subgraphs as they like. Each subgraph can have individual options, and all subgraphs can have global options denoted by the final options specified in the command.

Users must be careful with where they put options. For instance, the code `coefplot` *m1*`,` *options1* `||` *m2*`,` *options2* will treat *options2* as global, therefore applying to both subgraphs. This may not necessarily be desired. If users want to apply *options2* only to *m2* and apply other options globally to all subgraphs, then the correct setup is `coefplot` *m1*`,` *options1* `||` *m2*`,` *options2* `||,` *options3*. Here *options3* acts as the global option for all subgraphs, while *options1* and *options2* act as local options for the subgraphs *m1* and *m2*, respectively. This style is demonstrated below in figure 30.5, where the stored regression models are divided across two subgraphs and the global options `drop()`, `xline()`, and `byopts()` apply to all graphs. In each graph in this example, the constant is dropped, a reference line is drawn, and the x axis is rescaled for each subgraph to account for the scale difference between `price` and `turn`.

Creating figure 30.5

```
. * Plot coefficients of four saved regressions
. coefplot foreign domestic, bylabel(Price) || foreign_t domestic_t,  ///
> bylabel(Turn) ||, drop(_cons) xline(0) byopts(xrescale)
```

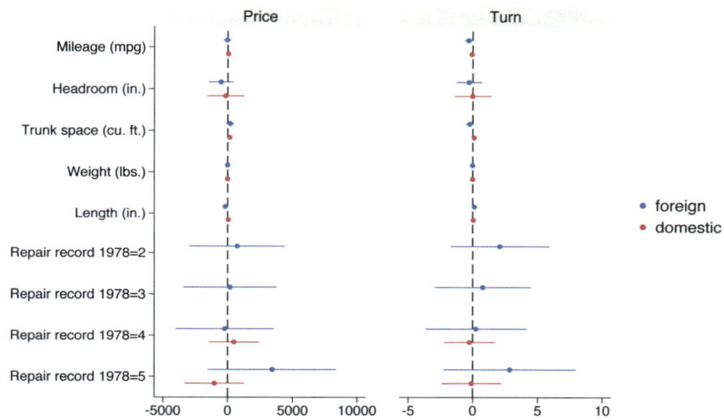

Figure 30.5. Graphing many regression coefficients with `coefplot`

Figure 30.5 depicts four groups of regression coefficients from four different regression models. It is relatively simple to extend this code to accommodate many different models to then compare visually. A key advantage is the fast and immediate acquisition of coefficient sizes and statistical significance. Differences are obvious, and this setup allows users to quickly visualize where major differences among various regression models lie.

30.5 Plotting coefficients from a regression with an interaction term

Users who specify interaction coefficients in their regressions using the on-the-fly expansion capabilities of Stata (for example, `c.`*var*`##c.`*var*) cannot use the **margins** and **marginsplot** commands to graph these coefficients easily. However, `coefplot` has no problems retrieving these coefficients and will plot them as they are. An example is shown below, where the variable **mpg** is fully interacted with the variable **headroom**.

```
. * A regression with an interaction term
. regress price c.mpg##c.headroom trunk weight length i.rep78

      Source |       SS           df       MS      Number of obs   =        69
-------------+----------------------------------   F(10, 58)       =      5.28
       Model |  274812359        10  27481235.9   Prob > F        =    0.0000
    Residual |  301984599        58  5206631.03   R-squared       =    0.4764
-------------+----------------------------------   Adj R-squared   =    0.3862
       Total |  576796959        68  8482308.22   Root MSE        =    2281.8

------------------------------------------------------------------------------
       price | Coefficient  Std. err.      t    P>|t|     [95% conf. interval]
-------------+----------------------------------------------------------------
         mpg |  -165.9601   254.8276    -0.65   0.517    -676.0529    344.1328
    headroom |  -1045.567   1911.661    -0.55   0.587    -4872.172    2781.038
             |
      c.mpg#|
  c.headroom |   17.48807   89.86018     0.19   0.846    -162.3866    197.3628
             |
       trunk |   101.1437   112.2375     0.90   0.371    -123.5242    325.8115
      weight |   5.246743   1.185113     4.43   0.000     2.874482    7.619004
      length |  -125.2814   45.09698    -2.78   0.007    -215.5528   -35.00998
             |
       rep78 |
           2 |   1721.304   1954.318     0.88   0.382    -2190.688    5633.296
           3 |   1579.583   1815.507     0.87   0.388    -2054.549    5213.715
           4 |   2644.181   1839.013     1.44   0.156    -1037.003    6325.365
           5 |   4092.876   1900.649     2.15   0.035     288.3138    7897.439
             |
       _cons |   15777.16   7844.607     2.01   0.049     74.47348    31479.85
------------------------------------------------------------------------------
```

Applying `coefplot` to the above regression output results in a graph that contains all relevant coefficients, including the interaction term. In the example below, the constant is dropped because of its magnitude; a reference line is included, and the option `cismooth` is used to add smoothed confidence intervals to the plot.

Creating figure 30.6

```
. * Plot coefficients vertically without constant
. coefplot, drop(_cons) yline(0) cismooth
```

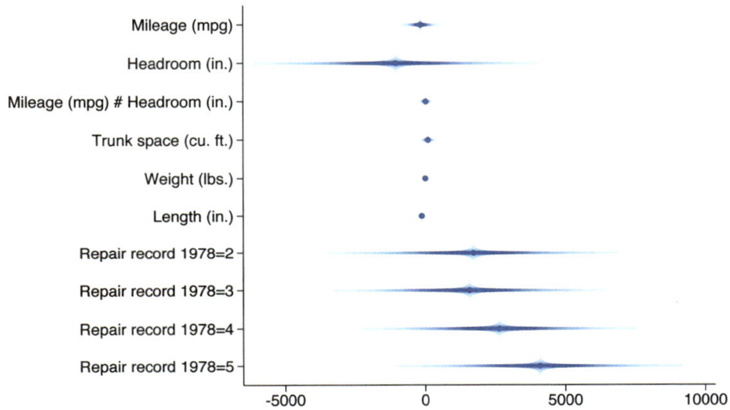

Figure 30.6. Graphing interaction regression coefficients with `coefplot`

Figure 30.6 demonstrates that `coefplot` will pick up all coefficients of a regression model, including complex interaction effects, and display these graphically. One is not limited to using `coefplot` only on ordinary least-squares regression (`regress`); it can access the estimates of many other estimation commands, including nonlinear models and panel-data models. Graphing regression output can be a valuable and convenient way to communicate complex models to others.

31 Maps

Advantages

1. Allow users to visualize geography within Stata
2. Can use color to provide choropleth visualizations of data at geographic levels

Disadvantages

1. Offer less options than dedicated mapping software
2. Require external map files
3. Take multiple steps and requires immediate coding experience

Required community-contributed packages

No additional installations are required to follow along with this chapter completely.

31.1 Introduction

Visualization of data within geographic maps has become much easier and more accessible in recent decades. Geolocation data are easier to obtain, and there are many different software packages that specialize in creating thematic maps. By default, Stata is able to use spatial data and has various commands available to help users visualize maps.

In particular, creating choropleth maps is possible through Stata. Choropleth maps use colors corresponding to a summary of geographic characteristics within spatial units. An example might be local unemployment rates, population density, or income. Such maps provide fast and easy ways to visualize how a continuous or categorical variable varies across a geography.

Choropleth maps are similar to heat maps but use geographic boundaries rather than predefined bins. The advantage of choropleth maps is that they are incredibly effective at transmitting information. They not only provide statistical information via colors but also add context by drawing geographic areas. Combining these two visual styles leads to a simple but powerful visualization that almost everyone understands.

A disadvantage is that colors lose statistical precision. It can be difficult to compare a light green area with another slightly lighter green area and understand the numerical difference between both regions. Moreover, the level of geography can matter greatly, just as bin sizes matter for heat plots. High-level geography can cause large areas to take one statistical value and thus one color. But realistically, there will be variation across such a large region because towns and cities differ from rural areas. Moreover, boundaries cause sharp color discontinuities while color (and hence statistical values) would most likely change gradually from one area to the next.

Finally, how color levels are derived can significantly change the visualization of the map. Are equal class intervals, quantiles, k means, or other custom classes used? Each can lead to different color variations across the map, leading to different narratives of what is happening.

A further disadvantage is that users must access geographic shapefiles from external sources to implement maps in Stata. Such shapefiles contain the geographic information to draw boundaries. Users must find such files from specialist websites, and their desired level of geography may not always be available. When obtained, shapefiles must be converted into a Stata format, to which users then must merge whatever information they wish to visualize. This operation can involve relatively complex code across multiple datasets and careful Internet searches.

31.2 A basic geographic map

To create a basic map, a shapefile must be obtained. In this chapter, an example of UK regions is used. A quick Google search of the terms "UK shapefile NUTS level 1" returns a desired result: a `.zip` file containing multiple files, including `.shp` and `.dbf` files (accessed via https://geoportal.statistics.gov.uk/). These files are unzipped and placed into the local directory that Stata is accessing.

The next step is to invoke the command `spshape2dta`, which translates shapefiles to Stata format. The command is relatively simple to use; specify the name of the `.shp` and `.dbf` files without the file name extension, for example, `spshape2dta` *name*. The option `replace` is used below to overwrite any current files. Once executed, two Stata files will be created: a *name*`_shp.dta` file, which contains all the boundary coordinates (this file will generally be large), and a *name*`.dta` file, which should be loaded and merged with any other data the user might have.

Once the *name*`.dta` file is loaded, users must activate the `grmap` command with the `activate` option. Now users are ready to generate maps in Stata. The below code demonstrates the aforementioned steps.

```
. * Convert shapefile to Stata file
. spshape2dta "NUTS_Level_1_(January_2018)_Boundaries", replace
  (importing .shp file)
  (importing .dbf file)
  (creating _ID spatial-unit id)
  (creating _CX coordinate)
  (creating _CY coordinate)

  file NUTS_Level_1_(January_2018)_Boundaries_shp.dta created
  file NUTS_Level_1_(January_2018)_Boundaries.dta     created

. * Load data and activate grmap
. use "NUTS_Level_1_(January_2018)_Boundaries", clear

. grmap, activate
grmap activated

. describe

Contains data from NUTS_Level_1_(January_2018)_Boundaries.dta
 Observations:            12
    Variables:            12                  21 Sep 2023 11:27
--------------------------------------------------------------------------------

Variable      Storage   Display    Value
    name         type    format    label      Variable label
--------------------------------------------------------------------------------

_ID             byte    %12.0g                Spatial-unit ID
_CX             double  %10.0g                x-coordinate of area centroid
_CY             double  %10.0g                y-coordinate of area centroid
objectid        byte    %2.0f                 objectid
nuts118cd       str3    %9s                   nuts118cd
nuts118nm       str24   %24s                  nuts118nm
bng_e           long    %6.0f                 bng_e
bng_n           long    %6.0f                 bng_n
v6              double  %24.15f               long
lat             double  %24.15f               lat
st_areasha      double  %24.15f               st_areasha
st_lengths      double  %24.15f               st_lengths
--------------------------------------------------------------------------------

Sorted by: _ID
```

The resulting dataset contains several important variables. The `_ID` variable is the spatial-unit identifier. This variable is numeric. The string data for each spatial unit (that is, the geographic name) is stored in the variables `nuts118cd` and `nuts118nm`. This will differ depending on the shapefile downloaded. The `_CX` and `_CY` variables provide the coordinates of the centroid of each geographic unit. These variables are used to add additional labels to maps. The remaining variables contain additional geographic information.

A simple map can now be drawn by calling on the `grmap` command. Note that data must be `spset` first; this declares data to be spatial data. The `spshape2dta` command automatically declares transformed data to be `spset`. No additional formatting commands are thus required. `grmap` is based on the community-contributed command `spmap` (Pisati 2007) and takes many of its options. The command has an extensive set of options, and users should examine the help file closely (see `help grmap`). To draw the underlying map without any additional information, no options or expressions must be specified. The command `grmap` is enough. Figure 31.1 demonstrates.

Creating figure 31.1

```
. * A basic map
. grmap
```

Figure 31.1. A basic geographic map

Figure 31.1 demonstrates how spatial data in the form of external shapefiles can be used to draw maps within Stata.

31.2.1 Custom labels

Users can also add custom labels to coordinates on each map. The centroid coordinates are convenient because they are automatically calculated when reshaping the data. Labels are added with the `label()` option. Several suboptions will need to be specified for the code to work. The suboption `label()` requires users to specify a variable that contains the relevant labels. This results in an odd-looking but correct `label(label())` code. The suboption `length()` should be specified for long labels to provide the maximum number of label characters. The suboption `size()` determines the label size, and the suboption `color()` determines color. The suboptions `xcoord()` and `ycoord()` must be specified and determine at which point the label should be inserted. Centroid values are used in the example below. Figure 31.2 presents a map with geographic labels.

Creating figure 31.2

```
. * A map with centroid labels
. grmap, label(label(nuts118nm) length(50) size(*0.7) color(maroon)  ///
> xcoord(_CX) ycoord(_CY))
```

Figure 31.2. Geographic map with custom labels

Figure 31.2 demonstrates how users can add custom labels to their spatial units.

31.3 Choropleth maps

The primary purpose of `grmap` is to create choropleth maps that color regions according to the values of another variable. Users must merge additional data files with the file created by `spshape2dta`. Alternatively, some basic information can be directly imputed. This is demonstrated below in figure 31.3, where the limited number of spatial units of a UK map suits itself to manual imputation of gross weekly wages. Then one simply calls `grmap` *varname*.

```
. * Impute geographic information
. generate wkgrosspay = .
(12 missing values generated)
. replace wkgrosspay = 546.8 if objectid == 1
(1 real change made)
. replace wkgrosspay = 578.0 if objectid == 2
(1 real change made)
. replace wkgrosspay = 568.5 if objectid == 3
(1 real change made)
. replace wkgrosspay = 573.4 if objectid == 4
(1 real change made)
. replace wkgrosspay = 581.8 if objectid == 5
(1 real change made)
. replace wkgrosspay = 628.6 if objectid == 6
(1 real change made)
. replace wkgrosspay = 728.4 if objectid == 7
(1 real change made)
. replace wkgrosspay = 660.1 if objectid == 8
(1 real change made)
. replace wkgrosspay = 577.3 if objectid == 9
(1 real change made)
. replace wkgrosspay = 570.6 if objectid == 10
(1 real change made)
. replace wkgrosspay = 622.0 if objectid == 11
(1 real change made)
. replace wkgrosspay = 575.0 if objectid == 12
(1 real change made)
```

Creating figure 31.3

```
. * A basic choropleth map
. grmap wkgrosspay
```

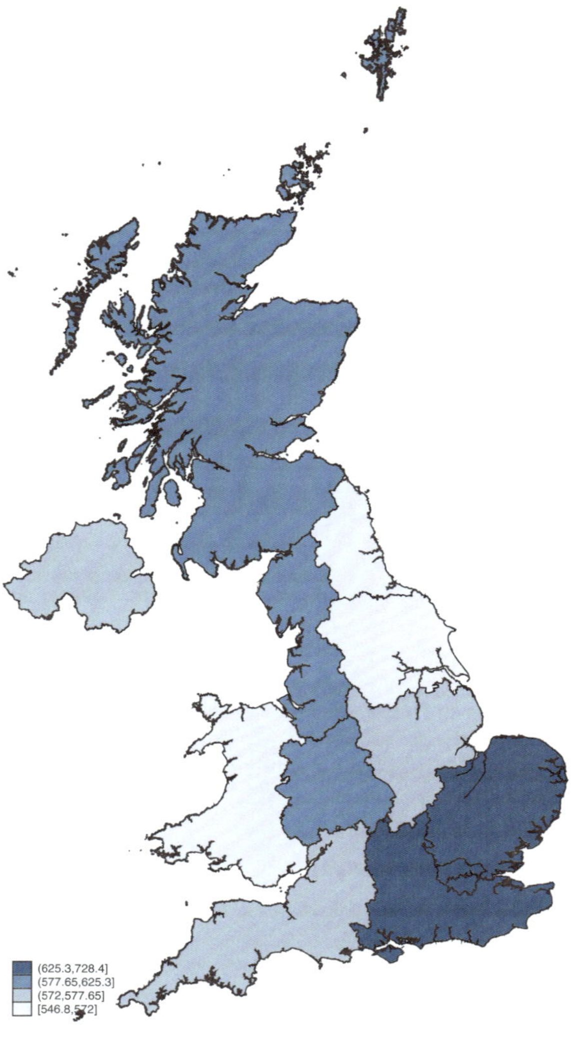

Figure 31.3. Choropleth map

Figure 31.3 presents a basic choropleth map of gross weekly wages in 2021 for UK regions. The default method for classifying the data and representing its spatial distribution is quantiles. It is relatively easy to identify that southern areas of the United Kingdom, including London, have high weekly gross wages. Northeastern areas and Wales have the lowest weekly gross wages.

31.3.1 Customization options

`grmap` offers users many options to customize their map looks. The legend position can be adjusted via the `legend()` option with the suboption `position()`. In the example below, it is put at the 10 o'clock position. The choropleth method is defined as a custom method via `clmethod()`. This, in turn, requires the `clbreaks()` option, which specifies custom breaks. In this case, different wage colors are defined in £50 increments. Finally, the fill colors of a region are adjusted by invoking the `fcolor()` option and specifying a red–yellow–green set of colors. Many colors are available, and users should examine the help file carefully (see `help grmap##color`). Figure 31.4 demonstrates a basic modified choropleth map with custom breaks and nondefault colors.

Creating figure 31.4

```
. * A custom choropleth map
. grmap wkgrosspay, legend(position(10)) clmethod(custom)   ///
> clbreaks(500 550 600 650 700 750) fcolor(RdYlGn)
```

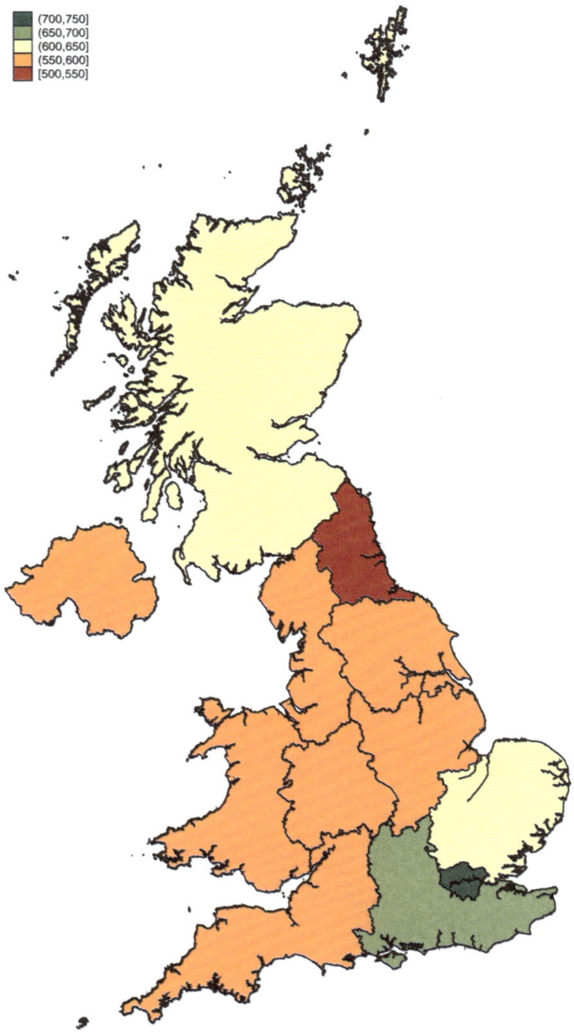

Figure 31.4. Custom choropleth map

Figure 31.4 demonstrates how choropleth map options can matter. Because of the different color cutoffs, the Northeast has now been identified as the "poorest" region in terms of gross weekly wages. London is identified as the highest region. Previously, the results were more mixed and identified multiple regions as the highest. As with heat maps, users should examine different choropleth classification methods to identify how different breaks can result in different map visualizations.

Figure 31.5 below demonstrates a different method for computing map colors. In this case, an equal interval method is specified that sets the number of intervals to 9 using the `clnumber()` option. Colors are changed to green.

Creating figure 31.5

```
. * A custom choropleth map with equal intervals
. grmap wkgrosspay, legend(position(10)) clmethod(eqint) clnumber(9)   ///
> fcolor(Greens)
```

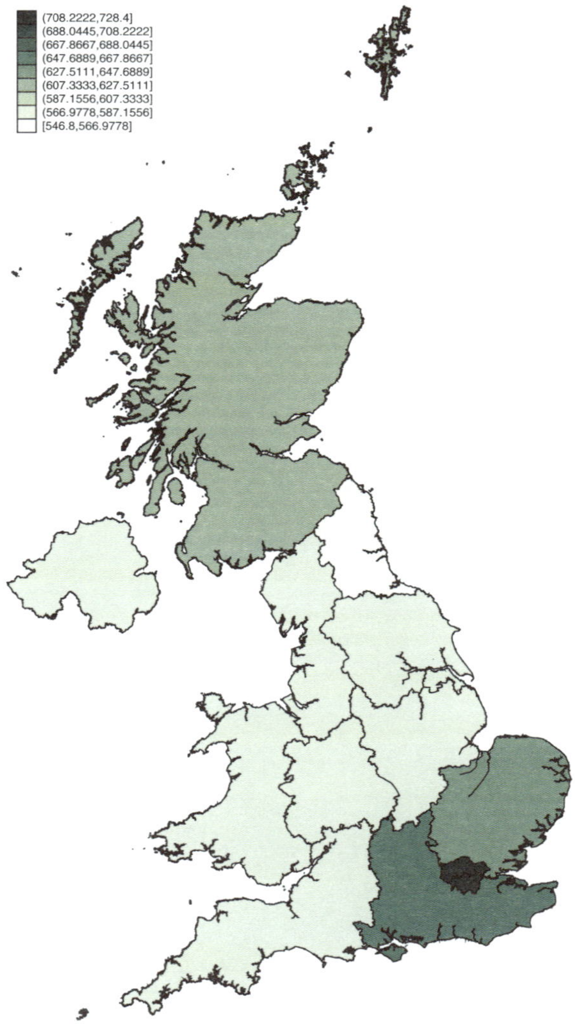

Figure 31.5. Equal interval choropleth map

Figure 31.5 demonstrates a different method of classifying gross weekly wages and also demonstrates another color palette. Users should experiment and find a color classification method that suits their needs.

31.4 Submaps

A unique function of `grmap` is that it will adjust the size of the regions displayed to fit the available graph area appropriately. Users are free to select subgroups of regions via `if` qualifiers or keeping or dropping observations via `keep` and `drop`. Figure 31.6 demonstrates an example where only regions from the south of the United Kingdom are drawn using an `if` qualifier.

Creating figure 31.6

```
. * A partial choropleth map
. grmap wkgrosspay if _ID > 5 & _ID < 10, legend(pos(10))    ///
> fcolor(YlGn)
```

Figure 31.6. Choropleth submap

Figure 31.6 shows how Stata will automatically zoom into the submap. This feature is particularly useful if users work with disaggregated spatial units where cities often contain tens or hundreds of subdistricts. On a national map, such small geographies get lost. However, by selecting only the relevant city or subarea, users can zoom in and better visualize small geographies.

32 Plotting equations

Advantages

1. Provides a visual depiction of statistical functions, distributions, and equations
2. Enhances comprehension of mathematical concepts
3. Can be used effectively to understand polynomials
4. Is useful for teaching and presentations

Disadvantages

1. Has limited applied use cases
2. Requires a good understanding of the underlying mathematics
3. Is often less effective than the `margins` command for representing polynomials

Required community-contributed packages

No additional installations are required to follow along with this chapter completely.

32.1 Introduction

Like other statistical and mathematical software packages, Stata is capable of plotting custom equations. Stata has many mathematical and statistical functions that users can visualize to help them better understand distributions, functional forms, or other mathematical concepts. Note that drawing such equations and functions is not reliant on data being loaded into Stata. One can use Stata interactively, independent of their data, and explore how various mathematical functions behave.

Users can also integrate theoretical functions with estimated results or data processes should they wish. Note that users cannot interactively adjust drawn equations via the Graph Editor. Each new equation or function will need to be coded and redrawn. When working with functions that relate to underlying estimation or data, one often should use the `margins` and `marginsplot` commands.

In many cases, drawing equations and functions helps users better understand particular data analysis processes. This may be related to specific theoretical requirements of their estimator or visualizing results, especially nonlinear relationships. While drawing custom functions is not frequently used in applied data analysis, it offers important learning and context capabilities.

32.2 A basic mathematical function

The basic problem is that users wish to visualize theoretical distributions or equations to help them better understand how variables might be related or distributed. Stata offers many statistical and mathematical functions that can be accessed via `help functions`.

Creating figure 32.1

```
. * Stata functions
. help functions
```

Description

Quick references are available for the following types of functions:

Type of function	See help
Date and time	datetime functions
Mathematical	mathematical functions
Matrix	matrix functions
Programming	programming functions
Random-number	random-number functions
Selecting time spans	time-series functions
Statistical	statistical functions
String	string functions
Trigonometric	trigonometric functions

Figure 32.1. Stata's functions

Users should further examine the functions available. Statistical and mathematical functions can be visualized via the `twoway function` command. The basic setup of this command is in the form of `y = ` $f(x)$, where $f()$ is a function from one of the aforementioned help files. Figure 32.1 below demonstrates this with the `sin()` function, where the code `y = sin(x)` is specified after the `twoway function` command. By default, the x axis is limited to the $[0, 1]$ range, so it often makes sense to extend the x-axis range using the `range()` option. Below, the range is extended to $[0, 10]$.

Creating figure 32.2

```
. * Draw a custom function y = sin(x)
. twoway (function y = sin(x), range(0 10))
```

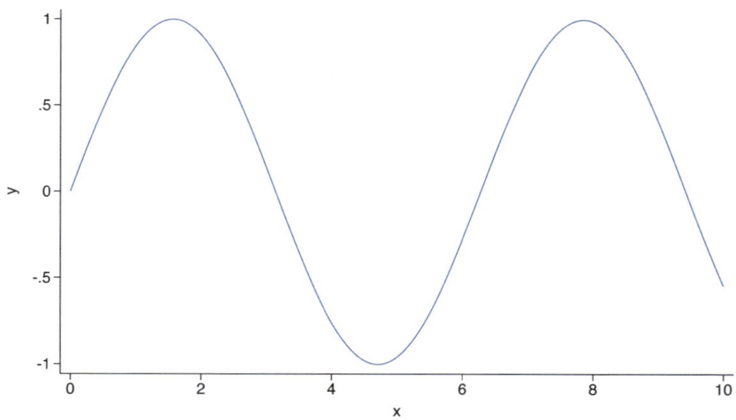

Figure 32.2. A sine wave

Figure 32.2 demonstrates a basic sine curve. Creating such a function can be useful if users wish to overlay a theoretical sine wave over observational data such as weather data or audio data.

32.3 A custom equation

Users can also specify custom equations that do not rely on any statistical or mathematical functions. A basic quadratic equation is demonstrated in figure 32.3 below. This is done by entering the x term twice, once as a linear x term and once as a squared term via x^2. A constant value of 1 and two coefficient values, 2 and 3, are also specified. Note the use of * before the x term to signify the multiplication factor.

Creating figure 32.3

```
. * Draw a quadratic equation
. twoway (function y = 1 + 2*x + 3*x^2, range(-5 5))
```

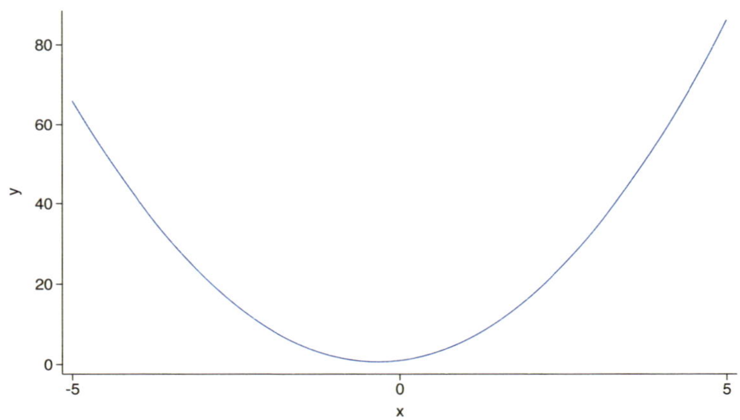

Figure 32.3. A quadratic equation

The result is a quadratic equation.

32.4 A statistical function

Stata has access to many statistical functions that users can call upon to visualize statistical distributions. In the below example, the cumulative logistic distribution with mean 0 and a default standard deviation of 1 is called upon via the `logistic()` function. Many functions also allow users to specify custom means, standard deviations, and other parameters.

Creating figure 32.4

```
. * Draw the logistic cumulative distribution function
. twoway (function y = logistic(x), range(-5 5))
```

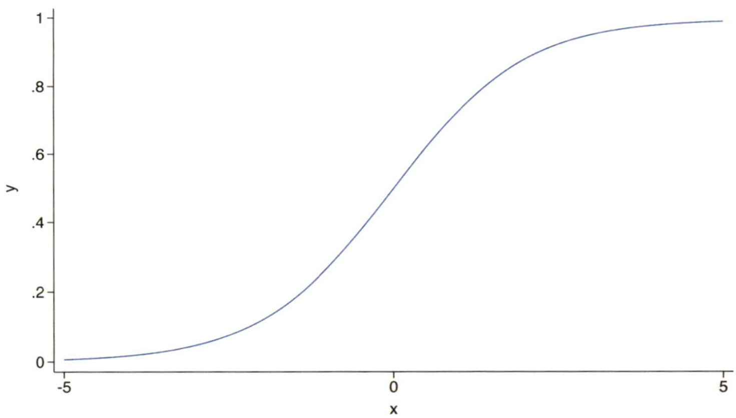

Figure 32.4. Logistic cumulative distribution function

Figure 32.4 presents a basic logistic cumulative distribution function that is often used to explain how logit regression models operate and why coefficients have complex nonlinear interpretations. For example, this kind of visualization can be used to show how marginal effect computation (the slope value of the function) depends on the value of x when interpreting logit coefficients.

32.4.1 Overlaying

The `function` plottype is part of the `twoway` graph command and can thus be overlaid with multiple `function` plots or other two-way plots. Figure 32.5 demonstrates how two `function` plots are overlaid to highlight the probability density function of the Cauchy distribution with location parameter 1 and scale parameter 1. The `recast()` option is used in the second `function` plot to shade the area under the graph, and the `range()` option is used to limit this area shading to between $[-5, 0]$.

Creating figure 32.5

```
. * Multiple overlaid functions
. twoway (function y = cauchyden(1,1,x), range(-5 5))     ///
> (function y = cauchyden(1,1,x), range(-5 0) recast(area) legend(off))
```

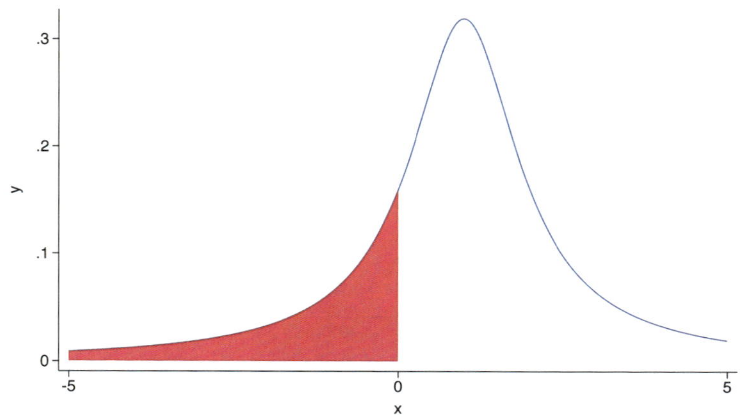

Figure 32.5. Cauchy probability density function with area shading

Figure 32.5 demonstrates how multiple `function` commands can be overlaid and how the `recast()` option can be specified to use different visual effects.

32.5 Polynomial regression terms

The previous examples all relate to theoretical concepts where users wish to graph pure mathematical equations and statistical functions. There are also applied uses for the `function` command: to overlay theoretical data distributions on observed data or to take estimates from models and graph them.

A simple example is how the `function` command can be used to graph regression polynomials. In the below example, a basic polynomial regression is estimated using `auto.dta`. The variable `price` is regressed against the terms `mpg` and `mpg`^{2}. After the `regress` command has been executed, results are stored in a matrix called `e(b)`. This matrix can be accessed by other commands, such as `matrix list`, because it is not deleted until another estimation command is executed.

```
. * Load auto data
. sysuse auto
(1978 automobile data)
. * Run regression
. regress price c.mpg##c.mpg
      Source |       SS           df       MS      Number of obs   =        74
-------------+----------------------------------   F(2, 71)        =     18.28
       Model |  215835615         2  107917807     Prob > F        =    0.0000
    Residual |  419229781        71  5904644.81    R-squared       =    0.3399
-------------+----------------------------------   Adj R-squared   =    0.3213
       Total |  635065396        73  8699525.97    Root MSE        =    2429.9

------------------------------------------------------------------------------
       price | Coefficient  Std. err.      t    P>|t|     [95% conf. interval]
-------------+----------------------------------------------------------------
         mpg |  -1265.194   289.5443    -4.37   0.000    -1842.529   -687.8593
             |
 c.mpg#c.mpg |   21.36069   5.938885     3.60   0.001     9.518891    33.20249
             |
       _cons |   22716.48   3366.577     6.75   0.000     16003.71    29429.24
------------------------------------------------------------------------------

. matrix list e(b)
e(b)[1,3]
                   c.mpg#
           mpg      c.mpg      _cons
y1  -1265.1942   21.36069  22716.476
```

To access stored estimates in the `e(b)` matrix, one can use system variables, which are built-in variables that are automatically created and updated after estimation commands are executed. The expression `_b[`*varname*`]` allows users to access the most recent estimated coefficients.

Inserting such system variables into `function` allows users to build a continuously updating function plot. In this case, figure 32.6 demonstrates the form of the estimated quadratic regression relationship by inserting system variables as coefficients before `x` in the `function` command. The `range()` option is also specified and contains the variable `mpg`, which means Stata will automatically extend to range from minimum to maximum values of `mpg`.

Creating figure 32.6

```
. * Plot regression coefficients as a custom function
. twoway (function y = _b[_cons] + _b[mpg]*x + _b[c.mpg#c.mpg]*x^2,  ///
> range(mpg))
```

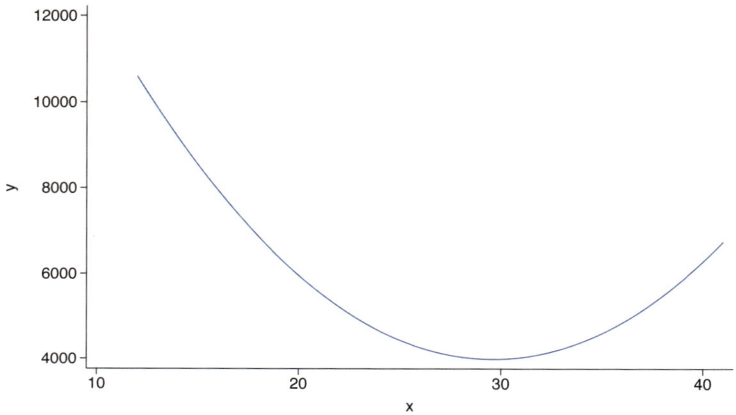

Figure 32.6. Quadratic regression relationship

The result is an accurate depiction of the relationship between `price` and `mpg` as estimated using the `regress` command. Many other estimation commands can be similarly illustrated.

Note that `margins` is often more suited to plotting complex polynomial shapes from regression modeling via its powerful prediction functions. However, `function` can call upon a comprehensive set of custom functions that allows users incredible flexibility when plotting equations.

References

Alexandersson, A. 2004. ellip: Stata module to graph confidence ellipses (version 8). Statistical Software Components S422603, Department of Economics, Boston College. https://ideas.repec.org/c/boc/bocode/s422603.html.

Buis, M. L. 2007. hangroot: Stata module creating a hanging rootogram comparing an empirical distribution to the best fitting theoretical distribution. Statistical Software Components S456886, Department of Economics, Boston College. https://ideas.repec.org/c/boc/bocode/s456886.html.

Cox, N. J. 1998a. distplot: Stata module to generate distribution function plot. Statistical Software Components S337502, Department of Economics, Boston College. https://ideas.repec.org/c/boc/bocode/s337502.html.

———. 1998b. triplot: Stata module to generate triangular plots. Statistical Software Components S342401, Department of Economics, Boston College. https://ideas.repec.org/c/boc/bocode/s342401.html.

———. 1999a. skewplot: Stata module to draw skewness plots. Statistical Software Components S388201, Department of Economics, Boston College. https://ideas.repec.org/c/boc/bocode/s388201.html.

———. 1999b. tabplot: Stata module to show table of frequencies as table of bars. Statistical Software Components S370601, Department of Economics, Boston College. https://ideas.repec.org/c/boc/bocode/s370601.html.

———. 2003a. eqprhistogram: Stata module for equal probability histogram. Statistical Software Components S432701, Department of Economics, Boston College. https://ideas.repec.org/c/boc/bocode/s432701.html.

———. 2003b. pairplot: Stata module for plots of paired observations. Statistical Software Components S431101, Department of Economics, Boston College. https://ideas.repec.org/c/boc/bocode/s431101.html.

———. 2003c. slideplot: Stata module for sliding bar plot for frequencies or percents. Statistical Software Components S432801, Department of Economics, Boston College. https://ideas.repec.org/c/boc/bocode/s432801.html.

———. 2003d. stripplot: Stata module for strip plots (one-way dot plots). Statistical Software Components S433401, Department of Economics, Boston College. https://ideas.repec.org/c/boc/bocode/s433401.html.

———. 2004. scat3: Stata module for crude three-dimensional graphics. Statistical Software Components S445801, Department of Economics, Boston College. https://ideas.repec.org/c/boc/bocode/s445801.html.

———. 2005a. beamplot: Stata module to draw horizontal dotplots using beams. Statistical Software Components S454001, Department of Economics, Boston College. https://ideas.repec.org/c/boc/bocode/s454001.html.

———. 2005b. Speaking Stata: Smoothing in various directions. *Stata Journal* 5: 574–593. https://doi.org/10.1177/1536867X0500500408.

———. 2007a. corrtable: Stata module to show correlation matrix as graphical table. Statistical Software Components S456893, Department of Economics, Boston College. https://ideas.repec.org/c/boc/bocode/s456893.html.

———. 2007b. spineplot: Stata module for spine plots for two-way categorical data. Statistical Software Components S456885, Department of Economics, Boston College. https://ideas.repec.org/c/boc/bocode/s456885.html.

———. 2007c. Speaking Stata: Turning over a new leaf. *Stata Journal* 7: 413–433. https://doi.org/10.1177/1536867X0700700308.

———. 2013. sparkline: Stata module for sparkline-type plots. Statistical Software Components S457583, Department of Economics, Boston College. https://ideas.repec.org/c/boc/bocode/s457583.html.

———. 2014a. crossplot: Stata module for scatter (or other twoway) plots for each y vs each x variable. Statistical Software Components S457819, Department of Economics, Boston College. https://ideas.repec.org/c/boc/bocode/s457819.html.

———. 2014b. sepscatter: Stata module to draw scatter (or other twoway) plots separated by a third variable. Statistical Software Components S457829, Department of Economics, Boston College. https://ideas.repec.org/c/boc/bocode/s457829.html.

———. 2016. Speaking Stata: Multiple bar charts in table form. *Stata Journal* 16: 491–510. https://doi.org/10.1177/1536867X1601600214.

———. 2018. fabplot: Stata module for plots for each subset with rest of the data as backdrop. Statistical Software Components S458494, Department of Economics, Boston College. https://ideas.repec.org/c/boc/bocode/s458494.html.

Dupont, W. D., and W. D. Plummer, Jr. 2002. sunflower: Stata module to generate density distribution sunflower plots. Statistical Software Components S430201, Department of Economics, Boston College. https://ideas.repec.org/c/boc/bocode/s430201.html.

Ender, P. B. 2002. histbox: Stata module to generate histogram with boxplot. Statistical Software Components S430101, Department of Economics, Boston College. https://ideas.repec.org/c/boc/bocode/s430101.html.

Jann, B. 2013. coefplot: Stata module to plot regression coefficients and other results. Statistical Software Components S457686, Department of Economics, Boston College. https://ideas.repec.org/c/boc/bocode/s457686.html.

———. 2017. palettes: Stata module to provide color palettes, symbol palettes, and line pattern palettes. Statistical Software Components S458444, Department of Economics, Boston College. https://ideas.repec.org/c/boc/bocode/s458444.html.

———. 2019. heatplot: Stata module to create heat plots and hexagon plots. Statistical Software Components S458598, Department of Economics, Boston College. https://ideas.repec.org/c/boc/bocode/s458598.html.

Kohler, U., and S. Kuehnel. 2000. hist3: Stata module to draws [*sic*] histograms for intervals of different size. Statistical Software Components S412001, Department of Economics, Boston College. https://ideas.repec.org/c/boc/bocode/s412001.html.

Luedicke, J. 2015. supsmooth: Stata module to perform Friedman's super smoother. Statistical Software Components S458030, Department of Economics, Boston College. https://ideas.repec.org/c/boc/bocode/s458030.html.

Mander, A. 2005a. cdfplot: Stata module to plot a cumulative distribution function. Statistical Software Components S456409, Department of Economics, Boston College. https://ideas.repec.org/c/boc/bocode/s456409.html.

———. 2005b. surface: Stata module to draw a 3D wireform surface plot. Statistical Software Components S448501, Department of Economics, Boston College. https://ideas.repec.org/c/boc/bocode/s448501.html.

———. 2005c. trellis: Stata module to create a trellis graph. Statistical Software Components S450802, Department of Economics, Boston College. https://ideas.repec.org/c/boc/bocode/s450802.html.

———. 2007. radar: Stata module to draw radar (spider) plots. Statistical Software Components S456829, Department of Economics, Boston College. https://ideas.repec.org/c/boc/bocode/s456829.html.

Naqvi, A. 2021. schemepack: Stata module providing ready-to-use graph schemes. Statistical Software Components S459012, Department of Economics, Boston College. https://ideas.repec.org/c/boc/bocode/s459012.html.

———. 2023. Stata package "joyplot". GitHub. https://github.com/asjadnaqvi/stata-joyplot.

Nichols, A. 2008. bihist: Stata module to produce bihistograms. Statistical Software Components S456983, Department of Economics, Boston College. https://ideas.repec.org/c/boc/bocode/s456983.html.

Over, M. 2020. superscatter: Scatter plot with marginal distributions. http://digital.cgdev.org/doc/stata/MO/Misc.

Pisati, M. 2007. spmap: Stata module to visualize spatial data. Statistical Software Components S456812, Department of Economics, Boston College. https://ideas.repec.org/c/boc/bocode/s456812.html.

Raciborski, R. 2009. Graphical representation of multivariate data using Chernoff faces. *Stata Journal* 9: 374–387. https://doi.org/10.1177/1536867X0900900302.

Rostam-Afschar, D., and J. Robin. 2014. graph3d: Stata module to draw colored, scalable, rotatable 3D plots. Statistical Software Components S457929, Department of Economics, Boston College. https://ideas.repec.org/c/boc/bocode/s457929.html.

Stepner, M. 2013. binscatter: Stata module to generate binned scatterplots. Statistical Software Components S457709, Department of Economics, Boston College. https://ideas.repec.org/c/boc/bocode/s457709.html.

Winter, N. 2014. combomarginsplot: Stata module to combine the saved results from multiple calls to margins into one marginsplot. Statistical Software Components S457804, Department of Economics, Boston College. https://ideas.repec.org/c/boc/bocode/s457804.html.

Winter, N., and A. Nichols. 2008. vioplot: Stata module to produce violin plots with current graphics. Statistical Software Components S456902, Department of Economics, Boston College. https://ideas.repec.org/c/boc/bocode/s456902.html.

Author index

Subject index